Technology of Empire

Telecommunications and Japanese Expansion in Asia, 1883–1945

Harvard East Asian Monographs 219

Technology of Empire

Telecommunications and Japanese Expansion in Asia, 1883–1945

Daqing Yang

Published by the Harvard University Asia Center
Distributed by Harvard University Press
Cambridge (Massachusetts) and London, 2010

Printed in the United States of America

The Harvard University Asia Center publishes a monograph series and, in coordination with the Fairbank Center for Chinese Studies, the Korea Institute, the Reischauer Institute of Japanese Studies, and other faculties and institutes, administers research projects designed to further scholarly understanding of China, Japan, Vietnam, Korea, and other Asian countries. The Center also sponsors projects addressing multidisciplinary and regional issues in Asia.

Library of Congress Cataloging-in-Publication Data

Yang, Daqing, 1964–

Technology of empire : telecommunications and Japanese expansion in Asia, 1883–1945 / Daqing Yang.

p. cm.

Includes bibliographical references and index.

ISBN 978-0-674-01091-8 (cloth : alk. paper)

1. Japan--History--1868– 2. Telecommunication systems--Japan--History. 3. Telecommunication systems--Asia--History. 4. Japan--History, Military--1868– 5. Japan--Armed Forces--Communication systems. 6. World War, 1939–1945--Communications. 7. Sino-Japanese War, 1937–1945--Communications. I. Title.

DS881.9.Y375 2010

384'.0952'09041--dc22

2004030493

Index by the author

Printed on acid-free paper

Last figure below indicates year of this printing

19 18 17 16 15 14 13 12 11

To my mother

Acknowledgments

A work that has taken a long time in the making naturally incurs too many intellectual debts to enumerate here. The greatest of all goes to Akira Iriye, a true scholar in the finest sense of the word. Since my sophomore year at Nanjing University he has been my source of inspiration for studying international history. John Dower, who has influenced me a great deal in thinking about war, race, and justice, is another. Mark Peattie has endowed with me not only with his book collection on Japan but also a lasting friendship. My other teachers, especially Albert Craig, Andrew Gordon, Bill Kirby, Philip Kuhn, John Stephan, and Ezra Vogel, all have nurtured me intellectually in countless ways.

At the George Washington University, my colleagues in the History Department and the Sigur Center for Asian Studies have been most supportive. Edward McCord read the entire manuscript and offered valuable suggestions. Chris Sterling deserves special thanks for keeping me interested in all things related to communication and transportation. Barney Finn at the Smithsonian has never failed to provide fellowship and support. Matt Zolotor produced early versions of the maps in the book. Outside Washington, Daniel Headrick, Laura Hein, and Tessa Morris-Suzuki have also offered excellent comments on the project. Over the years, a number of my good friends have helped with this project in various ways. A few of them have been extraordinarily generous with their time and my writing has benefited enormously as a result. My profound gratitude goes to each one of them.

Most of the research for this book was done outside the United States, especially in Japan. There many scholars generously shared their time and expertise with me: Fujii Nobuyuki, Hamashita Takeshi, Hatano Sumio, Hikita Yasuyuki, Ikei Masaru, Kobayashi Hideo, Nojima Yoko, Sugiyama Shinya, and Yanagisawa Asobu, to name just a few. In Copenhagen, Kurt Jacobsen has been a wonderful host when I visited. As a historian, I am indebted to too many librarians and archivists to name. Several must be mentioned, however. Without Nakano Michiko and Kosakai Masako in Japan, research for this book would not have been possible. The Interlibrary loan staff at Gelman Library at the George Washington University has been most patient with my many requests.

I have had the good fortune to work with John Ziemer and William Hammell, two superb editors at the Harvard Asia Center. I am extremely grateful for their high standards of professionalism and unfailing support throughout the years. I would like also to thank the two anonymous readers who provided very helpful comments on an earlier version. As this book underwent final preparations, Tabitha Mallory provided constant intellectual support and companionship. I wish one day I can pay her back adequately.

In the course of research and writing, I have received generous funding from Harvard University, Whiting Foundation, Japan Foundation, and George Washington University. For all these, I am very grateful.

I dedicate this book to my mother in China, who has waited patiently all these years, and to the memory of my father.

D-Q.Y.

Contents

Part IV Network, 1939–1945

Reference Matter

Figures, Tables, Maps, Photographs

FIGURES

TABLES

MAPS

PHOTOGRAPHS

Abbreviations

AT&T	American Telephone and Telegraph Company
CCTC	Central China Telecommunications Company
CEC	China Electric Company
DDJS	Nihon denshin denwa kōsha, *Denshin denwa jigyō shi*
DJ	*Denpa to juken*
DKS	*Denmu kenkyū shiryō*
DKSKS	*Daitōa kensetsu shingikai kankei shiryō*
DKZ	*Denshin kyōkai zasshi*
DT	*Denki tsūshin*
DTGZ	*Denwa denshin gakkai zasshi* (Tokyo), later renamed *Denki tsūshin gakkai zasshi*
DTJGKS	*Denki tsūshin jishu gijutsu kaihatsu-shi: hansō denwa-hen*
FER	*Far Eastern Review*
FRUS	United States, Department of State, *Foreign Relations of the United States*
GGK	Government-General of Korea
GGT	Government-General of Taiwan
GKDTS	Gaichi denki tsūshin shi hensan iinkai, comp., *Gaichi kaigai denki tsūshin shiryō*
GNTC	Great Northern Telegraph Company

GP	gutta-percha
ITC	International Telephone Company (1933–37) and International Telecommunications Corporation (1938–45)
JACAR	Japan Center for Asian Historical Records
JMFA	Japanese Ministry of Foreign Affairs
JSDB	Jiaotong shi bianzhuan weiyuanhui, *Jiaotong shi dianzheng bian*
JTTCC	Japan Telegraph and Telephone Construction Company
JWT	Japan Wireless Telegraph Company
KDD	Japan International Telegraph and Telephone Company
KDT	*Kachū denki tsūshin*
KDTKKS	Kokusai denwa kabushiki kaisha, *Kokusai denki tsūshin kabushiki kaisha shashi*
KSS	Ishikawa Junkichi, comp., *Kokka sōdōin shi*
KSSS	Ishikawa Junkichi, comp., *Kokka sōdōin shi shiryō*
MOC	(Japanese) Ministry of Communications
IMTFC	Inner Mongolia Telecommunications Facility Company
MTT	Manchurian Telegraph and Telephone Company
NCTA	North China Telecommunications Administration
NCTT	North China Telegraph and Telephone Company
NEC	Nippon Electric Company
NGNB	*Nihon gaikō nenpyō narabini shuyō bunsho*
NHK	Nippon hōsō kyōkai
NLC	non-loaded cable
PBX	private (telephone) branch exchange
RCA	Radio Corporation of America
SACS	Special Account for Communications Service
SCAP	Supreme Command of the Allied Powers

SMR	South Manchuria Railway Company
TDTZ	*Tōa denki tsūshin zasshi*
TKZ	*Teishin kyōkai zasshi*

Epigraph Sources

Part I (p. 15)

Fukuzawa Yukichi, "Denshin kaigyō no shukuji," reprinted in Wakai Noboru and Takahashi Yūzō, *Terekomu no yōake*, 97.

Shigemitsu Mamoru, in Teishinshō denmukyoku, *Nisshi tsūshin kaigi gijiroku*, 2.

Part II (p. 85)

Charles Bright, "Imperial Telegraph," *Quarterly Review* (April 1903), included in his *Imperial Telegraphic Communication*, 82–83.

Watanabe Otojirō, *Denki tsūshin kokusaku to denki tsūshin jigyō*, 283.

Part III (p. 207)

Sugitani Hidenosuke, "Nihon shinkō tsūshingaku taibō ron," *DT* 5.21 (1942): 12–15; also published in *TDTZ* 2.5 (1942): 19–23.

G. J. Mulgan, *Communication and Control*, 4.

Part IV (p. 315)

Matsumae Shigeyoshi, "Daitōa kensetsu to kōtsū tsūshin seisaku," *Senji seisan ron*, 142.

Johan Galtung, "Structure of Imperialism," 82.

Technology of Empire

Telecommunications and Japanese Expansion in Asia, 1883–1945

INTRODUCTION

Shortly after 12:00 o'clock Tokyo Time on August 15, 1945, the prerecorded voice of Japan's Emperor Hirohito was broadcast from a studio in downtown Tokyo. "After pondering deeply the general trends of the world and the actual conditions obtaining in Our Empire today," the emperor solemnly declared, "We have decided to effect a settlement of the present situation by resorting to an extraordinary measure." Observing that "the war situation has developed not necessarily to Japan's advantage, while the general trends of the world have all turned against her interests," the 44-year-old monarch called on his "100 million subjects" to "bear the unbearable."[1] The war had finally come to an end, in Japan's defeat.

Though lasting a mere 4 minutes 37 seconds, this unprecedented radio broadcast of the emperor's own speech was one of the defining moments in Japan's modern history. Even though many Japanese could not clearly hear the emperor's voice or fully understand his archaic language, nearly all listeners were overcome with profound emotions as defeat finally dawned on them.[2]

Understandably, much has been written about the political significance of the broadcast as well as its profound psychological impact on the Japanese people. Few, however, have ventured to comment on the

---

1. Text of Hirohito's Radio Rescript, *New York Times*, August 15, 1945. A program of the broadcast is included in Nippon hōsō kyōkai, *Hōsō 50-nen shiō*, 305–7. For a comparison of various English translations of the speech as well as the actual implementation of the broadcasting, see Kitayama Setsurō, *Zoku Taiyeiyō sensō mediya shiryō I*, vol. 2.

2. Only once before and then accidentally, in 1928, had the Japanese authorities allowed the emperor's voice to be broadcast over the radio. See Takeyama Akiko, *Gyo'on hōsō*, 13. For an analysis of the broadcast and its impact, see Takeyama, *Gyo'on hōsō*; and Sato Takumi, *Hachigatsu jūgonichi no shinwa*. In English, see Pacific War Research Society, *Japan's Longest Day* (Tokyo: Kodansha, 1973).

unprecedented geographical scale of the broadcast and the vital role of modern communications technology. The emperor's speech was not heard only on the four main Japanese home islands but also relayed and broadcast simultaneously in nearly the entire Asia Pacific region. From tropical jungles in Southeast Asia to rural settlements in Manchuria, from the colonies of Korea and Taiwan to metropolises in occupied China, hundreds of thousands of Japanese civilians and troops, like their compatriots on the war-ravaged home islands, heard the voice of their divine ruler at precisely the same moment.[3]

That tens of millions of Japanese scattered over such a vast area were able to experience those intense feelings simultaneously can only be considered a communication spectacle, made possible by Japan's vast communications network at home and throughout Asia. Japan's regular radio programs early that morning had announced an important speech upcoming at noon. Notifications were sent by telegraph or telephone to far-flung locations in advance to ensure that the broadcast reached its maximum audience. It is one of history's greatest ironies that the most impressive display of Japan's empire-wide communications network ushered in its own collapse as well as that of the Japanese empire.

Seen this way, the emperor's broadcast on August 15, 1945 serves as a reminder of the pivotal link between empire and communication in Japan's imperial expansion in Asia.

COMMUNICATION AND EMPIRE

Defined as a process that involves set arrangements and media that must be in place for any relay of information to occur,[4] communication is indispensable to any organization and thus often considered the "nerve system" of government and society. Since the dawn of human

3. According to Kitayama Setsurō's *Rajio Tōkyō*, the only exceptions were the Japanese military authorities in the Dutch East Indies, who refused to relay the broadcast for local stations. See also Hibi Tsuneaki, "'Gyo'on hōsō' o doko de kikimashitaka?" *Shinchō 45*, no. 220 (August 2000): 136–45; Takeyama, *Gyo'on hōsō*, 52–62; and Hanakada, *Shōwa 20-nen 8-gatsu*, 178–92. Some in the home islands also missed the broadcast.

4. Alleyne, *International Power and International Communications*, 3.

history, information has been transmitted over distances via humans, animals, or optical signaling. Together with transportation, advances in communications technology help reduce the barrier posed by sheer space. As such, communications have historically functioned as one of the most effective "space-adjusting" technologies.

Thanks to the discovery of electricity, the inventions of electric telegraph and telephone in the nineteenth century revolutionized point-to-point communication when "electricity freed communication from the constraints of space and from the limits of physical transport."[5] Even though the term "telecommunications" was not officially introduced until the International Communications Conference in Madrid in 1932, telecommunications technology played a key part in what sociologist James Beniger calls the "control revolution" of the late nineteenth century—a complex of rapid changes in the technological and economic arrangements by which information is collected, processed, and communicated, and through which formal or programmed decisions might effect social control.[6] As media historians increasingly acknowledge, telecommunications launched the information society we know today.[7]

Empire, which comes from Latin "imperium" meaning command, authority, rulership, power, has been the largest organization in human history. Empire is also a spatial construct. In other words, empire-building in a fundamental sense is a project of producing and controlling imperial space.[8] Communication scholars beginning from Harold A. Innis, the late Canadian economic historian, have pointed out what is perhaps obvious: that the geographical limits of empires are determined by the possibilities for effective communication, and that changes in the technology of transport and communications have permitted vast changes in the possibilities for the extension of empires.[9] It is in this sense that communications technologies become the technology of empire.

---

5. Mulgan, *Communications and Control*, 33. Throughout this study, the plural forms of "communication" and "telecommunication" are used except when describing the process and the act.

6. Beniger, *The Control Revolution*, vi.

7. Marvin, *When the Old Technology Was New.*

8. Colas, *Empire*, esp. 31–35; see also Maier, *Among Empires*, 24–25.

9. Melody, "Introduction," *Culture, Communication and Dependency*, 5–6; Innis, *Communication and Empire.*

The link between telecommunications and modern European empires has drawn much scholarly attention in recent decades. Starting with his *Tools of Empire*, historian Daniel Headrick has demonstrated telecommunications to be among the several key technological developments most crucial to European imperialism, "either by making imperialism possible where it was otherwise unlikely, or by making it suitably cost-effective in the eyes of budget-minded governments." As telecommunications gave value to a handful of mostly deserted islands in the most isolated parts of the world, and in many instances helped empires to expand, as Headrick puts it, the "web of power that tied the colonial empires together was made of electricity as well as steam and iron."[10] Recent works also debate the extent to which imperial rivalries had shaped the history of telecommunications. Bringing together geopolitics and technology, geographer Peter Hugill uses *geostrategy* to describe the ways a government mobilizes its resources—political, military, economic, and certainly technological—to overcome geographical limitations on the exercise of its power and to turn geography to its own advantage. Indispensable to the survival of the imperial powers, according to Hugill, control over telecommunications networks was often crucial in hegemonic struggles among them.[11]

More than 60 years after its demise, Japan's modern empire has generated a new wave of academic and public interest. The scope of scholarly explorations now embraces a much broader range of themes and subjects, and social and cultural studies of Japanese colonies in particular have dethroned earlier works that privileged the political or economic policies of the imperial center.[12] Although questions about the motivations for Japanese imperialism and its techniques of expansion have lost some of the earlier pre-eminence, they are by no means abandoned. Whereas many in the past saw Japan's empire-building as an un-

---

10. Headrick, *Tools of Empire*, 160–64; Headrick, *Tentacles of Progress*, 98. See also his later work, *The Invisible Weapon: Telecommunication and International Politics, 1851–1945*. See also Ferguson, *Empire*, 140–42.

11. Hugill, *Global Communications Since 1844*. In their 2007 book, *Communication and Empire*, Dwayne Winseck and Robert Pike caution against overestimating the influence of imperial rivalry in international telecommunications before 1930 and instead call for greater attention to globalizing capitalism and interdependence.

12. For surveys of recent scholarship, see Wilson, "Bridging the Gap: New Views of Japanese Colonialism, 1931–1945"; and Y. Tak Matsusaka, "The Japanese Empire."

folding of ultranationalist ideas, recent scholarship tends to emphasize Japanese discursive efforts to appropriate international norms or to locate "Japan's Orient" in Asia.[13] Japan's quest for security has been often cited as a leading factor behind Japan's overseas expansion,[14] but recent scholarship has broadened the inquiry to include considerations of strategic resources.[15] No longer slighting its economic roots, many recent English-language works on Japanese imperialism have explicitly rejected mono-causal explanations, including economic determinism.[16] As a result, Japanese overseas expansion is increasingly seen to involve much more than just open use of force or a quest for either security or profits; rather, it is viewed as a multifaceted project that also entails social and cultural mobilization.[17]

Such welcome trends notwithstanding, the material means of either building Japan's empire or holding it together are still largely taken for granted rather than being thoroughly investigated. Even the new scholarship on colonial science, urban planning, and medicine—an encouraging development—has not fully remedied the situation.[18] Writing

---

13. Some works are attuned to the more idealistic aspects; see, e.g., Berger, "The Three-Dimensional Empire," 355–83; and Lebra, "Postwar Perspectives on Japan's Greater East Asia Co-Prosperity Sphere." Recent examples include Stefan Tanaka, *Japan's Orient*; and Alexis Dudden, *Japan's Colonization of Korea: Discourse and Power.*

14. See, e.g., the multivolume studies by Japanese scholars in the early 1960s, *Taiheiyō sensō e no michi*. They were translated into English beginning in the 1980s, under the editorship of James William Morley.

15. "Since Japan had to pursue a raw-materials strategy," notes historian Ian Nish, "it may have been pushed into trying to establish control over lands which offered it the raw materials it needed and into drawing up guidelines for the Greater East Asia Co-Prosperity Sphere in the early 1940s" (Nish, "Some Thoughts on Japanese Expansion," in Mommsen and Osterhammel, eds., *Imperialism and After*, 83). Michael Barnhart (*Japan Prepares for Total War*) has emphasized the strategic planning for economic mobilization among military officers such as Ishiwara Kanji and their civilian fellow-travelers, largely to fulfill purposes of waging total war based on lessons drawn from the Great War in Europe.

16. Beasley, *Japanese Imperialism, 1895–1945*.

17. See, e.g., Duus, *The Abacus and the Sword*; and L. Young, *Japan's Total Empire*.

18. This trend is marked by the newly launched journal *East Asian Science, Technology and Society*. In Japanese, one volume of the recently published eight-volume *Iwanami kōza "Teikoku" Nihon no gakuchi* is devoted to science and technology. Yet one of its contributors, Japanese historian Iijima Wataru, laments the fact that scholarship on the Japanese empire has slighted material culture (ibid., 7: 39).

more than three decades ago, economic historian Kozo Yamamura remains perhaps the lone voice calling for greater attention to the link between Japan's technological capability acquired during the 1920s and its decisions to adopt an expansionist policy and wage war in the 1930s. Yamamura notes that although such decisions are made for many reasons—conflicting national interests, domestic politics, emerging militarism, human fragility, and irrationality—a nation's military capability is directly determined by its technological capability.[19] Few historians have taken up his suggestion of investigating such linkages in the crucial decade of the 1930s, let alone examining the role of technology in the broader context of Japan's empire-building in modern times. As a result, scholars of Japanese overseas expansion have occasionally marveled at the "sheer terms of square miles of territories occupied" during its half-century history,[20] but have taken for granted the technologies essential to conquering and controlling such territories.

To fully appreciate the critical importance of such technologies, it is helpful first to pause and consider the reality of Japan's imperial space. At the zenith of Japan's overseas expansion, Japan's imperium—a shorthand for its formal and informal empires—encompassed nearly all of Pacific Asia. In 1942, the so-called Greater East Asia Co-Prosperity Sphere combined home islands, colonies, and client states and occupied territories of immense proportions.[21] Just as important, the tempo of imperial Japan's spatial expansion since the early 1930s had been spectacular by any standard. As the semi-official *Japan Year Book* proudly declared in that same year, "The territories that have been occupied by the Imperial forces of Nippon, plus occupied areas on the Chinese continent, are about 10 times the size of Nippon Proper."[22] In other words,

19. Yamamura, "Japan's Deus ex Machina," 65–95. Interestingly, it is several political scientists that have included technology, together with population and access to resources, as one of the "master variables" in discussing Japan's national expansion; see, e.g., Choucri, North, and Yamakage, *The Challenge of Japan Before World War II and After.*

20. E.g., Nish, "Some Thoughts on Japanese Expansion," in Mommsen and Osterhammel, eds., *Imperialism and After*, 87; and Hata Ikuhiko, "Continental Expansion, 1905–1941," *Cambridge History of Japan*, 6: 314.

21. Throughout this book, the term "imperium," defined by *Random House College Dictionary* as an "area of dominion, sphere of control or monopoly," is used interchangeably with "wartime empire."

22. Foreign Affairs Association of Japan, *The Japan Year Book, 1943–44*, 297.

Table 1
Land and Population in Japan's Imperium, 1945

Region	Land area (000 km²)	Population (000s)	Under Japanese control since
Japan	382.6	71,420	
Taiwan	36.0	5,872	1895
Karafuto (Southern Sakhalin)	36.0	415	1905
Kwantung	3.5	1,367	1905
Korea	220.8	24,326	1910
Nan'yō	2.1	131	1918
SUBTOTAL IN 1931	681.0	103,531	
Manchukuo	1,303.0	43,203	1931
Mōkyō (Inner Mongolia)	615.4	5,508	1937
Republic of China (North China)	602.7	116,306	1937
Reformed government (Central & South China)	350.1	78,644	1937
Thailand	620.0	15,718	n/a
French Indochina	630.0	23,854	1945
British Malaya	136.0	5,330	1942
British Borneo	211.0	931	1942
Burma	605.0	16,119	1942
Dutch East Indies	1,904.0	60,727	1942
Philippines	296.3	16,000	1942
TOTAL	7,273.5	382,340	

SOURCE: Kobayashi Hideo, *Daitōa kyōeiken*, 43.

the decade after Japan's 1931 invasion of Manchuria saw the landmass under Japan's domination increase tenfold (see Table 1).

The spatial implication of Japan's rapid expansion was not lost on contemporaries in Japan. Nakayama Ryūji, president of Japan's Telecommunications Association, carefully made his own calculation and concluded that the Greater East Asia Co-Prosperity Sphere actually encompassed some 44.4 million sq km, or 10.7 percent of the Earth's land surface, a staggering 40 times the size of Japan's home islands. As Nakayama was quick to add, this was not empty space, but with a combined population of 674.6 million, or roughly one third of the population of the entire world. As he saw it, such a vast imperial space as well as the size and ethnic complexity of its population posed an immense challenge to an insular nation such as Japan.[23]

TECHNO-IMPERIALISM

This book examines modern Japan's endeavor to cope with this challenge, focusing on telecommunications. As becomes clear in the following chapters, telecommunications technology not only played a key role in Japan's state-building and economic development at home but also proved essential to its overseas expansion in modern times. As a technology of empire, telecommunications facilitated Japan's empire-building *strategically* in a number of ways. In times of military engagements, telecommunications facilities—whether in the form of a crucial submarine telegraph link or a wireless network—played a decisive role in gaining victory. In peacetime, telecommunications infrastructure such as a local telegraph office advanced Japan's overseas commercial interests and strengthened its ability to gather and disseminate news and information in Asia. In territories that came under Japanese control, communications technology was both a crucial tool of administration and suppression and also a harbinger of colonial modernity. More important, as Japan embarked on its quest for an autarkic imperium in Asia in the 1930s, Japanese communications engineers and administrative bureaucrats took the lead in reinventing technologies and techniques to cope with Japan's new spatial and other challenges. Major innovations in telecommunications technology in turn promised to create an integrated imperial space through an expanding imperial telecommunications network. This strategic practice of designing or using technology to advance empire-building goals can be best described as "techno-imperialism."[24]

---

23. Nakayama Ryūji, "Shin-Tōa kensetsu to denki tsūshin jigyō," later included in his *Sensō to denki tsūshin.* As the figures in Table 1 show, Nakayama might have underestimated the land area while overestimating the population under Japan's control.

24. This term has appeared in a number of recent works starting with Matsusaka's *Making Japanese Manchuria.* My definition is broader than the one used recently by David Wittner, who limits it to the "use of military technologies to facilitate territorial acquisition in an effort to find both the raw materials which support, and markets for, the aggressor's industries" (*Technology and the Culture of Progress in Meiji Japan*, 172n10). In his 2006

The technology of empire involves not only artifacts but also the body of skills, knowledge, and practice that make them work. Here technology is best seen as what historians of technology term "technological regimes," namely, "linked sets of individuals, practices, institutions."[25] To understand the dynamics of Japan's techno-imperialism, therefore, one must scrutinize its ideology and grand strategy, its political economy and organizational structure, as well as its human agents.

A case study of Japan's techno-imperialism in telecommunications provides an opportunity to explore the ideological context of technological development in modern Japan. As a cutting-edge technology, telecommunications is a perfect subject for studying changing Japanese attitudes toward modernity. Examining technology in Japan's overseas expansion allows us to build on but also go beyond what political scientist Richard Samuels has usefully called the phenomenon of "techno-nationalism"—the use of technology to enhance national security—and to probe Japan's discourse of technological leadership in Asia.[26]

Since bureaucratic institutions, financial structures, and modes of ownership have an enormous impact on the operations of telecommunications, a study of telecommunications and overseas expansion helps capture the broader dynamics of modern Japanese history. In particular, it reopens the question about the role of the state in Japan's economic development, an issue that preoccupied the first generation of postwar Japanese scholars of telecommunications history and has generated lively discussion outside Japan since the 1980s.[27]

---

work, Chinese scholar Liang Bo defined "technological imperialism" as imperialism that deploys science and technology for aggression and colonial rule, exploitation of wealth and resources, and control of ideas, culture, and even ways of life (*Jishu yu diguozhuyi*, 4).

25. For a succinct introduction to the difficulties in defining "technology," see Hughes, *Human-Built World*, 2–5, 175–77. On "technological systems," see Gabrielle Hecht, *The Radiance of France: Nuclear Power and National Identity After World War II* (Cambridge, Mass.: MIT Press, 1998), 56; and Hughes, *Network of* Power, 465. A valuable work that studies Japan's technological development through a social-network approach is Morris-Suzuki, *The Technological Transformation of Modern Japan.*

26. Samuels, *"Rich Nation, Strong Army,"* 42. Samuels largely focuses on military technologies such as the armament and aircraft industries. The recent work by Hiromi Mizuno, *Science for the Empire*, is an attempt to analyze the discourse of what she calls "scientific nationalism."

27. Johnson, *MITI and the Japanese Miracle.* More recently, see Meredith Woo-Cumings, ed., *The Developmental State* (Ithaca: Cornell University Press, 1999). In this

Moreover, a study of techno-imperialism in modern Japan sheds light on its human face—the Japanese engineers, technicians, and others who championed the cause of technology for empire. After all, it was their vision and action that shaped the contours of Japan's technological development and, to a significant extent, its imperial expansion. Equally important, telecommunications also provide an interface to explore the complex relationships between the Japanese and other peoples in the empire. The native population in Japan's colonies, like the Chinese and others in Asia during wartime occupation, were not simply passive objects of Japanese domination. As customers, employees, and sometimes technical assistants involved in Japan's empire-wide telecommunications operations, their experience opened up new venues to examine issues of colonial modernity and collaboration, as well as technological transfer in the age of imperialism.

An examination of Japan's imperial telecommunications network brings technology back to the study of Japanese imperialism. Moreover, it also adds a much-needed systemic perspective to the study of Japanese empire. Since the empire has been largely examined as separate units rather than as a dynamic whole, how its different parts related to one another remains poorly understood.[28] As its overseas empire not only became more spatially extended but also organizationally more complex, Japan sought to transform existing colonial and international telecommunications links into an integrated imperial network with Japan at the center. A study of Japan's evolving imperial telecommunications system thus yields a better appreciation of the inner dynamics of Japan's empire.

This book consists of ten chapters, subsumed under four sections that are largely chronological but also thematic. Part I is a schematic overview of the early history of telecommunications in the emergence of modern Japan and its nascent empire. It situates Japan's later quest for regional hegemony in its historical, political, and technological contexts, focusing on the role of the state; the dualistic aspects of Japan's dependency on the West for technology and its quest for autonomy; and

sense, my use of the term "technology" is also similar to Kenneth Pyle's in "The Technology of Japanese Nationalism."

28. Many edited volumes take up the empire as a whole, but few monographs do so. One exception is Jennings, *The Opium Empire*.

the importance of submarine telegraphy in overseas expansion. Ending with the seminal events in Manchuria in 1931, Chapter 2 demonstrates that Japan's experience with wireless communications reflects both the success and the failure of Japan's imperial strategy as well as the limits of its institutional adaptation.

Part II places telecommunications technology in the larger context of Japan's renewed continental expansion in the 1930s and emphasizes the technological underpinnings of Japan's new empire. Chapter 3 analyzes the creation of Japan's first semi-public, full-service telecommunications enterprise in the newly created puppet state of Manchukuo as well as its influence on Japan's telecommunications operations in occupied areas in China after the outbreak of war in 1937. Chapter 4 explores the process and wider implications of a key innovation in long-distance telecommunications technology in the context of Japan's evolving imperial agenda in Asia. Chapter 5 links evolving designs of telecommunications networks to the government's expanding imperial vision of an integrated East Asian sphere under Japan's leadership. Although technology did not drive history, advances in communications technologies did provide the crucial milieu in which the new strategic visions of Japan's imperium took shape.

Whereas Part II deals with innovations and blueprints, Part III examines their implementation as well as organizational adaptations from the perspective of asserting control both at the imperial center and in the periphery. Chapter 6 discusses how the bureaucracy attempted to reshape the communications service in response to the new demands as well as to the new frictions in the old empire. This led to the creation and reorganization of the International Telecommunications Company (ITC) as well as further control over remaining foreign communications interests at home. As Chapter 7 shows, managing telecommunications in occupied China involved coping not only with Chinese resistance and competition from Western firms but also with deeply entrenched Japanese interests. Japan's further expansion into present-day Southeast Asia before and during the Pacific War, examined in Chapter 8, brought new kinds of challenges to Japan's technological and organizational capabilities as well as its political influence.

With the vast area of China and Southeast Asia incorporated into the imperium, the vision of the Greater East Asia Co-Prosperity Sphere finally found its ultimate expression in an expanded imperial communications network spanning the entire region. Part IV examines the formation and operations of Japan's wartime imperial telecommunications network as a single system. The increasing organizational complexity of Japan's imperium and the divergent interests of some of its constituents are the subject of Chapter 9. Added to this organizational stalemate was overreach of technological capacity, which further hastened its gradual meltdown. But as Chapter 10 reminds us, the telecommunications network did function as an imperial "nerve system" and "artery," and not all was lost with the demise of Japan's empire. Parts of the imperial infrastructure as well as aspects of the empire-building experience continued to exert influence on the postwar world, both at home and in Asia. The book concludes by situating Japan's experience with deploying communications technology for overseas expansion in the broader discussion of technology of empire as well as of communication imperialism. It also suggests implications for understanding communications and control in the present.

The nexus of communication and empire in modern Japan has multiple facets that cannot be adequately examined in a single study. This book focuses on point-to-point, two-way communication via electronic means and treats other aspects of communication such as radio broadcasting and military and postal communication in a cursory manner.

Using essentially the same technology as wireless telephony, radio broadcasting primarily serves the purpose of one-way dissemination of information and plays a key role in mass communication and propaganda. Organizationally and functionally, it remained largely separate from telecommunications, their dramatic convergence on August 15, 1945, notwithstanding.[29] Therefore, broadcasting is discussed as part of the broadly defined telecommunications network in this book, but the

29. There were exceptional occasions when radio broadcasting was used primarily to send a message to a particular party. For instance, after the Combined Fleet began sailing toward Hawaii, Japan's Imperial Headquarters relied on overseas broadcasting to inform the fleet of action plans. And in August 1945, both the Japanese and the Allied governments used overseas news broadcasting to announce their decisions about acceptance of the Potsdam Declaration. Broadcast on the evening of August 10, Japan's reply reached President Truman ahead of the diplomatic telegrams sent via Switzerland and Sweden (Tsūshinsha shi kankōkai, *Tsūshinsha shi*, 554–56). The emperor's speech blurred that distinction for the last time in the war.

all-important issues of broadcast programming and of audience must be left to another study.[30] This book addresses military communications insofar as they had a strategic dimension. Tactical military communications, certainly a vital part of Japan's overseas expansion, are a subject more appropriate for military historians studying particular combat operations. Similarly, communications security, including issues such as ciphers, code-breaking, and other "software" of the communications system, has been the subject of numerous studies[31] and is addressed here only briefly. Postal communication is obviously an essential part of the information system in modern Japan, affecting more people directly than the use of telegraph and telephone. Technologically, there is a fundamental difference between the postal service, which is completely dependent on means of transportation—rail, shipping, or aviation—and electric communication, which has taken a revolutionary break from that constraint. In this sense, telecommunications are considered more effective in annihilating space and time, and thus more appropriately regarded as modern Japan's technology of empire.

30. In English, one may turn to Jane Robbins's *Tokyo Calling* for a start.

31. See, among others, Drea, *MacArthur's Ultra.* For an argument that errors and distortions in the U.S. translations of intercepted Japanese diplomatic communication contributed to the misunderstandings in crucial negotiations, see Keiichiro Komatsu, *Origins of the Pacific War and the Importance of Magic* (New York: St. Martin's Press, 1999).

PART I

Genesis, 1853–1931

Although there have been many inventions in recent years, nothing is greater than the telegraph. . . . When the telegraph serves as the nerve system of a country, the Central Telegraph Office is like the brain, and branch offices elsewhere are like nerve ends. As Japan sharpens its new nerve system, its body gains new vitality.

—Fukuzawa Yukichi, 1875

The present world is entering the era of "speed." One can speak of neither politics nor economy without the concept of time. Especially in relations between countries, both politically and economically, communication and transportation are essential. That is to say, the speedier the communication and transportation, the more thorough the exchange of ideas and closer the relationship.

—Shigemitsu Mamoru, 1930

CHAPTER I

An Emerging Empire in the Age of Submarine Telegraphy

In February 1854, an American squadron of eight all-black warships led by Commodore Matthew Perry returned to Edo Bay in Japan. Over the next two months, Perry and associates pressed the Shogun's government for negotiations to open the island-country to trade. In the meantime, the commodore presented gifts carefully selected to impress the Japanese. Among them were two telegraph sets, complete with wires and batteries. The official record of the expedition contained a vivid description:

> [Posts] were brought and erected for the extension of the telegraph wires, the Japanese taking a very ready part in all the labors, and watching the result of arranging and putting together the machinery with an innocent and childlike delight. The telegraphic apparatus, under the direction of Messrs. Draper and Williams, was soon in working order, the wires extending nearly a mile, in a direct line, one end being at the treaty house, and another at a building expressly allotted for the purpose. When communication was opened up between the operators at either extremity, the Japanese watched with intense curiosity the *modus operandi*, and were greatly amazed that in an instant of time messages were conveyed in the English, Dutch, and Japanese languages from building to building. Day after day the dignitaries and many of the people would gather, and, eagerly beseeching the operators to work the telegraph, watch with unabated interest the sending and receiving of messages.[1]

---

1. Matthew Galbraith Perry, *The Japan Expedition, 1852–1854* (Washington, D.C.: Smithsonian Institution, 1968), 357. Placing the encounter in the broader context of American belief in technological superiority are Yakup Bektas, "Displaying the Ameri-

The first public demonstration in Japan of a working electric telegraph and other technological gadgets, which would inscribe Japan into a position of technological inferiority in the minds of both Westerners and Japanese,[2] seemed to have produced its intended effect. Weeks later, the Shogun's government signed the Treaty of Kanagawa and formally ended two centuries of Japan's self-imposed seclusion. Interestingly, just as the broadcasting of the emperor's speech in 1945 signaled the end of an era in Japan's modern history, a starting point in that history nearly a century earlier coincided with the introduction of modern communications technology.

If telecommunications in modern Japan seem to have acquired a definite origin in this encounter, the course of its subsequent development was by no means predetermined. As we shall see, a decade and a half would pass before the telegraph service was started in Japan, a consequence less of technology and more of politics. Moreover, in contrast to the United States, where the telegraphic service was left to the private sector, the Japanese government would establish a comprehensive and long-lasting monopoly over domestic telegraph services. Lastly, while initially dependent on foreign telegraph companies for international communications, Japan successfully sought to enhance its autonomy and even secure advantageous positions in neighboring countries in a few decades. None of these developments was harmonious. On both the domestic and the international front, tension and conflict characterized the development of telecommunications service in modern Japan, reflecting the contested nature of the state's role in the domestic economy as well as Japan's new place in the world.

---

can Genius: The Electromagnetic Telegraph in the Wider World," *British Journal for the History of Science* (June 2001): 199–232; and Adas, *Dominance by Design*, 1–31. For Japanese accounts of this first encounter with the telegraph, see Watanabe Masami, *Nihon denshin denwa sōgyō shi*, 1–22; and especially, Kawanobe Tomiji, *Teregarafu komonjo kō*, 1–172.

2. As the Americans duly noted, the Japanese were not completely ignorant; a number of them had read about the new device in books imported from the Netherlands. For a detailed discussion of the Japanese encounter with the telegraph via Dutch publications, see Kawanobe, *Teregarafu komonjo kō*, 173–280.

TELECOMMUNICATIONS AND THE RISE OF MODERN JAPAN

Beginning of Telecommunications in Japan

After the Shogun's government gave in to Western pressure to set up diplomatic representation and settlements in its ports and to open Japan to foreign trade, reliable and efficient international communication became a necessity. As part of the so-called unequal treaties, which typically included extraterritoriality and reduced tariffs for imports, the United States, Britain, and France insisted on maintaining their own postal services in their settlements in Japan, as they had done earlier in China, on the grounds that the Japanese service was inadequate. Through their government-subsidized steamship routes, Western shipping companies linked the once-closed island country to the global postal communication system.[3]

In the meantime, the nearly instantaneous telegraphic communication was extending its tentacles throughout the hemisphere. In the decades after Samuel Morse's 1844 telegraph experiment between Washington, D.C., and Baltimore, Maryland, the world experienced what can only be described as a global communication revolution. In 1851, the first operational submarine telegraph cable was laid across the English Channel between Calais and Dover. By the end of the decade, the trunk-line telegraph network in Europe was complete. The 1860s saw rapid expansion of telegraphic lines outside Europe. In 1869, the Indo-European Telegraph Company completed the line between Britain and its Crown Colony of India. A year later, the British Indian Company connected Alexandria and Bombay. Competition was rampant, not only between actual lines in operation but also between companies seeking concessions from foreign governments to land submarine cables or to build land lines. As a result, mergers or pooling agreements among the cable companies became common. In 1869, the Great Northern Telegraph Company (GNTC) was founded in Copenhagen following the merger of three companies that engaged in telegraphic communication

3. Takahashi, *Tsūshin*, 46–51; Shinohara Hiroshi, *Gaikoku yūbin kotohajime* (Tokyo: Yūkyo saabisusha, 1982).

in northern Europe. In 1873, three British companies operating in Asia joined forces to become the Eastern Extension, Australasia, and China Telegraph Co., Ltd. Extending telegraph communications to China and Japan by way of Siberia and India, respectively, these two telegraph companies were to dominate international communication in East Asia for decades to come.[4]

Even before the Meiji Restoration of 1868, the Shogun's government and a few domains had shown interest in the use of telegraph and placed an order for telegraph machines in Europe.[5] A few Japanese and foreigners even demanded telegraphic service. Soon after its establishment, the new Meiji government received a similar proposal from Terashima Munenori, governor of Kanagawa prefecture, whose jurisdiction included the important port of Yokohama. Terashima had studied the workings of telegraphy in his native Satsuma domain and had toured Europe in 1861 in a Bakufu-sponsored mission, where he had observed the telegraph in action. Keenly aware of the need for swift communications between his own office in Yokohama and government authorities in Tokyo, he suggested first setting up telegraph lines between these two cities. In December 1868, the Meiji government decided to begin telegraph service in Japan and put Terashima in charge of the proposed Tokyo–Yokohama line.[6] On January 26, 1870, Japan's first telegraphic line went into operation. Initially, telegrams in Japanese and in European languages could be sent only between two telegraph offices—one in Tsukichi in Tokyo, and one at the Court House in Yokohama—every day between 8:00 A.M. and 8:00 P.M. Those not in the vicinity of these two offices had to rely on messengers to deliver the telegrams.[7] Spanning 32 km and supported by some 600 poles, the telegraph line became the first modern public facility introduced by the Meiji government, two years before the first railroad between Tokyo and Yokohama or the first gaslights in Japanese cities.

---

4. On the activities of these two companies in East Asia prior to World War I, see Ahvenainen, *The Far Eastern Telegraphs.*

5. Wakai and Taahashi, *Terekomu no yoake,* 46–60.

6. On Terashima's role in the establishment of the telegraph in Japan, see Takahashi, *Nihon denki tsūshin no chichi*; and Terashima Munenori kenkyūkai, *Terashima Munenori kankei shiryōshū,* 1: 453–629.

7. Watanabe, *Nihon denshin denwa,* 57–65; Takahashi, *Tsūshin,* 160–61.

Demand also arose for international telegraphic communication. As the Meiji government began an ambitious program of modernization modeled after that of the advanced Western nations, it dispatched officials and students to the West and set up diplomatic representations overseas. Foreign trade continued to expand. All this called for more rapid and reliable communication than that afforded by the postal service relying on steamships. In September 1870, the two-year-old Meiji government entered into an agreement with a Danish diplomat representing the Danish firm GNTC. The company was to construct submarine telegraphic cables linking Nagasaki with Shanghai and Vladivostok, as well as a cable between Nagasaki and Yokohama. The agreement allowed the GNTC to land cables on Japanese soil and to establish a telegraph office in Nagasaki. A year later, the GNTC completed the first two cables, 909 km and 1,430 km respectively, thus linking Japan to the expanding worldwide telecommunications network for the first time.

Perry's generous gifts notwithstanding, modern technologies such as telecommunications came to Japan at a price. Like nearly all other modern enterprises, Japan had to rely on foreign technical know-how in construction and maintenance of its first telegraph. The Japanese government hired British engineers to build the land line between Yokohama and Tokyo.[8] Japan's dependency was even more pronounced in international communications. The new Meiji government had neither the financial resources to invest in costly submarine cables nor the technological ability to build and repair them. This gave the GNTC of Denmark considerable bargaining power. Although the Danes initially attempted to secure a total monopoly on Japan's telegraphic communication with the Asian mainland, they failed, partly due to strong opposition from the British Foreign Office, which sought to protect the interests of the British company. As a result, the 1870 agreement between the Meiji government and the GNTC stipulated that the latter would "not make any complaint even if the Japanese government gave permission to others to begin the same service." However, if Japan

8. Nihon kagakushi gakkai, comp., *Nihon kagaku gijutsu shi daikei*, 53–55; Takahashi Zenshichi, *Oyatoi gaikokujin* (7).

granted more benefits to companies from other countries, the GNTC would be entitled to the same.[9]

As a result, Japan's very access to the international telecommunications network came at considerable cost of compromise to its autonomy. The Japanese government was more concerned with foreign influence over its domestic communications. Partly to prevent the GNTC from laying the Nagasaki–Yokohama cable in Japanese waters, the Japanese government rushed to construct its domestic telegraph network and did so at an impressive pace in the early 1870s. The 1,340-km Tokyo–Nagasaki trunk line on land was completed in 1873, followed two years later by the completion of the 825-km Tokyo–Aomori trunk line. Telegraph lines were constructed in Hokkaido and Kyushu simultaneously. It took only six years for the Meiji government to link all major Japanese cities, from Sapporo in the north to Kumamoto in the south, in a nationwide telegraphic communications network. In 1875, Japan's telegraph lines extended a total of 1,760 *ri* (6,912 km, or 4,294 mi) and carried 612,000 telegrams a year.[10]

Although the public was initially suspicious of the unfamiliar structures being erected all over the country, Japanese elites, both inside and outside the government, were quick to emphasize the great promise of telegraphic communications. Itō Hirobumi, then in charge of the Ministry of Engineering overseeing the telegraph construction, memorialized the Meiji emperor in 1875 that "telegraph lines are most effective in facilitating public administration and promoting human intelligence. Now that land lines extend from Tokyo to both Nagasaki and Aomori, from Hakodate to Kotaru in the north, even people in remote areas can benefit from our prosperity."[11] Itō's enthusiasm was shared by Fukuzawa Yukichi, the leading Japanese intellectual on modernization issues in the Meiji era. "When we think about the function of the telegraph," Fukuzawa observed with apparent excitement at the opening of the Tokyo Central Telegraph Office in 1878, "it is the nerve system of the country. The Central Office is like its head and branch offices are its nerve ends."

---

9. Nagashima Yōichi, "Taihoku denshin kaisha no Nihon shinshutsu to sono haikei." The text of the treaty is reprinted in Nihon kagakushi gakkai, comp., *Nihon kagaku gijutsu shi daikei* (19), 49–50.

10. Wakai and Takahashi, *Terekomu no yōake*, 76–78.

11. Quoted in Watanabe, *Nihon denshin denwa*, 71.

Exuding confidence, Fukuzawa claimed that "the distance of 1,700 *ri* has been reduced to naught. In other words, the body of the Japanese has been extended to all locales. Since there are also telegraph lines in foreign countries, not only Japan but the entire world will be shrunk and made more manageable."[12] Thus by more than 80 years Fukuzawa predated Marshall McLuhan, who famously characterized media as "the extensions of men." Confidence in the benefits of science and technology was indeed the hallmark of Japan's modern transformation.[13]

Modern Communications and National Transformation

The importance of modern communications networks to the transformation of modern Japan cannot be overstated. To be sure, the telegraphic operation was part of the larger modern information infrastructure then taking shape in Japan. Pre-industrial Japan had already developed courier services known as *hikyaku.* Dating from the Kamakura period (1192–1333), this system was widely used by the Shogun, daimyo, and even private business in the Tokugawa period (1603–1868). By relying on relay stations along the way, couriers could bring letters or money from Osaka to Edo in as little as three days, though at considerable expense.[14]

In 1871, the Meiji government replaced the existing courier services with a modern postal service modeled on that of Britain, which operated as a government monopoly with a nationwide flat rate.[15] The new post offices were initially concentrated in urban areas. As local business and commerce grew, the government extended the postal network to rural areas by establishing a nationwide third-class post office system that employed local landowners as postal employees. In the early 1880s,

---

12. Fukuzawa Yukichi, "Denshin kaigyō no shukuji," reprinted in Wakai and Takahashi, *Terekomu no yōake*, 97.

13. McLuhan, *Understanding Media: The Extensions of Men.* On the role of science and technology in Meiji Japan, see Low, *Science and the Building of a New Japan.*

14. Fujimura Jun'ichirō, "Jōhō dentatsusha hikyaku no katsudō," 311–48. For a study of domestic travel in premodern Japan, see Constatine Nomikos Vaporis, *Breaking Barriers: Travel and the State in Early Modern Japan* (Cambridge, Mass.: Council on East Asian Studies, Harvard University, 1994).

15. D. Eleanor Westney, "Building the National Communications System: Adopting and Adapting Western Organizational Models in Meiji Japan," 39–59; Takahashi, *Tsūshin,* 45–144.

there were 300 telegraph offices throughout the country, compared to over 4,000 post offices. Following unification of the postal and the telegraphic administrations in 1888, telegraphic service became available in many local post offices, and the next three years saw the number of telegraph offices increase by about 750. Beginning in 1903, due to the combined effect of a tight state budget and lobbying by local politicians, villages and towns could petition to have a telegraph office set up by paying the construction cost and part of the maintenance. As a result, by 1910, the number of telegraph offices in Japan had increased to nearly 2,000.[16] Beginning in 1885, a nationwide flat rate for domestic telegrams was adopted, essentially eliminating the factor of distance in telecommunications within the entire country.[17] Although still considerably more expensive than the postal service, telegraphic service was widely available and generally efficient.

Examining Meiji developments in education and conscription as well as in infrastructure, communication scholar Katō Hidetoshi has concluded that by the late 1880s a nation-wide "communication market" had emerged in Japan for the first time. Telecommunications as well as roads and railway, in his view, not only expanded this market but also deeply integrated its organization.[18] Economic historian Syndey Crowcour likewise has pointed out that, together with the postal system, telecommunications services had initiated an information-based society in Japan by the beginning of the twentieth century.[19] As Japan pursued the twin goals of "rich country, strong army," telegraphic communication came to play a unique and indispensable role in the country's politics and military affairs, economy, and cultural life.

The telegraphic network had a major impact on the economic development in Meiji Japan. Japan is probably the only nation to have had a

---

16. Fujii Nobuyuki, *Terekomu no keizai shi*, 43–47.

17. This was partly in response to business petitions. For a business petition, see "Denshin toriatsukaikata ni gokaisei o yosurugi ni tsuki kengi" (April 22, 1885), in *Shibusawa Eiichi denki shiryō*, 18: 282–83. Tsushima was not brought into the flat rate system until 1891, when the Japanese government purchased the portion of the cable owned by the Great Northern.

18. Katō, "Meiji 20-nendai nasonarizumu to komyunikeshon," in idem, *Bunka to komyunikeshon*, 153–77.

19. Sydney Crowcour, *Cambridge History of Japan*, vol. 5, *The Nineteenth Century*, ed. Marius B. Jansen, 399. For an evaluation of the postal network in early Meiji, see Sugiyama Shin'ya, "Meiji zenki ni okeru yūbin nettowaku."

complete domestic telegraphic network in place from the outset of its industrialization. As Fujii Nobuyuki has shown, by the end of the 1880s, commercial and transportation businesses became particularly dependent on an expanded telegraphic network. The extension of telegraphic service in Japan, as Fujii demonstrates, gave merchants all over the country a strong incentive to expand existing business relationships and open new ones. At the same time, as regional price differences began to disappear, some merchants were threatened by the new information network. International business, too, increasingly came to rely on overseas telegraphy. As Ishii Kanji and others have shown, the use of international telegraphy enabled trading companies such as the Mitsui Bussan to establish their own global information network, thus breaking the monopoly of foreign trading companies and gaining an advantage over their domestic rivals as well.[20] To economize on the relatively high cost of sending telegrams, Japanese business increasingly adopted commercial codes that had been used in the West.[21] As a result, the initial preponderance of government telegrams was soon replaced by an overwhelming number of private telegrams, most of which related to business activities. By 1879, for instance, 90 percent of domestic telegrams were paid for by private customers. A survey in 1886 and 1887 showed that, of some 4 million telegrams sent in Japan, 43 percent were for the purpose of commerce and industry, 8 percent were devoted to the exchange market, and 39 percent were for miscellaneous purposes. In 1892 and 1893, of 10 million telegrams, 63 percent were devoted to agriculture, industry, or commerce, 2 percent were for banking, and 34 percent were for miscellaneous purposes.[22]

The telecommunications network was closely intertwined with the development of mass media, contributing directly to the emergence of modern news agencies in Japan. The telegraph had been used to transmit news between the capital of Tokyo and the business center of

20. For case studies of use of the telegraph by individual businesses in Japan, see Fujii Nobuyuki, *Terekomu no keizai shi*, 127, 175; and Ishii, *Jōhō tsūshin no shakai shi*, 85–90; see also Suzuki Kunio, "Nihon zaibatsu no jōhō nettowaku: Mitsui zaibatsu o chūshin ni," unpublished paper presented at the Rikkyo University symposium on Globalization as History, March 14, 2004.

21. For one such internal code used by the Mitsubishi Steamship Company, see Nihon kagakushi gakkai, comp., *Nihon kagaku gijutsu shi daikei* (19): *denki tsūshin*, 61–63.

22. Ishii Kanji, *Jōhō tsūshin no shakai shi*, 111.

Osaka from about 1880. Seven years later, the country's premiere newspaper, the Osaka-based *Ōsaka Asahi shimbun*, proudly announced that "telegraph would replace postal service in transmitting news reports from Tokyo." On February 11, 1889, when the Meiji Constitution was promulgated, the Osaka-based newspaper was able to publish its entire text in an extra on the same day, thanks to the telegraph. The advantage over the postal service, which required at least three days to cover the distance between Tokyo and Osaka around that time, was obvious.[23] The international telegraphic network also gave Japan access to news from around the world. Through submarine cables, foreign news agencies such as the London-based Reuters began supplying news to Japan after 1887. During the Sino-Japanese War (1894–95), many Japanese correspondents on the continent or in Hiroshima sent reports to their respective news agencies by way of emergency telegrams. The Russo-Japanese War (1904–5) a decade later saw another sharp increase in news-related telegrams. Since news reports tended to be long, newspapers in Japan demanded a special rate for news telegrams, as instituted at the 1896 Budapest International Telegraphic Conference. In 1906, the government introduced the reduced press-rate telegram in Japan, making it 60 percent cheaper than a regular telegram of the same length and allowing newspapers and news agencies to take full advantage of electrical communications.[24] In 1908, although domestic news telegrams sent to 236 news outlets made up only 0.6 percent of the total number of telegrams, they accounted for 5 percent of all words.[25]

As Meiji elites like Itō and Fukuzawa clearly understood, the purpose of telecommunications was as much to inform as to control. It was no accident that Japan's domestic telegraphic network took shape with such amazing speed, because establishment of the telegraph coincided with intense state-building efforts. The telegraph served as the "nerve of government" as various state bureaucracies ranging from the police to the military signed up for service. As Yoshimi Shun'ya has shown, frantic construction of telegraph lines often preceded the frequent imperial

23. Tsūshinsha shi kankōkai, *Tsūshinsha shi*, 887–88; Wakai and Takahashi, *Terekomu no yōake*, 101. The U.S.-based United Press (UP) began supplying news to Japan in 1907.

24. Nippon denshin denwa kōsha, *Denpō ryōkin no enkaku*, 119–21.

25. Wakai and Takahashi, *Terekomu no yōake*, 103.

tours by the Meiji emperor and his entourage.[26] Citing Western newspapers published in Yokohama, one Japanese newspaper called attention to the use of the telegraph in military operations during the Franco-Prussian War of 1870–71.[27] Soon, telegraphic communications would play a critical role in consolidating the Meiji government's control over all of Japan in the face of incessant rebellion and internal turmoil during its first two decades in power. The embryonic telegraphic network proved a timely weapon for the Meiji government when disaffected ex-samurai under Saigo Takamori rose in arms in Kyushu in 1877. In early February of that year, government officials in Nagasaki prefecture reported the disturbance in Kagoshima by telegraph to Tokyo. By the middle of the month, the governor of Nagasaki had confirmed the rebellion in Kagoshima and requested expeditionary troops from Tokyo.[28] Upon receiving these reports, the Meiji government quickly sent reinforcements to Kyushu. Before launching its suppression, the government made Kyoto, where the emperor happened to be visiting, a command nerve center by installing telegraph facilities there. It also rushed to build 850 km of additional telegraph lines around the rebellious area and extended existing lines well beyond the government stronghold of Kumamoto, thus covering the entire island of Kyushu. The rebel troops made little attempt to use modern electric communications, relying instead on old-fashioned couriers.[29]

The successful suppression of the Satsuma rebellion no doubt confirmed Meiji leaders' belief in the importance of a government monopoly of telegraphic communications. In the first two decades after the beginning of telegraph service in Japan, the government introduced various measures to consolidate its control over the new media. The government decision to impose a state monopoly on the telegraph was based, above all, on national security. As early as 1872, the Executive Council expressed concern that since "the nations of the West are

---

26. Yoshimi Shun'ya, *"Koe" no shihonshugi*, 142–45.

27. Nihon kagakushi gakkai, comp., *Nihon kagaku gijutsu shi daikei* (19), 50–52.

28. Many such government telegrams exchanged in February 1877 and later found in the Nagasaki Municipal Library are included in Watanabe, *Nihon denshin denwa sōgyō shi*, 179–224.

29. Nakamura Fumio [pseud. Habaki Jūn'ichirō], "Seinan sensō to tsūshin," *Denpa to juken* (hereafter *DJ*) 32.7 (July 1982): 59–64.

thoroughly versed in our telegraphic codes," the establishment of a private telegraph company would jeopardize official secrecy and interfere with the conduct of foreign relations.[30] The government also took economic considerations into account, reasoning that leaving telegraphic operation in private hands could pose serious competition to the government-run postal service. Together with laws protecting telegraph lines against vandalism, the Meiji government issued regulations for the telegraph service that had far-reaching consequences in modern Japan. In 1874, the government adopted the Telegraphic Code, which, among other things, reaffirmed the principle of state monopoly of telecommunications in Japan. In 1885, when Japan launched the new cabinet system, the Ministry of Communications (MOC) was established and charged with operating telecommunications and postal service, in addition to oversight over beacons and shipping.

The Meiji state's preoccupation with control over telecommunications helps explain the different fate of the telephone in Japan. In contrast to the relatively swift deployment of the telegraph, telephone service developed much more slowly. The nature of the technology had as much to do with it as the state's perception of its usefulness. In 1877, the year after the telephone patent was granted in the United States, two telephone sets were brought to Japan and their use demonstrated in Tokyo for officials and other dignitaries. Government bureaucracies such as the Home Ministry adopted the telephone for internal use rather quickly.[31] Public telephone service, however, was a different matter. In 1882, after a business trip to Shanghai, Director of the Telegraph Ishii Tadaakira reported to the minister of engineering that major commercial firms in Shanghai were using telephones as indispensable daily necessities. If installed in commercial centers such as Tokyo and Osaka, he suggested, the telephone would contribute much to the business and economy. But due to the government's financial retrenchment policy in the mid-1880s as well as internal disagreements over the form of ownership, it was not until 1889 that the first public telephone line was built between Tokyo and the resort town of Atami. In 1892, tele-

---

30. *Meiji zaisei keizai shiryō shūsei* (Tokyo, 1931–36), 17: 215, quoted in Thomas C. Smith, *Political Change and Industrial Development in Japan: Government Enterprises, 1868–1880* (Stanford, Calif.: Stanford University Press, 1955), 45.

31. Watanabe, *Nihon denshin denwa*, 226–32.

phone service also began in Osaka and Kobe.[32] Despite an initial lack of public interest, the total number of private telephone subscribers more than tripled in the next two years. Telephones remained largely for local use until the early twentieth century, however. To the Meiji state, there was simply not the same urgency that existed with the telegraph, where national interests and political control were at stake. For a modern bureaucracy that relied on recordkeeping and written communication, the telegraph also had an advantage over the telephone. As a result of government monopoly and chronic lack of funding, public access to telephone service in Japan continued to lag far behind that of postal and telegraph communication well into the twentieth century.[33]

OVERSEAS EXPANSION IN THE AGE OF SUBMARINE CABLES

The Korea Cable and the Logic of Dependent Expansion

The quest for autonomy and the pursuit of expansion have always been intertwined in modern Japan, and nowhere was this clearer than in Japan's relations with Korea in the last decades of the nineteenth century. It was also in Korea that Japan first attempted to exert influence beyond its shores via modern communications. The 1876 Treaty of Kangwha, which followed Japan's retaliation when a Japanese navy vessel surveying Korea's waters was fired upon, allowed Japan to open a post office in the southern Korean port city of Pusan. Still, the only way to communicate from Korea to Japan (and vice versa) was by ship, either to Shimonoseki in western Japan or via the Chinese port of Tianjin, whence messages could be sent by telegraph to Nagasaki via Shanghai. Neither alternative was satisfactory to Japan. Already, some Japanese were calling for the construction of a submarine cable to Korea. In 1882, political tensions in

---

32. Westney, "Building the National Communications System," pp. 39–59; Takahashi, *Tsūshin*, 178–79.

33. This led to the emergence of what one American survey called "the most unique system of telephone charges," with private telephone connections being traded like an expensive commodity (Dilts, *The Telephone in a Changing World*, 56–57).

Korea grew into a crisis when anti-Japanese sentiment led to expulsion of the Japanese minister and an attack on Japan's legation. A telegraph link to Korea became an absolute necessity.[34]

In March 1883, the Japanese government entered into an agreement with the GNTC whereby the latter would lay the first submarine telegraph cable across the Korea Strait. After further negotiations with the Korean government, Japan was allowed to build a short land line from the submarine cable's landing site to the Japanese settlement in Pusan, where Japan would establish a telegraph office. From there, Japanese telegrams could be delivered to as far as Seoul by postal service for free. As a result, Japanese-language telegrams could be received and sent outside the Japanese islands for the first time, an important boost to the small but growing Japanese presence in Korea. To protect this vital communications link on strategic and business grounds, the Japanese government secured a 25-year monopoly during which Korea would not build competing telegraph lines itself or allow a foreign government or company to do so.[35] Meiji Japan thus made good use of the advanced West in facilitating its own expansion in Asia.

In the 1883 agreement, Great Northern agreed to add new cables between Nagasaki and Shanghai and Nagasaki and Vladivostok. In addition, the company promised that it would not raise cable rates without the consent of the Japanese government and that it would keep all submarine cables in working order. All this came at a price to Japan's own autonomy, however. In exchange for the GNTC's timely service, the Japanese government had to grant the Danish company a twenty-year monopoly over Japan's international cable communications.[36] While solving Japan's immediate communication needs with Korea, the arrangement meant prolonged foreign control of Japan's overseas communications, with serious economic and strategic consequences. Japan's subsequent attempts to establish an independent link with Asia would encounter persistent opposition from the Danish firm. The deal would receive much criticism decades later from Japanese writers who blamed

34. Nippon denshin denwa kōsha, Kaiteisen shisetsu jimusho, *Kaiteisen hyakunen no ayumi*, 110–15.

35. For the complete text, see Ishikawa Junkichi, comp., *Kokka sōdōin shi* (hereafter *KSS*), 3: 717–20.

36. For the compete text of the agreement, see ibid., 696–701.

near-sighted bureaucrats for placing Japan's international communications under the yoke of the GNTC monopoly for decades to come.

Japan had little choice at the time. The technological challenge of laying the cable in deep water was formidable. Moreover, the cable between Japan and Korea was estimated to cost 300,000 yen—amounting to 25 percent of all expenditures on telegraph construction in Japan between 1869 and 1882. The government was not in a financial position to engage in such a costly undertaking due to the Matsukata Deflation policy adopted in the early 1880s. Moreover, the benefit seemed worth its price. The Korea cable not only gave Japan a strategic advantage but also proved useful in meeting the demands of public communication. As Japan's economic presence in Korea expanded, telegraphic traffic between Japan and Korea increased from a trickle of 3,800 telegrams in 1884 to some 100,000 each year a decade later, 98 percent of them in Japanese.[37]

Alarmed by events in 1882 as well as by Japan's new access to information in Korea via the Korea cable, the Qing government quickly decided to build a telegraph line from the Chinese border to the Korean capital of Seoul. After Japan protested against it as a violation of its monopoly, the Korean government agreed to build its own telegraph line linking Seoul and Pusan.[38] Rivalry over telecommunications in Korea thus contributed to the brewing conflict between Japan and China.

The Korea cable proved an essential strategic asset to Japan when it started the war with China in 1894, the first by Japan's modern military machine against a numerically superior foreign force. Because of its easy access to land and marine transportation, the city of Hiroshima in western Honshu served as the imperial headquarters throughout the eight-month-long conflict. To oversee the military operation and to boost the morale of Japanese troops, the Meiji emperor was moved temporarily to that city. The crucial role of telegraphic communication

---

37. Nippon denshin denwa kōsha, Kaiteisen shisetsu jimusho, *Kaiteisen hyakunen no ayumi*, 82–84, 171; Ahvenainen, *Far Eastern Telegraphs*, 186.

38. Jiaotong shi bianzuan weiyuanhui, *Jiaotong shi dianzheng bian* (hereafter *JSDB*), 45; Nippon denshin denwa kōsha, Kaiteisen shisetsu jimusho, *Kaiteisen hyakunen no ayumi*, 112; on Qing China's competition with Japan over telegraph communications in Korea, see Larsen, *Tradition, Treaties, and Trade*, 135–40; see also Chiba, *Kindai kōtsū taikei to Shin Teikoku no henbō*.

for the conduct of the Sino-Japanese War was reflected in the sharp increase in the telegraphic traffic of the city. The total number of telegrams sent and received in the Hiroshima area increased almost threefold, from 114,000 to 308,000. The number of overseas telegrams sent and received in Hiroshima prefecture jumped from a negligible 50 per year before the war to over 160,000 in 1894 and 227,000 in 1895. Had the imperial headquarters stayed in Tokyo, one Japanese official reasoned, such a staggering number of telegrams would have been a huge burden on the incipient telegraph lines in central Japan, likely disrupting business activities in cities like Osaka and Kobe. To help ease the unprecedented traffic, the Japanese government installed the then-cutting-edge duplex automatic telegraph on the existing telegraph lines between Hiroshima and Tokyo, doubling the capacity of this important route.[39]

After the war, Japan retained control over the telegraph line linking Pusan and Seoul, built by its military during the conflict. As Japan further expanded its influence in the Korean Peninsula, it instituted several reductions of telegram rates between Japan and Korea. By 1905, some 430,000 telegrams were transmitted through the Korea cable. Before Japan's final annexation of Korea in 1910, telegrams to and from Korea made up 60 percent of Japan's international traffic. Nearly all the traffic with Korea traveled on a cable that was completely foreign-owned until 1891, when the Japanese government purchased the portion between Nagasaki and Tsushima. During the Sino-Japanese War, the Japanese government sought to buy the entire cable, but Great Northern refused. It was not until after Japan annexed Korea that it was finally able to complete its control of the Korea cable by purchasing the remainder of the cable for 160,000 yen.[40] After annexation, Japan also paid off the loans Korea had obtained from China for construction of its western-route telegraph line between Seoul and Sinuiju on the Chinese border.[41]

---

39. Nakamura, "Daihon'ei to tsūshin I," *DJ* 32.9 (September 1982): 106–7.

40. Chōsen sōtokufu, Teishinkyoku, *Chōsen tsūshin jigyō enkakushi*, 107–9, 133; Nippon denshin denwa kōsha, Kaiteisen shisetsu jimusho, *Kaiteisen hyakunen no ayumi*, 171; Ahvenainen, *Far Eastern Telegraphs*, 186.

41. *JSDB*, 45.

The Taiwan Cable and the Beginning of Technological Autonomy

Japan's military and diplomatic victories over Qing China brought many rewards. The large indemnity China was forced to pay helped finance, among other things, the first great telecommunications expansion in modern Japan. Over the next ten years, Japan's domestic telecommunications network saw a threefold increase in the number of telegraph offices, the total length of telegraph lines, and the total number of incoming and outgoing telegrams. The length of new lines added from 1904 to 1905 was 201,390 km, compared to 1,989 km from 1894 to 1895. Cutting-edge technologies such as multiplex telegraphy and teleprinters were installed on the trunk lines. This was done to meet the surging demand created by the postwar business boom and to strengthen the country's military preparedness for the next war. Partly thanks to these and other improvements in domestic communications, it was no longer necessary to establish imperial headquarters in a forward position in western Japan to fight another war on the continent.[42]

Thanks to its victory in the war, Japan acquired a major overseas colony, Taiwan. As the embryonic colonial empire was born, Japanese leaders recognized communications facilities as important infrastructure. No sooner had the two governments concluded the Shimonoseki Treaty than Japan's Ministry of War set up a provisional agency for constructing telegraph facilities and navigation beacons for the new colony. Headed by General Baron Kodama Gentarō, one of its top priorities was establishment of a telegraphic link between the home islands and Taiwan via Okinawa, a once-independent kingdom Japan annexed in 1879.

This was not the first telegraphic link for Taiwan, however. While still a remote province in the Chinese empire, Taiwan had been connected with mainland China by a submarine cable in 1887 shortly after the Sino-French War, at the suggestion of a reform-minded Chinese official. The Taipei–Fuzhou (also known as Danshui–Chuanshishan) line was completed under the supervision of Danish and British engineers. The Chinese government was the main customer, although the

42. Nakamura, "Daihon'ei to tsūshin," *DJ* 32.10 (October 1982): 107–8; "Nisshin, Nichiro no koro no tsūshin," *DJ* 33.10 (October 1983): 99.

telegraphic link to Taiwan was also used by British and German merchants in the sugar and tea business, and even by some Chinese merchants from Shanghai.[43] Because there was no direct telegraphic link between Taiwan and Japan, however, the Japanese expedition forces sent to occupy the island in 1895 had to communicate with Tokyo by way of this link, via Chinese cities such as Fuzhou and Shanghai.

Although the large indemnity received from China made it possible to start several ambitious projects in telecommunications expansion, Japan had just embarked on its own industrialization and lacked the capacity to manufacture submarine cables itself. Laying the cable was no easy task either. Initially, opinion was divided within the government as to whether to have an experienced British engineer supervise the laying of the cable to Taiwan. Among those who objected were two MOC engineers who argued that British supervision would keep Japan dependent on foreign skills in future submarine cable expansions. General Kodama agreed, and their view prevailed. As a result, in addition to submarine telegraph cables, Japan ordered its first cable-laying ship, named the *Okinawa maru*, from a shipyard in Britain. Laying of the cable began in July 1896 and reached Okinawa a month later. The entire 1,608 km–long cable to Taiwan was completed a year later, entirely by Japanese hands. In his report to the army minister, General Kodama praised Japanese technicians for overcoming adversities ranging from rough seas and terrain (land lines were built in hard-to-reach mountains for security reasons, to be out of sight of ships) to interruptions by "local bandits" in Taiwan.[44]

The significance of the completion of the Taiwan–Japan submarine telegraph cable was manifold. The Okinawan islands, whose precarious independence ended in 1879 when Okinawa became a prefecture, were connected to Japan proper by telegraph for the first time. For many of Japan's colonial theorists, rapid means of communication between Japan and its first overseas colony seemed to favor adopting the French model of colonial assimilation. Hara Takashi, a bureaucrat in the Tai-

43. *Taiwan nenkan* (Taihoku, 1925 edition), 184; Taiwan yinhang, Jingji yanjiushi, *Taiwan jiaotong shi*, 11–12.

44. Funatsu Shigetarō, "*Okinawa maru* to taiyō kaiteisen," *Teishin shiwa* 1: 277–81; Nippon denshin denwa kōsha, Kaiteisen shisetsu jimusho, *Kaiteisen hyakunen no ayumi*, 129–45.

wan Affairs Bureau in 1895 and a future prime minister, listed relative geographical proximity and ethnic and cultural similarity among the reasons for integrating Taiwan into the homeland. He then noted that the distance would be "shortened" even further with the completion of the submarine cable and the development of shipping routes.[45]

Technologically, the completion of the Taiwan cable marked Japan's ascendance as a world-class player in submarine telegraphic communications, a status enjoyed by only a handful of nations. As the project gave Japan valuable experience in laying submarine cable in the open sea, the cable ship acquired for this project would play an active role over the next several decades. Hence Japan's first colonial telecommunications project became a major boost to its overall technological capacity. There were other benefits as well. The exact amount of submarine cable ordered from the British manufacturer was kept a military secret, but with remarkable foresight, General Kodama had ordered 1,648 nautical miles of cable to cover the 1,045-nautical-mile route between Japan and Taiwan. As a result, the Army was left with some 600 nautical miles of submarine cable as a strategic reserve. Entrusted to the MOC for storage, this reserve would be a most critical asset less than a decade later. When tensions on the Korean peninsula began to mount, this spare cable was immediately put to use for a new submarine telegraphic link connecting Japan with Korea and the Liaodong peninsula in southern Manchuria.[46] Clearly, warfare and colonial expansion not only benefited from progress in telecommunications technologies but stimulated their development as well.[47]

The Shanghai Cable and Continental Geostrategy

The war with Qing China proved telegraphic communications to be indispensable in an international conflict. However, the Russo-Japanese War of 1904–5 was in many ways the first full-fledged war of modern communications. Even before the initiation of the conflict, Japan's

---

45. Quoted in Chen, "The Attempt to Integrate the Empire," 250–51.

46. Nippon denshin denwa kōsha, Kaiteisen shisetsu jimusho, *Kaiteisen hyakunen no ayumi*, 141–42, 152–57; Funatsu, "*Okinawa maru* to taiyō kaiteisen," 279–80.

47. For a stimulating essay on the general theme of military expansion and industrial development in modern Japan, see Yamamura, "Success Ill-gotten?"

communications presence in Korea, which included a number of post offices in major cities, gave it a strategic advantage over Russia, its arch-rival on the peninsula at the beginning of the twentieth century. During the Sino-Japanese War, advancing Japanese troops not only took over the existing western (between Sinuiju and Seoul) and northern telegraph routes (between Wonsan and Seoul) in Korea but also constructed military telegraph lines in the south following destruction of the previous lines in the peasant uprisings. After the war, Japan had to return all telegraph lines to the Korean government, but it managed to retain control over its military line linking Pusan and Seoul.[48] Although all telegraph lines north of the capital of Seoul were still administered by the Korean government, Japan's Army General Staff was able to exchange telegrams with Japanese officers dispatched to northern Korea by sending telegrams through Japanese post offices in Chinnamp'o and P'yongyang. In anticipation of the military operations on the continent, Japan set out to lay a new submarine cable between Sasebo in northern Kyushu and the Liaodong peninsula.[49]

Japan's communication warfare began before the first shot was fired. Immediately after the two countries broke off diplomatic relations in February 1904, the Japanese Navy seized a Russian merchant ship off the southern port city of Pusan. Acting under the order of a young Japanese consul by the name of Shidehara Kijurō, Japanese consular policemen in Pusan threatened the Korean post office staff and prevented them from sending a Russian diplomatic report about the start of hostilities, which would have tipped off the Russian fleets in Inch'on and Port Arthur. Soon afterward, Shidehara ordered Japanese policemen to cut the Pusan-Seoul telegraph line altogether, although it belonged to the Korean government and the act was a clear violation of international law. In the meantime, the Japanese minister in Seoul also ordered all telegraph lines out of Seoul except Japanese-owned lines to be severed for three days.[50] Japan thus successfully imposed

48. Duus, *The Abacus and the Sword*, 121.

49. Nakamura, "Nisshin, Nichiro no koro no tsūshin," *DJ* 33.10 (October 1983): 103.

50. Shidehara, *Gaikō gojūnen*, 16–19; Shidehara heiwa zaidan, *Shidehara Kijūrō* (Tokyo: Shidehara heiwa zaidan, 1957), 40–43. It is indeed ironic that Japan's championing of international order and cooperation in the interwar period contributed to Japan's war effort by violating international protocol.

a "communication blackout" on Russian diplomatic and military personnel in Korea on the eve of the conflict and gained a critical advantage over its adversary. The Japanese navy launched a surprise attack and sunk most of the Russian fleet two days later. On February 10, the two countries formally declared war.[51]

As soon as the Russo-Japanese War broke out in 1904, Japanese troops stationed in Korea quickly seized control of Korea's telegraph network despite Seoul's pledge of neutrality.[52] The war saw the deployment of new technologies and improvement of telegraphic service. The debut of field telephones in land battles and wireless sets in naval engagements, for instance, would usher in changes in tactics. Approximately 10,000 telegrams were exchanged each day among the imperial headquarters in Tokyo and various Japanese Army units in Manchuria. The average time it took a telegram to travel between Manchuria and Tokyo was reduced from 4 hours, 57 minutes in 1904 to 3 hours, 5 minutes in 1905. The fastest telegrams took less than 1 hour, compared to about 4 hours during the Sino-Japanese War.[53]

Ironically, the war with Russia also exposed serious weaknesses in Japan's overseas communications capabilities. Before a Pacific submarine cable connected Japan and the United States in 1906, much of Japan's overseas communication had to travel through the submarine cables owned by GNTC. There was widespread suspicion from the beginning that Japan's overseas communication security had been compromised by the Danish company, which maintained close relations with the Russian government.[54] Japan had good reason to be concerned

---

51. See the reminiscences by Nagai Jūtarō and others in Chōsen sōtokufu, Teishinkyoku, *Teishin shūi*, 42–43, 49–50. At the end of January 1904, when it became apparent that war with Russia was imminent, Seoul Post Office Chief Tanaka Jirō undertook what he vaguely described as "resolute measures" at this critical juncture. The exact nature of the measures is unknown, because at the request of the discussant, this episode was not recorded in the minutes of the 1935 roundtable discussion (ibid., 79–81). Whatever he did, Tanaka was promoted to director of the Communications Bureau of MOC in 1911 (Nakayama Ryūji, *Sensō to denki tsūshin*, 20–21).

52. Duus, *Abacus and Sword*, 184–86.

53. Nakamura, "Nisshin, Nichiro no koro no tsūshin," *DJ* 33.10 (October 1983): 100, 103.

54. The Japanese government ordered an investigation into the allegation that GNTC's Shanghai office had passed copies of Japanese diplomatic telegrams to the Russians, an allegation flatly denied by the company (see Japanese Ministry of Foreign

about its own communication security. In fact Japan had been an early beneficiary of communication espionage. Shortly before the conflict with Qing China broke out in 1894, the Japanese Foreign Ministry succeeded in breaking the Chinese diplomatic codes by analyzing telegrams sent and received by the Chinese legation in Tokyo. As Takahashi Hidenao has shown, access to such highly sensitive Chinese government communication had a direct impact on Japan's decision for war. It also gave Japan an enormous advantage on the battleground as well as during subsequent diplomatic negotiations between the Chinese and Japanese leaders, conducted conveniently in the Japanese city of Shimonoseki at Japan's insistence.[55]

Not surprisingly, it was the military that raised the greatest alarm over Japan's overseas telegraph connections. Already in 1894, Japan adopted a Military Telegraph Law, setting up a separate military communications system. Because telecommunications in Japan were already in government hands, military communications facilities initially included only a limited number of fixed lines between military bases and ports, thus avoiding redundancy and unnecessary intrusion onto Ministry of Communications turf.[56] As Japan's military now established a permanent presence in the colonies and in China following the Boxer Rebellion, it understandably sought to shape Japan's overseas communications policy. After the war with Russia, however, some government officials felt that wartime military lines should be withdrawn, since their maintenance was costly, their value seemed negligible, and they posed diplomatic complications. But as the military saw it, this would amount to ruining the "hundred-year blueprint of the country":

Affairs [hereafter JMFA] Archives 7.1.4.18). Recent research has shown that the Russian government indeed had access to Japanese diplomatic telegrams in Paris from April 1904 to March 1905, with the help of the French authorities (Inaba, "International Telecommunications During the Russo-Japanese War").

55. On Japan's success in deciphering Chinese diplomatic telegrams as well as the impact of the temporary breakdown of telegraphic communications between Tokyo and Seoul, see Takahashi Hidenao, *Nisshin sensō kaisen katei no kenkyū*, esp. 56–68, 89–96. See also Mutsu Munemitsu, *Kenkenroku: A Diplomatic Record of the Sino-Japanese War, 1894–95*, trans. Gordon Mark Berger (Princeton: Princeton University Press, 1982), 266*n*2.

56. Nakamura, "Rikugun tsūshinhei yomoyama hanashi," *DJ* 31.7 (July 1981): 45.

All countries of the world consider autonomy in communications a basic tenet in national defense. All aspire to possess their own communications systems. In the early years of Meiji, Japan gave in to the Great Northern Company, which represented Russian interests, and until today Japan has been unable to break away from its grip. At present, with the newly established Anglo-Japanese Alliance, Japan is looked upon as a formidable Power. However, it could not even construct a submarine cable for the purpose of joint military operations. Isn't that a disgrace to the reputation of Japan? Leaving aside the issue of reputation, anyone who knows of the attitude of Great Northern in the course of this war must realize the precarious position of Japan's overseas communications. Both land and underwater telegraph lines set up during the war gradually expanded with the progress of military operations, and the magnitude of its organization is truly spectacular in East Asia. However, this is nothing more than a temporary phenomenon based on belligerent rights. Whether or not they can be maintained in peacetime depends on the resolution of Great Northern's privileges.[57]

Following Japan's victory over Russia in 1905, Japan made important progress in improving its position in international communications. In 1906, a Japanese government cable linked with the cable of America's Pacific Commercial Cable Company at Guam, thus opening the first Pacific route that bypassed the Great Northern lines. After its annexation of Korea in 1910 Japan purchased the remainder of the Korea cable from the GNTC in 1911. That same year, the Qing dynasty in China was overthrown, drastically increasing uncertainties as well as opportunities for Japan.[58] It was against such a backdrop of geopolitical change in East Asia that the Japanese government began to formulate a comprehensive international telegraph policy commensurate with its national strength and its new objectives.

In China, Japan remained at a disadvantage in telecommunications. To begin with, it was the European cable companies that handled traffic between China and Japan. This was particularly evident in Shanghai, by far the most important city for Japan's communication link with China. Here, submarine cables from Danish, British, American, and German-Dutch companies made the city the international hub of news,

57. "Sengo ni okeru teikoku denshin seisaku" (n.d.), JMFA Archives 1.7.4.36–2.

58. On the Army's role in Japan's continental policy during this period, see Kitaoka Shin'ichi, *Nihon rikugun to tairiku seisaku.* In English, see Matsusaka, *The Making of Japanese Manchuria*, 149–60.

finance, and commerce in the Far East. In fact, setting up Japan's own cable link to Shanghai had been Japan's objective since the dawn of its international communications. Terashima Munenori, who oversaw the start of telegraphic service in Japan, had advocated laying such a cable when he was still a self-described "hot-blooded" young man in the early Meiji period.[59] However, at the time Japan possessed neither the technological nor the financial capacity to do so. As the city's importance—and the Japanese presence within it—grew, the Japanese government became painfully aware of the disadvantage of lacking a link to Shanghai despite Japan's geographic proximity, and made a Japanese cable to Shanghai a high priority.

Acutely aware of the presence of Western interests in neighboring China under various treaty protections, the Japanese government saw the monopolies of British and Danish cable companies as the main obstacle. The fact that Great Northern's ten-year monopoly in China was to expire at the end of 1912, however, provided a perfect opportunity. In early 1912, the Ministry of Communications set up a Committee for Investigating Overseas Telegraphy. The committee called for a general reduction of international cable rates; it also recommended setting up a new link with Russia and reopening the Taiwan–Fuzhou cable for international traffic. The centerpiece of the new policy, however, was an independent telegraphic route to China.[60]

The fall of the Qing dynasty gave an added sense of urgency. In 1912, Kodama Kenji, president of the Japanese Chamber of Commerce in Shanghai, petitioned the Japanese consul general to install wireless sets in ships sailing between Japan and Shanghai. Kodama did so in part because the political uncertainties in China required additional communications routes with Japan, but also because installation of wireless would pressure the GNTC to reconsider its monopoly over telegraphic communications between Japan and China, which Japanese residents had endured for many years.[61] The military in Japan also demanded that a separate Japanese cable to China be given top priority in upcoming

59. Ahvenainen, *Far Eastern Telegraphs*, 187; Machida Itsuyoshi, "Taigai denshin ryakushi," *Denki tsūshin gyōmu kenkyū* 19 (July 1951): 60.

60. Hanaoka Kaoru, *Kaitei densen*, 174–75.

61. "Nihon Shanhai kan kisen ni musen denshin sochi ni kansuru ken," MOC Records I.

negotiations. Minister of Communications Gotō Shinpei seemed to agree: "The trend among the Great Powers was the extension of a telegraph network all over the world, and in particular, the attempt to open telegraphic links with the great market of China."[62]

As Japanese officials were fully aware, the greatest obstacle to Japan's aspiration was the Danish GNTC, which would be threatened by a potential loss of its monopoly over communication between Japan and China. Captain J. J. Bahnson, the company's general manager for the Far East, was known for his arrogance and proved to be a tough negotiator. He steadfastly refused to relinquish GNTC's monopoly on Japan's overseas telegraphic communication. The company maintained that existing cables were sufficient and even offered to handle Japanese-language telegrams. However, it eventually gave in to Japan's demand to lay a new cable between Shanghai and Nagasaki, after Japan promised that the new cable would not cause financial damage to the GNTC. The new Shanghai cable would handle Japanese-language telegrams between Japan and Shanghai as well as European-language government telegrams between Japan and China. In return, Japan agreed to set up a Joint Purse and share the revenue of *all* telegraphic traffic between Japan and China with the GNTC until 1930, turning over as much as 64.5 percent of the revenue from the new Shanghai cable. Between 1914 and 1930, this payment alone would amount to over 25 million yen.[63] Once again, financial sacrifice was the price paid for strategic advantage.

Once Great Northern gave in, negotiations with Chinese representatives in Tokyo proved much easier than they had been. In exchange for China's consent to the new Japanese cable, Japan granted all Chinese government telegrams a 50 percent discount and agreed to pay

62. Memo by Gotō Shinpei, February 6, 1912, in *KSS*, 3: 734–38; "Daihoku denshin kaisha ni ataetaru tokkyo kigen manryōgo ni okeru teikoku gaishin seisaku ni taisuru gunjijō no yōkyū," Vice Minister of Communications to Vice Minister of Foreign Affairs, September 18, 1913, in JMFA Archives 1.7.4.36–1, vol. 1.

63. For the minutes of these negotiations, see JMFA Archives 1.7.4.36–3, vol. 1. See also *JSDB*, 1: 173–77. See also Nippon denshin denwa kōsha, Kaiteisen shisetsu jimusho, *Kaiteisen hyakunen no ayumi*, 180–81; and Matsunaga Tadao, "Shin'yu mudentai no fukkō to Nisshi kan denki tsūshin no konseki," *Teishin kyōkai zasshi* (hereafter *TKZ*) 370 (June 1939): 38. For a recent Japanese study emphasizing the importance of monopoly, see Kishi, "Nagasaki Shanhai kan 'Teikokusen' o meguru takokukan kōshō to kigyō tokkyoken no igi."

China a 5-cents-per-word terminal rate for all telegrams sent over the cable. In a secret agreement, both governments pledged that they would not extend privileges to a third party in telecommunications between the two countries without first consulting the other country. In particular, Japan insisted that China not extend landing rights to Great Northern after these rights expired in 1930. Interestingly, the definition of "electric communications" caused considerable disagreement between the Japanese and Chinese negotiators. The Chinese preferred to use "submarine telegraphy" so as to eliminate any future prospect of including wireless and telephone in this exclusionary agreement. Since they had lost much of their submarine cable communication to foreign interests, the Chinese were determined not to repeat the same mistake with wireless facilities. The Japanese negotiator insisted that Japan had no such intention but refused to put it in writing. In the end, the Japanese prevailed.[64] Given Japan's own objective in augmenting its influence over China's communications, such a clause would pave the way for Japan's ascendance in that field.

The first completely Japanese-owned Nagasaki–Shanghai cable opened to public service on the first day of 1915, just as the Great War began to engulf Europe. The inauguration of Japanese-language telegraphic communication between Japan and China, together with the establishment of a Japanese government telegraph office in Shanghai, was just in time for the rapid expansion of Japan's presence on the continent. In a report submitted to the prime minister on Japan's international telegraphic communications shortly after the agreement with China, then Minister of Communications Motoda Hajime proudly enumerated the major interests (*riken*) that Japan had recently obtained. The new cable to Shanghai, despite costing 1.62 million yen, "had profound implications in terms of the military, commerce, and diplomacy and had been an age-old objective for Japan." As Minister Motoda put it, the agreement with China would serve as "a major guarantee for Japan and a check on China's telegraph communications (*tai-Shi denshin*

---

64. Chinese Ministry of Foreign Affairs to Japanese Minister to China, January 13, 1912; *JSDB* 1: 177–78. See also the memorandum of a meeting between Chinese Foreign Minister Lu Zhengxiang and Japanese Foreign Minister Ijuin, January 14, 1913, in Zhongyang yanjiuyuan, Jindaishi yanjiusuo, comp., *Zhong-Ri guanxi shiliao*, 2: 52–53. For minutes of the negotiations in Tokyo, see JMFA Archives 1.7.4.36–4.

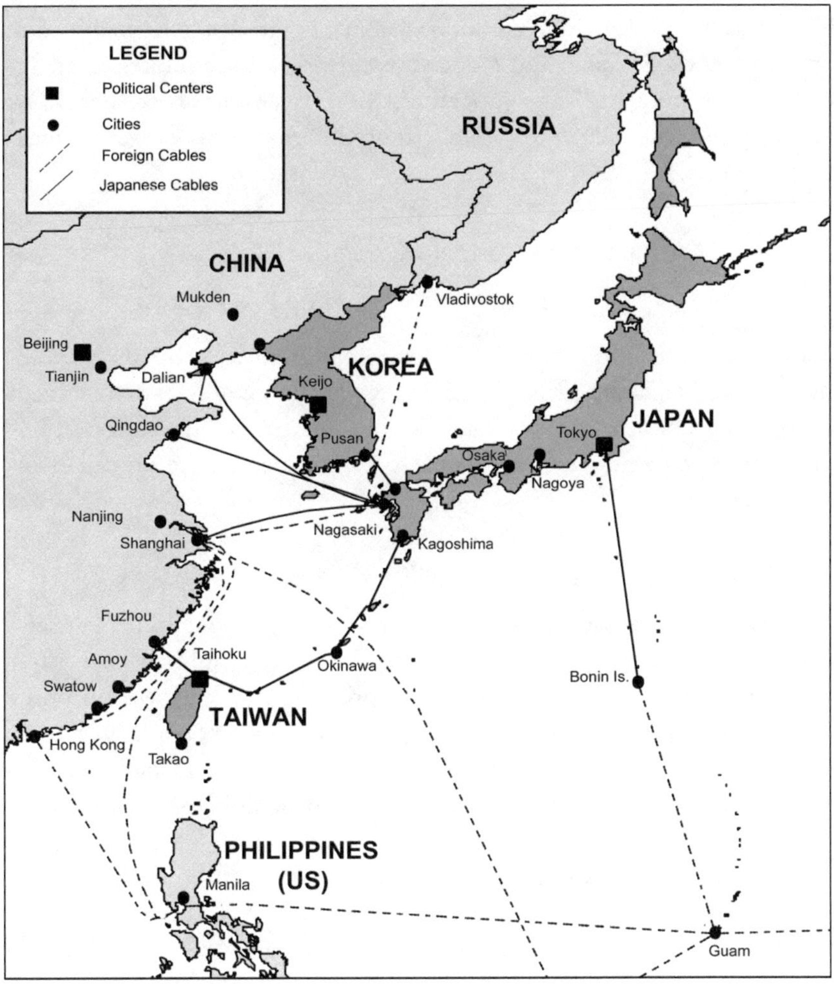

Map 1 Japanese submarine cables in East Asia, 1915.

no seichū)."[65] Indeed, the construction of a submarine cable from Nagasaki to Shanghai was a crowning achievement in Japan's overseas communications expansion (see Map 1). By the early 1930s, more than half of the telegrams exchanged between Japan and China proper went through this Shanghai cable, at an average of more than a thousand

65. "Taigai denshin mondai ni kansuru sho kyōyaku ketsuryō ni tsuki hōkoku" (October 7, 1913), reproduced in Ishikawa Junkichi, comp., *KSS*, 3: 740–44.

telegrams each day.[66] In addition, Japan acquired a telegraphic link with Russia via Sakhalin and reduced worldwide telegram rates by an average of 26 percent. Taken together, Motoda emphasized, these new gains lay the foundation of Japan's future international telecommunications policy.[67]

TELECOMMUNICATIONS AND MANAGEMENT OF THE EMPIRE

To be sure, telecommunications technology such as submarine cables did not only facilitate Japan's overseas expansion; by the early twentieth century, telecommunications had also become part of the fabric of imperial management, both in formal colonies and in the so-called informal empire.

Telecommunications and Colonial Control in Korea

Although from the outset Japanese authorities made much of their mission to develop the Korean economy, the modern information infrastructure in Korea did not take shape primarily for business purposes until the 1930s. The bulk of the telecommunications network in Korea, especially the inter-city telephone links, was developed to help meet Japan's urgent political and military goal of consolidating its control over the Korean population.

Japan's defeat of Russia in the war assured its position as the unchallenged power in postwar Korea and as a major power in the world. Korea was made a Japanese protectorate, and Japan was also to control Korea's telegraph, telephone, and mail systems. Ikeda Jūsaburō, an experienced MOC bureaucrat then in charge of the Tokyo Post and Telegraph Office, was dispatched with a team of Japanese technicians and officials to take over Korea's communications facilities. The Koreans did not give up without last-ditch resistance to the Japanese attempt,

---

66. Iino Takeo, "Tai-Ka denshin mondai no keii ni tsuite," *TKZ* 281 (January 1932): 45. The author, a MOC official, reiterated that given Shanghai's role in influencing world opinion about the Far East, the absence of a Japanese telegraph link to the city was a grave disadvantage for Japan's information policy.

67. "Taigai denshin," *KSS*, 3: 741–44; Hanaoka, *Kaitei densen*, 177.

but Ikeda succeeded after applying considerable coercion, by his own admission.[68]

The first open challenge to Japan's dominance came in July 1907, when its forceful dissolution of what remained of Korea's own army ignited large-scale armed resistance across Korea. The telegraph and telephone lines under Japanese control were heavily damaged by Korean fighters known as "righteous soldiers." Although the Japanese quickly repaired the damaged lines, at the considerable cost of 130,000 yen, the general lack of efficient means of communication severely handicapped Japan's response during the initial stage of the rebellion. One example given in an official Japanese report was a campaign in Kang'won-do where, due to poor communication, Japanese troops often moved independently of one another, once even marching off in opposite directions.[69] Since swift communication between all locations in Korea was considered vital not only to suppress the Korean resistance but to maintain order and security in the future, Vice Resident-General Sone Arasuke proposed a peninsula-wide police telephone network linking all police stations, military barracks, and postal and other government offices, so that all sites could be reached by telegraph or telephone within 24 hours. Upon completion, such a network would considerably strengthen Japan's control of Korea's vast rural areas.[70]

Given the urgent need, this police telephone network had to be constructed as quickly as possible. Telephone lines were added to some existing telegraph lines; elsewhere, they were set up from scratch. Between June 1908 and September 1910, more than 4,000 km of new lines spanning some 3,000 km were constructed, and 45,000 new telephone poles were erected, significantly altering the landscape of much of the peninsula. By the end of 1910, more than 800 police telephone sets had been added to security posts and government postal offices.

---

68. See the recollection by Okamoto Keijirō at a roundtable discussion in 1935. Okamoto was one of the seven Japanese officials sent to Korea in May 1905 to take over its communications administration (Chōsen sōtokufu, Teshinkyoku, *Teishin shūi*, 10–16).

69. Chōsen chūsatsugun shireibu, *Chōsen bōto tobatsu shi*, 50.

70. Kankoku keibi denwa kensetsubu, *Kankoku keibi denwa kensetsubu jigyō gaiyō hōkoku*, 1–2, 13.

Although this police telephone network, costing 376,000 yen, was not a sophisticated one, the significance of its completion by 1910 cannot be overestimated.[71] That same year, Japan forced a treaty of annexation on the Korean government, turning the country into an effectual Japanese colony. Japan's control of Korea's telecommunications network and its subsequent expansion of that network were an important precondition of the annexation.

Control of the telecommunications system in Korea could not eradicate Korean resistance, but it gave Japanese authorities an effective means of dealing with it. The famous March First Movement in 1919, when Koreans staged demonstrations throughout the peninsula demanding independence, took the Japanese by such surprise that it led to another wave of urgent expansion of Japan's police communications network in Korea. In addition to equipping local governments with automobiles, the new Superintendent General Mizuno Rentarō stressed the need to install telephone connections in every district (*do*).[72] Within two years' time, more than 1 million yen were spent to extend the network in 35 districts between Pusan and Sinuiju. The three years after March 1919 registered a 40 percent increase, totaling nearly 2,000 km, in telephone lines in Korea. As a result, the police telecommunications network was further strengthened.[73]

The effect of swift communication in consolidating control soon became obvious. Thanks to the telegraph and telephone network, the Government-General of Korea (GGK) was able to prevent a member of the Korean royal family from fleeing to Shanghai in 1919. After discovering that Prince Yi Gang had been spirited out of his Seoul residence, Japanese authorities immediately alerted the Japanese police throughout Korea and in neighboring northeastern China. Japanese policemen arrived at the Andong Railway Station, across the border in China, just as the prince was disembarking from the train from Korea. The prince was thus apprehended and brought back to Seoul, sparing

71. Ibid., 21–29, 53–54; Nihon denshin denwa kōsha, *Denshin denwa jigyō shi* (hereafter *DDJS*), 6: 341; Chōsen sōtokufu, Teshinkyoku, *Teishin shūi*, 16–18.

72. Chōsen gyōsei henshūkyoku, *Chōsen tōchi hiwa*, 70–71.

73. *DDJS*, 6: 333. The total budget for communications for the same year was 7 million yen (*Chōsen sōtokufu tokei nenpō 1930*, 320).

the Japanese a major embarrassment. This was but one of many examples of the GGK's use of telecommunications facilities.[74]

What started as a police network in the colony was opened to civilian use after the Japanese consolidated control of Korea. Firmly controlled by the colonial government, this information network also served economic and cultural purposes. The influx of Japanese business and agricultural immigrants, together with the gradual economic development and urbanization in Korea, all contributed to increasing use of electric communications. It was a paradox of Japan's colonial rule in general that domination and development went hand in hand.[75]

Telecommunications and Informal Empire

Even before World War I, telecommunications was an important component of Japan's informal empire in China.[76] This was certainly the case in the northeastern provinces of China, better known as Manchuria, where Japan had had a distinctive advantage since 1905. An area more than four times the size of the Japanese home islands, Manchuria was rich in resources and thinly populated before the twentieth century. It also had a contentious history, as reflected in its telecommunications facilities. The first telecommunications facilities in the region were built by the Qing government in the late nineteenth century as military threats mounted on Chinese borders. A few enlightened Chinese officials recognized the importance of swift communications as a vital component of national defense. During the 1883 Korean crisis, as a court official named Chen Jigong pointed out, China's interests were gravely harmed as a result of the fact that information from Korea had

---

74. Ibid., 58–59, 119, 220–37. I thank Young-Key Kim-Renaud and Christine Kim for clarifying the background of the prince and the significance of the incident.

75. See Daqing Yang, "Colonial Korea in Japan's Imperial Telecommunications Network." For an overview of the so-called Cultural Policy during the 1920s, see Michael Robinson's chapter in Carter Eckert et al., *Korea Old and New* (Cambridge, Mass.: Harvard University Press, 1991), 276–304.

76. On applying the concept of informal empire to China, see, among others, Jürgen Osterhammel, "Semi-Colonialism and Informal Empire in Twentieth-Century China," in Mommsen and Osterhammel, eds., *Imperialism and After*, 290–314; Duus, "Japan's Informal Empire in China, 1895–1937: An Overview," xi–xxix.

to travel via Japanese telegraph lines. Therefore, he suggested, China must promote the spread of telegraph on its own. Shortly after the Sino-French conflict broke out in southern China in 1885, the Qing government heeded Chen's advice and extended the Shanghai–Tianjin telegraph line to Yingkou, Mukden, and Lüshun, three cities in Manchuria. A year later, it was furthered extended from Liaoyang to Inchon, Korea. Meanwhile, to strengthen defense against Russian encroachment in the north, telegraph lines were extended from Mukden toward and along the boarder with Russia during the 1880s.[77]

Soon after, Imperial Russia became the first foreign power to build telecommunications in Manchuria. By then, Russia was already extending the China Eastern Railway (CER) across the northern Manchurian plains to reach its Pacific Coast. As elsewhere, it built telegraph lines along these railways to ensure their safe operation. Russia's right to operate the telegraph in Manchuria was recognized in a treaty signed with the Chinese government in 1897. Following the successful Triple Intervention in the wake of the Sino-Japanese War, together with Russia's lease of the Liaodong peninsula, Russian-built telegraph lines linked Dalian and Port Arthur (Lüshun in Chinese) with the Imperial Russian Telegraphs at home. For the next decade, Russia enjoyed uncontested supremacy in Manchuria.

As a result of its defeat in the Russo-Japanese War, itself largely fought in Manchuria, Russia handed over its special privileges in southern Manchuria to the emerging victorious power of Japan. They included the railway trunk line that became Japan's highly successful South Manchuria Railway (SMR).[78] Japan also took over the lease of the Kwantung Territory at the tip of the Liaodong peninsula as well as the jurisdiction over areas adjacent to the South Manchuria Railway including their telecommunications facilities, marking the beginning of Japanese telecommunication operations in Manchuria. A series of treaties with China signed in 1908 and 1909 formally recognized these rights.[79] By 1927, a total of 354 Japanese communications service facili-

---

77. Manshū teikoku yūsei sōkyoku, *Manshū teikoku yūsei jigyō gaiyō*, 94–95.

78. For an authoritative work in English, see Matsusaka, *The Making of Japanese Manchuria.*

79. For the texts of these agreements, see Mantetsu, Harbin jimusho, *Manshū no densei*, 1: 388–97. Russia sold telegraph facilities outside the railway zone to China (Zhonghua minguo dianxin zongju, *Zhongguo denxin jiyao*, 21).

ties existed in Manchuria, consisting of 216 postal offices, 176 telegraph offices, 30 telephone exchanges, and 74 telephone offices. Japan's special position in Manchuria was also clearly reflected in telecommunication links with Japan. Till the end of the World War I, Manchuria was linked with Japan through two telegraph routes. One of them, the Dalian–Sasebo submarine cable, was laid during the Russo-Japanese War, and another linked Mukden with Keijō (present-day Seoul). As the telegraph traffic increased steadily, from one million relays in 1909 to three million ten years later, a new Dalian–Tokyo cable was laid in 1919. Due to the frequent problems on the Dalian–Sasebo cable, the South Manchuria Railway Company invested 2.7 million yen in a new Dalian–Nagasaki cable in 1921 and leased it to the MOC.[80]

The administration of its special rights in Manchuria gave Japan an opportunity to demonstrate its "continental management." The South Manchurian Railway was hugely successful. Dalian also became Japan's colonial showcase. One leading example was the installation of automatic telephone exchanges in the Japanese-administered city in 1923, which predated the installation of such facilities in the home islands by two years. Long-term economic consideration was a factor to be considered. World War I brought unprecedented economic growth in Japan as well as in Japan's leased territories in Manchuria. The subsequent rising labor costs as well as labor unrest had some impact on the telecommunications industry as well. In the meantime, telephone subscriptions increased rapidly during the wartime economic boom. In the colonial city of Dalian, these increases were compounded by high wages paid to Japanese employees for working outside the home islands and by the ethnic mix of the population in the city. The many Chinese and Russian residents among the subscribers made manual telephone exchange an extremely demanding task. The Dalian-based Kwantung Government General proposed the adoption of the latest automatic exchange system that had just debuted in the United States. Despite its high cost—some 2 million yen was needed for the conversion—imported equipment was installed starting from 1920 and service began in 1923. Unlike the telecommunications expansion plans in Japan, which were constantly curtailed by financial constraints, Japanese telecommunications in Manchuria enjoyed strong financial backing. This achieve-

80. Mantetsu, Harbin jimusho, *Manshū no densei*, 397–98.

ment made Japan almost the first to install automatic telephone exchanges in East Asia.[81] Although all equipment as well as technical supervision was provided by the British firm Automatic Telephone Manufacturing Company, it proved a valuable experience to Japan's adoption of automatic telephone switching after the 1923 Kantō Earthquake. Even MOC sent many technicians to Dalian to learn about such facilities firsthand.[82] Thus, telecommunications in a city like Dalian came to symbolize Japan's "colonial modernity."

Though much smaller in number compared to those in Manchuria, a chain of Japanese telegraph offices in China proper, together with the submarine cables, formed part of Japan's strategic assets. Unlike those of other foreign countries, these were government telegraph offices under the control of Japan's Ministry of Communications. In this sense, the establishment of Japanese postal and telegraph offices in China also extended the tentacles of Japan's Ministry of Communications to these areas in China and allowed the ministry to observe developments there.[83] Varying both in size and volume of traffic, these offices were deemed of critical importance to Japan's multifaceted activities on the continent. Although operating costs could be high, even an office with little peacetime traffic could become busy overnight. This was the case of the Japanese telegraph office in Chefoo (present-day Yantai) in Shandong, which was linked with Dalian by a submarine telegraph cable. Established in 1909 and consisting of five Japanese and two Chinese employees, the office did not handle much telegram traffic in a port city with only 400 Japanese residents. World War I changed everything literally overnight. When Japanese troops launched an attack on the German stronghold on the Shandong peninsula, Chefoo became the only landing port of entry for Japanese forces. Added to the flood of the Japanese evacuated from the German-occupied city of Qingdao were many Japanese journalists who rushed to Chefoo from Japan,

---

81. Matsuo Matsutarō, "Nihon de saisho ni jidokashita Dairen denwa," in Teishin gaishi kankōkai, *Teishin shiwa*, 1: 403–9. See also *Kantō teishin jigyō 30-nen shi.* Russia had adopted automatic telephone exchanges in Harbin a year earlier.

82. Nihon denshin denwa kōsha, *Jidō denwa kōkan 25-nen shi*, 3: 434–42. MOC had installed a 300-circuit ATM automatic exchange the previous year as an experiment inside the ministry, but it was destroyed in the 1923 earthquake (ibid., 1: 17).

83. For a book on Japan's postal presence in China, written from a philatelist's perspective, see Mosher, *Japanese Post Offices in China.*

often with nothing more than a dozen straw shoes and several pounds of rice on their back. As a result, telegraphic business, particularly in the category of lengthy press telegrams, increased dramatically.[84]

Japanese influence in China also came in the form of advisors to the Chinese government on telecommunications. To be sure, as recipients of foreign advice only decades earlier, the Japanese were latecomers in China. In 1902, former MOC Director of Telecommunications Yoshida Masahide was invited by the Chinese government to offer advice on telecommunication matters. Two years later, MOC engineer Tsujino Sakujurō joined him and would remain in China as the chief engineer of the Beijing Telephone Office for over three decades. In terms of personal influence and ambition, however, few would match that of Nakayama Ryūji.

Born in rural Niigata in 1874, Nakayama went to Tokyo at the age of ten after his father's weaving business failed. While working as a servant boy in the homes of American missionaries, he acquired a good command of English and later entered the newly established Tokyo Post and Telecommunication School. After graduation, he joined MOC as a junior technician. Between 1895 and 1903, Nakayama went on several study trips to Europe, including one year of study in Germany and England. When Japan took over the telecommunications facilities in Korea, Nakayama was sent there twice. He had already become a promising official well versed in international developments, when he was called upon to serve as an advisor to the new Chinese government on telecommunications matters in 1913. These extremely well-paid positions with little formal duties were designed by Chinese President Yuan Shikai to enlist Japanese support for his imperial ambitions. Nakayama was initially reluctant to accept a position that offered little excitement. He changed his mind after Mitsui Bussan's general manager Fujise Seijirō, widely known for his knowledge

---

84. Sugiyama Etsuzō, "Chefoo denshinkyoku no omoide," in Teishin gaishi kankōkai, *Teishin shiwa*, 1: 95–98. The Japanese had severed the Russian cable during the Russo-Japanese War and built a new cable after lengthy negotiations with China following the war. All but seven nautical miles from Chefoo of the cable belonged to Japan. However, due to the existing agreement with GNTC, the cable was limited to local traffic between Dalian and Chefoo (*Kaiteisen no hyakunen*, 168–70).

of Chinese affairs, convinced him that China after the Republican Revolution of 1911 offered great new opportunities.[85]

During the next decade and a half, Nakayama Ryūji became a major voice calling for the modernization of Chinese telecommunications as well as the expansion of Japanese influence there. To Nakayama, the two were synonymous. Well informed about developments in Western countries and fluent in both English and German, Nakayama was an unwavering champion of extending Japanese influence in China in competition with the Western powers. The second year in his post, he proposed to the Chinese government that it build a great wireless station in China, in direct opposition to a similar proposal by the British Marconi Wireless Company to link China to its worldwide chain of wireless stations. The following year, Nakayama drafted another proposal on the improvement of the Chinese telegraph administration and expansion of the telephone service, both to be financed by Japanese loans. Nakayama also devoted much effort to education in China and gave lectures in English at the Jiaotong (Communications) University. Under his auspices, several Chinese students were sent to study telecommunications in Japan.[86]

Japan's presence in China's telecommunications grew rapidly during the war in Europe. After Japan's attack on the German stronghold of Qingdao in the Shandong peninsula, Japan's cable-laying capabilities enabled it to sever Germany's Shanghai–Qingdao cable and, in May 1915, to connect it between Qingdao and its own naval base in Sasebo.[87] Thanks to the combination of personal efforts and good timing, as well as progress in Japan's telecommunications industry, Japan also gradually gained a foothold in the telecommunications market in China. In 1915, the Chinese government planned an expansion of telephone facilities

---

85. For an authorized biography of Nakayama, see Matsuoka Yuzuru, *Nakayama Ryūji*. For a personal reminiscence of his years as an advisor in China, see Nakayama Ryūji (as told to Watanabe Otojirō after World War II), "Shina seifu komon no koro," in Teishin gaishi kankōkai, *Teishin shiwa*, 1: 261–65.

86. Matsuoka, *Nakayama Ryūji*. Many of Nakayama's proposals and suggestions to the Chinese government were published by him in Chinese as *Zhongguo dianzhen yijianshu* (Beijing, 1919). For his numerous reports to the Japanese government on Chinese developments, see Nakayama Ryūji, *Nakayama Ryūji gishi shokanshū*.

87. Nippon denshin denwa kōsha, Kaiteisen shisetsu jimusho, *Kaiteisen hyakunen no ayumi*, 195–99.

in the city of Wuhan. Three Japanese companies, represented by the joint venture China-Japan Industrial Corporation, entered the fierce competition in bidding against three British and American companies for the project, which was worth nearly a million yen. Thanks to Nakayama's efforts, the Japanese companies won the bid. To avoid competition among Japanese companies in the China market, these three cable and wire manufacturers formed a cartel headquartered in Shanghai. The work was completed in two years under the supervision of an ex-MOC engineer, marking the first successful export of Japanese telecommunications products and skills to China.[88]

The war in Europe kept most Western powers preoccupied and exhausted, while providing an excellent opportunity for Japan to bolster its influence in China. The Twenty-One Demands imposed by the Ōkuma cabinet on China's Yuan Shikai government in 1915, signaling Japan's intention to assume substantial influence over political and economic life in China, was the most notorious example. Incensed by the strong opposition to such high-handed diplomacy both at home and abroad, the Terauchi cabinet in Japan switched to a "yen loan" diplomacy by extending to China's central and local governments various large loans, often with many strings attached.[89] This was made possible by the fact that Japan had become a creditor country for the first time in modern history. In 1918, the China-Japan Industrial Corporation signed an agreement with the Chinese government, providing the latter with 10 million yen for telephone expansion in China. The agreement specified a seven-year period during which Japanese companies would be given preference in providing telephone equipment. If the telephone expansion in China proceeded smoothly, the Japanese expected a demand for 30 to 40 million yen worth of equipment. The agreement also stipulated that a Sino-Japanese joint venture would be established for the manufacturing of cables and wire, which would receive various protections from the Chinese government. China Electric Company, capi-

88. Sumitomo denki kōgyō kabushiki kaisha, *Shashi Sumitomo denki kōgyō kabushiki kaisha*, 251–52.

89. For a detailed discussion of Japanese policies, see Frederick R. Dickinson, *War and National Reinvention: Japan in the Great War, 1914–1918* (Cambridge, Mass.: Harvard University Asia Center, 1999).

talized at 1,500,000 yen, was founded by the Chinese Ministry of Communications and the Japan-China Industrial Corporation in 1919.[90]

By the end of the Great War, Japan had greatly expanded its overall in the telecommunications field in China, beginning with the strategic cable links and telegraph offices on Chinese soil. In addition, Japan placed advisors in important posts in the Chinese government, extended several high-profile loans to China in the telecommunications field, as well as secured contracts to supply telecommunications equipment. By any standards, Japan had not only caught up with the West in China but increasingly seemed to be winning the game of informal imperialism.

In the half century after Commodore Perry's visits, the novel communication technology once greeted with "childlike delight" had been studied, domesticated, and fully utilized in Japan. The revolutionary means of communication—both telegraph and telephone—was thus woven into the very fabric of modern Japan, which had transformed itself from a closed island-country to a centralized nation-state, an industrializing and trading nation and a rising military power in the region.

As this chapter illustrates, telecommunications also played a crucial role in the creation and management of Japan's colonial empire in East Asia. Rapid means of communication facilitated overseas military operations and helped consolidate colonial control in Taiwan and Korea. By the time of World War I, Japan also made significant inroads in the informal empire in China, as reflected in the Nagasaki–Shanghai submarine cable and numerous other communications assets. The new communications technology was not an unproblematic tool of empire for Japan, however. While submarine telegraph cables enabled Japan to take advantage of the global telecommunications network and extend its influence over its neighbors, its technological dependency on the West meant such benefits came at a cost to its own autonomy.

The Japanese state loomed large in this early history of telecommunications, as many Japanese historians long ago pointed out.[91] Indeed,

---

90. Nihon keieishi kenkyūjo, *Sōgyō 100-nen*, 138–39.

91. Takahashi Tatsuo, *Nihon shihonshugi to denshin denwa sangyō*. Amazawa Fujirō and others also emphasize the "state character" of telecommunications in prewar Japan; see Amazawa, ed., *Gendai Nihon sangyō hattatsu shi XXI*.

from the beginning, the Japanese state exerted monopolistic control over telecommunications operations at home, on the grounds of national interest. However, as we shall see, state control was neither omnipotent nor static. As new communication technologies emerged on the horizon after the turn of the century, more assertive business interests challenged the principle of state control at home, just as Japan searched for alternatives to bolster its communication influence abroad.

CHAPTER 2

Wireless and the Crisis in the Informal Empire

In the early morning of May 26, 1905, an armada of a dozen or so Russian warships sailed in thick fog in the East China Sea toward the Korea Strait. They were part of the Russian Baltic Fleet on a long journey around the world to rescue the besieged Russian troops in the Far East, then locked in a bitter battle with the empire of Japan. The armada was soon spotted by the *Shinano maru*, one of several Japanese merchant ships serving as an early warning system against the approaching Russian fleet. Immediately, short messages indicating the exact locations of the Russian warships were sent from the rudimentary wireless set on board the Japanese ship and relayed to Admiral Tōgo Heihachirō's flagship *Mikasa*. These critical reports confirmed the intent of the Russian fleet to sail through the Korea Strait instead of the Tsugaru Strait to the north and gave the Japanese Combined Fleet just enough time to prepare for the ambush at Tsushima.[1] Just as Japan's resounding naval victory sealed the outcome of the war with Russia, that single wireless transmission would acquire a legendary place in Japan's naval and communications history.

The wireless was one of the most dynamic technologies of the early twentieth century. In less than a decade after its invention, the spark technology used in the earliest machines was largely replaced by the continuous wave. The invention of the vacuum tube, a key component

1. Nakamura, "Nichi-Ro kaisen to tsūshin—Nihonkai kaisen no musen tsūshin," *DJ* 32.4 (April 1982): 35–40; Jolly, *Marconi*, 148.

of wireless transmission, made possible the crucial transition from the cumbersome long wave to the highly efficient shortwave transmission. If technological change can create new opportunities, it can also challenge existing balances of power, threaten the stability of established institutions, and even exacerbate tensions.[2] As the Japanese soon discovered, the Age of the Wireless created both new opportunities and new challenges at home and abroad. Despite Japan's hard-earned success with land and submarine telegraphy, rapid technological development and a growing domestic demand for telecommunications service would pose a fresh dilemma for the Japanese government. It was also in the Age of the Wireless that Japan's impressive inroads into telecommunications in neighboring China encountered fierce challenges—technological and political—from other powers as well as the Chinese. Japan's failure to meet such challenges in its informal empire resulted in the military takeover of northeastern China, an event that greatly changed the contours of Japan's overseas expansion.

HARNESSING TECHNOLOGY FOR AUTONOMY AND INFLUENCE

First tested across the English Channel in 1897 by the Italian Guglielmo Marconi, the wireless brought about a revolutionary change in communications. This time, Japan responded to the technological innovation with remarkable speed. As the wireless enabled ships at sea to communicate with shore stations or other ships, the Imperial Japanese Navy in particular showed a keen interest in the potential of the new technology almost immediately. As early as 1900, the Navy set up a research committee and produced a prototype wireless set in 1901, named "Type 34" to mark the thirty-fourth year of the Meiji era. It had a maximum range

2. For an excellent discussion of the technological progress and the geostrategic significance of the wireless technology, see Hugill, *Global Communications Since 1844*, 83–138. For an early perspective on wireless in international affairs, see Schreiner, *Cable and Wireless and Their Role in the Foreign Relations of the United States*, especially chap. 5. See also Headrick, *The Invisible Weapon*, 116–37. For a discussion of its wider effect, see Kern, *The Culture of Space and Time, 1880–1918*, 65–67.

of 18.5 nautical miles (34 km). By the end of 1903, Navy engineer Kimura Toshikichi improved the set into a new "Type 36," which had a maximum range of 80 nautical miles (150 km).[3] Having succeeded in manufacturing small wireless sets, the Navy ordered their installation on all Japanese navy vessels on the eve of the Russo-Japanese War. The early detection of Russia's Baltic Fleet by the Japanese, relayed to Japan's Combined Fleet through these wireless sets, contributed to the latter's decisive victory at Tsushima. Japan's adroit use of this new telecommunications technology—the first time in any naval battle—would go on to acquire legendary status in subsequent histories.[4]

Harnessing Wireless for Autonomy and Influence

As the new technology gradually but steadily improved, the wireless was transformed from a tool used primarily for navigation communication into one of land-based long-range communication. Wireless communications effectively ended the monopoly of submarine cables and land lines over telegraphic communications. Britain, the world's leader in submarine cable communications, was quick to detect the strategic dimension of the new wireless technology. Already in 1908, the Marconi Wireless Company proposed an Imperial Wireless Communications System for the British empire, consisting of a chain of wireless stations around the world. In 1913, the British government signed an agreement with Marconi for construction of such an empire-wide network.

Although the Imperial Japanese Navy immediately recognized the tactical value of wireless communication even before the Russo-Japanese War, appreciation of its multifaceted strategic importance in Japan was mixed. Having made use of those simple wireless sets on its ships with great success, the Navy went on to build a land-based wireless station in Funabashi near Tokyo. The Army also set up a Wireless Telegraphy Research Committee in 1910, but still placed its faith in

---

3. Nakamura, "Know-How," *DJ* 31.9 (September 1981): 40; "Nichi-Ro kaisen to tsūshin," *DJ* 32.4 (April 1982): 36; Nihon musen shi hensan iinkai, *Nihon musen shi*; Nihon kagakushi gakkai, comp., *Nihon kagaku gijutsu shi daikei* (19): *Denki tsūshin*, 181–90.

4. Nakamura, "Nichi-Ro kaisen to tsūshin—Nihonkai kaisen no musen tsūshin," *DJ* 32.4 (April 1982): 35–40; Jolly, *Marconi*, 148.

wired communications.[5] The Ministry of Communications found wireless a highly useful supplement to fixed-line telecommunications facilities that were broken or overburdened. When Japan's cables to Taiwan were interrupted in 1912, the government managed to transmit public telegrams from wireless stations for the first time. Wary of any erosion of its control, the MOC drafted a Wireless Telegraphic Communications Law, passed in the Diet in 1915, to ensure the state monopoly of this new form of communication.

World War I was a wake-up call. As one postwar Japanese proposal described, the new wireless technology could perform important functions in addition to communicating between one's own units or ships at sea. They would include reception and deciphering of enemy military communications; detection of the direction and location of enemy troops; monitoring enemy propaganda communications and responding to propaganda; completion of communications routes for intelligence-gathering; and perfection of diplomatic and business communications.[6]

Government leaders were not the only ones in Japan who embraced the long-range wireless communications. Rapid economic growth in Japan also rendered existing telecommunications inadequate. The Great War in particular had created an unprecedented business boom in Japan, sharply increasing the traffic on the existing telecommunications infrastructure. Whereas in 1913 some 33 million telegrams were sent in Japan, 75 million were sent in 1919. The inadequate facilities caused much chaos in telegraph and telephone services in Japan during 1918–19. Japan's international communication increased even more sharply. Between 1914 and 1920, as trade soared between Japan and the United States, telegraphic traffic between the two countries increased fivefold. Because of the limited capacity of cable connections between the two countries, delays in telegraphic service—sometimes as long as a week—were frequent.[7] In 1921, the MOC built a high-power wireless station in Iwaki in Fukushima. Two years later, the Great Kantō Earthquake dramatically demonstrated the effectiveness of wireless: when wired

---

5. For a contemporary report, see H. C. Huggins, "Radio Development in Japan," *Far Eastern Review* (hereafter *FER*) 18 (July 1922): 431–34.

6. "Gunji jō yori mitaru min'ei musen denshin jigyō" (January 1923), reprinted in *Shibusawa Eiichi denki shiryō* 52: 31–36. Although unsigned, the document was most likely written by the military.

7. Kokusai denwa kabushiki kaisha, *Kokusai denki tsūshin kabushiki kaisha shi*, 1–4.

telegraph connections out of Tokyo were put out of operation, a lone wireless operator at the Iwaki Wireless Station was credited with first breaking the news to the world.[8]

And if adding wireless to the existing wired connections was the desirable solution at home, wireless acquired crucial importance in international communications for Japan. The MOC came to view the advent of long-range wireless communications as a blessing in its effort to end foreign dominance of its international cable routes: traffic over wireless facilities operated by Japan could reduce payments to the foreign cable company (the GNTC). Wireless communications required fewer set-up costs than submarine cables and would encounter less difficulty regarding landing rights in foreign territories, so they were deemed most desirable in establishing direct communications with countries that were of great importance to Japan.[9]

To many contemporary observers, the war in Europe was a propaganda war, with Great Britain widely viewed as the victor in manipulating information.[10] Japan was particularly concerned with Germany's relentless propaganda activities in China at the beginning of the war. The propaganda measures of the belligerents as well as what some called Japan's "bitter experience" in being unable to initiate countermeasures during and after the war once again reminded the Ministry of Foreign Affairs that "the future of our diplomacy depended on whether the governments and peoples of the Powers can properly understand our position and policies."[11]

In early 1924, the government set up a Committee for Investigating International Information and Communication.[12] The committee brought together officials from the Army, the Navy, and the Ministries of Foreign Affairs, Communications, and Finance. Many experts in information and intelligence matters—for example, section chiefs in the Foreign Ministry's Information Department—hosted the monthly

8. "Radio's Part in Japan's Reconstruction," *FER* 19 (October 1923): 647.

9. Nomura Yoshio, "Taigai musen denshin no kakuchō ni tsuite," *TKZ* 332 (April 1936): 145–49; Kobayashi Takeji, "Nihon musen denshin kabushiki kaisha-hō no kaisei ni tsuite," *TKZ* 345 (May 1937): 44–49.

10. The classic study of this subject is Harold D. Lasswell, *Propaganda Technique in the World War* (New York: A. Knopf, 1927).

11. Quoted in Gaimushō hyakunenshi hensan iinkai, *Gaimushō no hyakunen*, 1030.

12. "Kokusai jōhō denshin chōsa iinkai kiroku," MOC Records II, 297.

meetings. The Navy and the Army sent their section chiefs in charge of telecommunications and intelligence; their delegates often outnumbered those from the ministries.[13] The establishment of such an interministerial committee indicates widely shared awareness on the part of the government of the great potential of the new communications technology as well as its challenges. Given these promises of the new technology, the government began studying measures to boost Japan's communications capability via wireless. An Army-Navy-Communications Joint Committee investigated high-power wireless for overseas communications, reaffirming the need to "quickly establish Japan's superior position in international communications in East Asia." The joint committee proposed that wireless telegraphy be given precedence over cable in the future expansion of Japan's international telecommunications facilities and called for direct wireless communications with foreign countries without relays in third countries.[14]

The challenge was to find the financial resources to put Japan on the map of international wireless communication, a highly dynamic and competitive field after the Great War. Although the Imperial Diet agreed to a seven-year expansion of telecommunications facilities at a cost of 73.8 million yen, this was soon scaled back due to the Minseitō Cabinet's postwar retrenchment policy and subsequent budget cuts. The devastating Great Kantō Earthquake further complicated the picture as recovery programs drained the government's finances and forced further drastic cuts.[15]

The rapid pace of technological development added a sense of urgency for Japan. As Japan considered expanding its wireless operations, international competition was intensifying as other countries—notably Germany, France, and especially the United States—all joined in. Since wireless technology was far from perfect, and the problems of static and interference remained to be overcome, the long-wave frequencies

---

13. "Kokusai jōhō denshin chōsa iinkai dai-3-kai kiroku" (June 13, 1924), MOC Records II, 293.

14. "Rikugun daijin, kaigun daijin, teishin daijin ni teishitsu taigai dai musen keikaku ni kansuru sanshō kankei kanri no chōsa hōkokusho" (received on August 12, 1924), MT 3.6.11.23, JMFA Microfilm, Library of Congress.

15. Teishinshō, *Teishin jigyō shi*, 3: 751; Nihon musen shi hensan iinkai, *Nihon musen shi* (5): *Kokusai musen shi*, 155.

then suitable for wireless communication were limited to the 8,000- to 30,000-meter range. As a result, the 130 or so suitable frequencies had to be allocated to operating stations by the International Telegraphic Union. The scramble for the frequencies best suited for long-distance communications was a fierce one. By one account, there were only 69 frequencies in the most efficient category, the 10,000- to 20,000-meter range; 46 of them were already being used by existing stations. The "Five Powers"—Britain, France, Italy, Japan, and the United States—demanded a total of 111 at the 1920 preliminary telecommunications conference in Washington.[16] Seeking to secure a sufficient share of radio frequencies to fulfill its ambitions of further expansion, a financially constrained Japanese government was faced with the stark possibility of losing out on the most desirable channels if it waited too long.

A compromise solution had to be found to meet these challenges. The Army-Navy-Communications Joint Committee recommended that private enterprise in Japan be allowed to build long-distance wireless facilities and even submarine cables while operation and management remained in government hands. The Ministry of Foreign Affairs also found incentives for such a formula. Since Japan planned to "operate wireless in East Asia, especially in China, on a large scale," as the ministry realized, it would be convenient to have a nongovernment company that was also involved with wireless in Japan. The actual operation and delivery must remain in government hands to ensure the security of diplomatic and military communications. This would also enable the government to exercise control over news sent and received by news agencies, both Japanese and foreign. As wireless stations around the world became more powerful and began to broadcast news that contained propaganda, the ministry argued the government must be able to select the news to be broadcast in order to avoid damage to its goals in foreign affairs.[17]

---

16. Teishinshō, "Nihon musen denshin kabushiki kaishahō seitei riyū" (March 1925), in *Shibusawa Eiichi denki shiryō*, 52: 59; Anazawa Chūhei, "Chōba kakutoku ni doryoku suru kakkoku to sanshō kyōgikai," in Teishin gaishi kankōkai, *Teishin shiwa*, II: 315–24. Another source gives 67; see "Developing Wireless in Japan," *FER* 24 (October 1928): 476.

17. Memo of the five-ministry meeting on the private operation of the wireless (May 29, 1924), MT 3.6.11.23, JMFA Microfilm.

In 1925, the Japan Wireless Telegraph Company (JWT) was founded by a group of businessmen for the sole purpose of constructing and maintaining wireless facilities in Japan. The service operation remained in government hands. This required a novel reinterpretation of the Telegraph Law so that the government monopoly was understood to apply to telecommunication (traffic) but not to telecommunications (facilities). To consolidate limited resources, the Japanese government turned over its Iwaki Wireless Station and other sites in Mie and Aichi prefectures, valued at 2.3 million yen, to the JWT as capital-in-kind. However, the company did not have to pay dividends on the government's shares for ten years unless dividends paid to private shares exceeded 8 percent. With an authorized capital of 20 million yen, the JWT was expected to open communications routes both with Europe and with Southeast and East Asia.[18] Faced with the imminent external danger of losing valued frequencies and under financial distress, on the basis of broad domestic agreement, the Japanese government was flexible enough to modify the principle of a complete state monopoly of telecommunications service for the first time. The JWT became the first special company engaged solely in constructing and maintaining communications facilities for the government. This public-private partnership in telecommunications operations would develop further in the years to come.

If the 1920s can be considered the Age of the Wireless, this had much to do with the beginning of radio broadcasting. Begun in 1925, broadcasting boasted nearly 780,000 subscribers in Japan by 1930. Based on the continuous-wave wireless technology and occasionally referred to as wireless telephony, radio broadcasting started in Japan with certain similarities in public-private partnerships. As Gregory Kasza has shown, the Ministry of Communications recognized the state's limited capability and opted for a mixture of civil and bureaucratic controls of the new medium. Instead of taking on a new commitment of resources, the ministry issued permits to private groups to build transmission facilities in Tokyo, Osaka, and Nagoya in early 1925. Relying on the same Wireless Communications Law, the ministry was quick to assert state

18. Kokusai Denki tsūshin kabushiki kaisha shashi hensan iinkai, *Kokusai denki tsūshin kabushiki kaisha shi* (hereafter cited as *KDTKKS*), 4–20. For a Japanese study, see Sunaga Noritake, "Senzenki Nihon no taigai tsūshin jigyō no tokushitsu to tsūshin jishuken."

control and stipulated ministerial approval over finance, organization, and programming. Barely had a year passed before the ministry forced a merger of the three broadcasting companies into the Japan Broadcasting Corporation (NHK). While still replying on private capital, the state enhanced its control by appointing the top executives of the new institution.[19] Despite the financial restraints, the Ministry of Communications again made it possible to take advantage of the new opportunities afforded by the wireless technology without compromising control. Limited institutional adjustment was the key.

MITSUI WIRELESS FIASCO AND THE CRISIS IN THE INFORMAL EMPIRE

Given the heightened awareness of its importance during the Great War, it was no coincidence that telecommunications became an important part of the agenda at the postwar settlement. At the Paris and Washington conferences, the deposition of German submarine cables and the status of Yap Island, where important German cables in the Pacific landed, were hotly contested by the United States and Japan.[20] And because wireless transmission had been most strongly affected by the war, the biggest postwar drama in international communications had to do with wireless concessions in China. In the words of one American observer, "Today the greatest interest in the international aspects of electrical communications in the Pacific area, if not in the world, focuses on chaotic China."[21]

A Pyrrhic Victory

As international rivalry over wireless communication in postwar China intensified, Japan's endeavor to retain influence in telecommunications in China would meet new kinds of challenges. Nowhere was this better illustrated than in Mitsui Bussan's wireless enterprise. In 1916, the Danish advisor to the Chinese Ministry of Communications signed a secret

---

19. Kasza, *The State and Mass Media in Japan, 1918–1945*, 72–101, esp. 72–88.
20. Headrick, *The Invisible Weapon*, chap. 10, especially 173–77.
21. Tribolet, *The International Aspects of Electric Communications in the Pacific Area*, 86.

deal with the Chinese government to build a high-power wireless station near the capital, Beijing. Rumored to be linked to German interests, the agreement was strongly protested by Britain, whose Marconi Wireless Company had plans to build smaller stations in China, and from its ally Japan. Tokyo had long resented the foreign cable companies for their control over China's international communications. Now it saw an opportunity to secure influence over wireless communications in China.[22] In 1918, Mitsui Bussan, Japan's leading trading company, acted on behalf of the Japanese government and took over the deal in an agreement with the Chinese Navy Ministry. According to this agreement, Mitsui was to raise the 536,067 pounds needed to build a great "radiotelegraphic station with transmitting power and special receiving apparatus capable of direct radiotelegraphic communication with Japan, America, and Europe." The loan would be paid back in 30 years; during this time Mitsui would be in charge of the station's operation.[23] In the following month, a secret clause was added specifying that Japan would be given a 30-year monopoly over building wireless stations for the purpose of overseas communication in China.[24]

At first, Japan seemed to have scored a resounding victory, thanks to close government and business cooperation. From the start, Japan's Foreign Affairs, Finance, Army, Navy, and Communications ministries were in regular consultation. They agreed that the funds for the proposed loan from Mitsui were to come from the government's Emergency Military Budget in order to avoid questions in the Imperial Diet, which might reveal the secret agreement to the media and to foreign countries. The Japanese government would not only supervise Mitsui's operation of the wireless station but could assume direct control of the station from Mitsui at any time. A special office was established inside the Ministry of Communications in Tokyo to supervise the project in

---

22. Gaimushō, *Mitsui musen mondai keika gaiyō (zōho)*, 2. Accessed through JACAR. A recent study of the Mitsui wireless issue based on Japanese and Chinese sources is Kishi, "Tsūshin tokkyo to kokusai kankei," 229–48. However, the author fails to mention Japan's larger strategy to bolster its influence over China's communications.

23. MacMurray, *Treaties and Agreements with or Concerning China, 1894–1919*, 1519. See also "Text of Sino-Japanese Wireless Installation Agreement," reprinted in *FER* 15 (May 1919): 399–401.

24. Nangyō Jūtarō, "Mitsui musen to tōji no Chūgoku," in Teishin gaishi kankōkai, *Teishin shiwa*, 1: 268–71.

both administrative and technical matters. Anticipating protests from the Great Northern and the Great Eastern companies, Japan insisted that these cable companies' monopolies did not extend to wireless communications in China.[25]

Technologically, the project posed no small challenge to Japan. As a Japanese-language weekly in Beijing pointed out, there was widespread skepticism in Chinese and foreign circles that Japan could build the largest wireless facility in East Asia.[26] Japan sought to modify the high-cycle transmitter in the original Danish design, which had been chosen only because the Danes had copied all designs from Germany, and no other countries were capable of manufacturing the equipment at the time. When the Japanese government engineer tried to present a different design blueprint, changing the relatively new high-cycle transmitter to the older arc type and reducing the number of antennas from six to two, the Chinese immediately rejected it. Mitsui also attempted to reduce the height of the masts from 250 to 210 meters, which became a subject of prolonged disagreement with the Chinese Naval Ministry. Other alterations, such as replacing the AC generator with a simple turbine engine, although allowed by the agreement, caused additional concern among the Chinese and other radio engineers that the Japanese simply could not complete the original task. They were proved wrong, however. Except for the boilers, all equipment for the wireless station, including the 500-kW radio frequency generators and the 1,000-horsepower steam turbine, were manufactured in Japan. Installation was completed in early 1923, and the station made successful contact with wireless stations in Europe for the first time. The Japanese were understandably proud.[27]

---

25. "Shina ni okeru musen denshin ni kansuru keiyaku jisshu hōhō gaiyō," *Kobun ruiju 42-hen 25-kan*, cited in Sunaga Noritake, "Chūgoku no tsūshin shihai to Nichi-Bei kankei," 173; Nangyō Jūtarō, "Mitsui musen to tōji no Chūgoku," in Teishin gaishi kankōkai, *Teishin shiwa*, 1: 272.

26. "Tōyō ichi no musen denshin naru," *Pekin shūhō* (August 18, 1923), reprinted in Fujiwara Kamaashi, *Pekin nijūnen*, 173–75.

27. *JSDB*, 3: 182–84. An unsigned article in the English-language *Far Eastern Review* welcomed the Mitsui effort as a remarkable attempt "to give China the best possible radio installation for the money to be expended." The original plans were changed, according to the article, because there were difficulties in obtaining necessary structural materials for the construction of the masts and because Beijing had no power plant to

Economically, the agreement exerted a heavy price on Japan. Initially, Mitsui's managers justified the deal by evoking national interest. "As a leading Japanese enterprise," they argued, "we cannot think just about our own profit. It is at least necessary to pay considerable attention to the position of our country beside the issue of profit."[28] However, a combination of flawed design and miscalculations, not to mention rising material and labor costs due to the wartime inflation in Japan, increased the total cost of the project by over 60 percent. To avoid "disgrace to Japan's prestige" and loss of control of the wireless rights, the Japanese government finally paid the bill, totaling more than 8 million yen, in 1925.[29]

The Mitsui Wireless Station encountered other problems. In mid-1919, not long after the Mitsui agreement, Britain's Marconi Wireless Company signed an agreement with the Chinese Ministry of War to provide several smaller wireless stations for domestic military communications. The first three were set up on China's north and northwestern frontier, and many more portable wireless telephone installations were exported to China for use by troops in the field. In addition, the Chinese National Wireless Company, capitalized at 700,000 pounds, was formed between the Marconi Company and the Chinese government to "manufacture and deal in wireless telegraph and telephone apparatus, material, and supplies, and to repair and maintain wireless installations now existing and hereafter to be established."[30] Although Japan's monopoly over Chinese international wireless communications (as opposed to domestic communications) was not immediately affected by this agreement, its ambition to expand into China's vast wireless market was significantly threatened.

Another blow to the Mitsui project came in October 1921. While its construction was still going on, the Chinese Ministry of Communications signed a secret agreement with the Federal Wireless Company of

---

deliver the requisite amount of energy to the station; see "The Shuang-Chiao Wireless Station," *FER* 20 (July 1924): 347. See also Fujiwara, *Pekin nijūnen.*

28. *Mitsui Bussan dai-6-kai shitencho kaigi gijiroku* (Tokyo, 1918), quoted in Sunaga, "Chūgoku no tsūshin shihai," 174.

29. See *JSDB*, 3: 182; and Nangyō Jūtarō, "Mitsui musen to tōji no Chūgoku," in Teishin gaishi kankōkai, *Teishin shiwa*, 1: 271–73.

30. "The Chinese National Wireless Co.," *FER* 15 (July 1919): 509; "China's New Wireless Telegraph and Telephone Installations," *FER* 15 (December 1919): 750–51.

the United States to build a powerful wireless station near Shanghai, China's commercial center. This precipitated one of the most difficult diplomatic problems between Japan and its two Pacific neighbors during the 1920s. The Japanese government strongly protested the American contract as a violation of the Mitsui agreement, while Washington resorted to the axiom of "Open Door," considering the Japanese claim a monopoly and therefore a violation of this time-honored principle.[31] Given the strong opposition of these countries and new emphasis on cooperative diplomacy, Japan sought compromise. In fact, at the Washington Conference dealing with the wireless problems in China, the Americans found the Japanese delegate "to have no definite proposal in mind and appeared willing, providing a use could be found for the Mitsui (Japanese) Station erected near Peking, not to press claims to monopolies or special privileges."[32]

In addition to the Mitsui Wireless, other of Japan's gains in China during World War I turned out to be short-lived. At the end of the war, the United States, France, Italy, and Japan maintained unauthorized telecommunications facilities in China. Japan had by far the largest number of them.[33] The Washington Conference between 1921 and 1923 raised the issue of the widespread practice of setting up unauthorized foreign wireless facilities in China. Japan not only faced pressure to withdraw its wireless facilities in China, but was also called upon to restore Chinese jurisdiction over the Shandong peninsula, including control of telegraph and telephone facilities in that area.[34] Sino-Japanese negotiations took place in Beijing after the Washington Conference and proved a protracted process. The Japanese agreed to return telephone

31. W. W. Willoughby, quoted in Tribolet, *International Aspects*, 104. For a contemporary account written from an American perspective, see Tribolet, *International Aspects*, 86–104. For a good summary, see H. O. Kung, "The Future of Radio in China," *FER* 24 (October 1928): 463–68.

32. The Technical Expert (Rogers), American Delegation at the Conference on the Limitation of Armaments, to the Secretary of State, in United States, Department of State, *Foreign Relations of the United States* (hereafter *FRUS*), 1922, 1: 839.

33. For a list of wireless stations in operation or under construction in China, see "The Wireless Monopoly in China," *FER* 22 (October 1926): 461–63; and Nihon musen shi hensan iinkai, *Nihon musen shi*, 9: 192–95.

34. For the text of "Resolution Regarding Radio Stations in China and Accompanying Declarations," see Willoughby, *China at the Conference*, 385–87; and Nihon musen shi hensan iinkai, *Nihon musen shi*, 9: 192–95.

offices in Qingdao and Jinan to China, but demanded that the Japanese language continue to be used in telephone exchanges in Qingdao as well as in several telegraph stations in Shandong. The use of the Japanese language inside China became one of the most sensitive issues. Since English had been used in telegraphy in China along with Chinese, the Japanese representatives insisted, there was no reason why Japanese-language telegrams could not to be accepted as well. In the end, both sides compromised by allowing Japanese to be used for another six months in the telephone exchanges.[35] Japan also agreed to a joint operation of the Qingdao–Sasebo submarine cable line, which had been converted from the German submarine cable linking Yantai (Chefoo), Qingdao, and Shanghai, and was considered by Japan as war booty. The agreement specified that until all foreign companies relinquished their telegraph rights in China, the Chinese government would allow Japan to operate the Qingdao end of the cable.[36]

Although the technical and financial difficulties associated with Mitsui wireless were eventually overcome, the victory turned out to be a hollow one for Japan. The work was severely delayed, and although the Japanese could blame this on the civil strife among Chinese warlords, the delay was all the more damaging from a technical standpoint. Due to the rapid progress in wireless technologies, high-frequency, longwave transmitters became dated within the matter of a few years, as countries shifted to shortwave wireless.[37]

Setbacks in the Informal Empire

The Mitsui wireless dispute was emblematic of the bigger problem Japan faced in its informal empire in the postwar decade. Several factors worked against Japan. First, itself dependent on Western countries for many advanced technologies, Japan was at a disadvantage in commercial competition with those countries in China. Unlike the textile industry, where Japan enjoyed a lead over its Western competitors by the early twentieth century, telecommunications was a capital-intensive field in

35. For minutes of the negotiations, see Duban lu'an shanhou gongshu bianjichu, comp., *Lu'an shanhou yuebao tekan*, 299–501.

36. *JSDB*, 2: 39–40.

37. Zhonghua minguo dianxin zongju, *Zhongguo dianxin jiyao*, 54.

which Japan was still trying to catch up.[38] Second, after the Twenty-One Demands episode of 1915, when Japan pressured the Chinese government into accepting existing concessions and granting new ones, Japan's actions in China were increasingly seen to harbor suspicious political motives and met Chinese resistance. Despite its justification on business grounds, Japan's insistence on a 30-year monopoly served only to strengthen Chinese belief that Japan was aiming to control China's political destiny. Third, the chaotic domestic politics in China also worked against Japan. When an American writer described this episode in the late 1920s, he noted certain irony in the "peculiar" fact that "Japan[,] which has the most unquestioningly loyal army in the world," would make the agreement with the Chinese Navy Ministry, while England, "who has boasted of her Navy for centuries," had negotiated with China's Ministry of War. Only an American company negotiated with the Chinese Ministry of Communications, which "seemed most logical."[39] Given Japan's deep involvement in Chinese domestic politics, the constant shifts in the power struggle in China during the civil war meant that Japan often sided with the wrong party. The Beijing government that the Japanese had once supported would lose power to the Nationalists in the south. The telecommunications loans Japan had extended to the warlord government to gain political favor, now in default, failed to deliver the political returns they had once promised.

The final nail in the coffin of the Mitsui wireless project was struck when the Chinese Nationalist government established its new capital in Nanjing in 1928. As Beijing lost much of its political clout and potential government traffic, Shanghai quickly became the logical location for meeting the communications needs of the new government as well as of China's commercial interests. Considered by the Japanese as "the test case for the solution of various unresolved issues in Japan-China relations," the Mitsui wireless fiasco ended up poisoning the atmosphere of international cooperation in East Asia.

To make things worse, Japan was also losing the advantage in telecommunications export it briefly enjoyed over its Western competitors

38. On Japan's textile operations in China, see Peter Duus, "Zaikabō: Japanese Cotton Mills in China, 1895–1937," in Duus et al., eds., *The Japanese Informal Empire in China*, 65–100.

39. Tribolet, *International Aspects of Electric Communications in the Pacific Area*, 90.

in China. Despite the civil war among various warlords, the following decade saw a rapid expansion of China's domestic telecommunications, especially with local and long-distance telephone facilities. The equipment installed in these new facilities was imported from Europe or America.[40] Japan was left out in the cold in all the new telecommunications ventures in China. This setback was particularly painful for Japan's new electronic industry, as it had counted on the Chinese market to recoup the heavy investment into costly machinery.

Last but not least, the Nationalist government in China that came to power in 1927 sought to harness modern technology for state-building and national unification, often by brute force. In August 1928, the Nationalist government promulgated its Telecommunications Law, placing the nation's telecommunications operation under the jurisdiction of the Chinese Ministry of Communications. After a year of internecine rivalry, the wireless administration under the National Reconstruction Committee was also turned over to the ministry.[41] In September 1929, the Nationalist government convened a national telecommunications conference that discussed domestic and international telegraph and telephone matters.[42]

As in Japan, modern technology such as the wireless had also become a powerful tool for gaining communications autonomy in China. In the field of international wireless communications, the Chinese government sought to break the deadlock by entering into contracts for the erection of totally new wireless stations near Shanghai, and assumed responsibility for breaching former contracts with foreign companies. At the beginning of 1928, a 500-kW shortwave built by the National Reconstruction Committee began communications with Manila, despite Japan's loud protest against what it saw as a violation of Mitsui's wireless monopoly. By the end of 1930, the Radio Corporation of America (RCA) and the

40. For telephone expansion, see the following *FER* articles: "Canton Automatic Telephone System," 25 (October 1929): 467–69; "The Automatic Telephone in Shanghai," 24 (January 1928): 20–22; "Shanghai Telephone Company Expands Service," 27 (September 1931): 541–46; "Hupeh's Long-Distance Telephones," 27 (July 1931): 446–47; and "Telephones in Chekiang," 27 (August 1931): 481–82.

41. Zhonghua minguo dianxin zongju, *Zhongguo dianxin jiyao*, 54–55; Youdianbu, *Zhongguo jindai youdian shi*, 176–77.

42. "Telegraph Conference Meets" (September 10, 1929), *North China Herald News*. Clipping found in MOC Records I, 163.

Société Française Radio-électrique had constructed an international wireless station outside Shanghai. Known as the Chinese Government Radio Administration, the new station operated from its central office in the Sassoon House inside the International Settlement. From early 1931, it was in constant communication with San Francisco (North and South America), Manila, Hong Kong, Java, Berlin, and Paris.[43]

In hindsight, a turning point had already been reached in late 1928. On an autumn day that year, more than a thousand people gathered at China's new international radio (wireless) station outside Shanghai. The opening ceremony began with an exchange of telegrams between Generalissimo Chiang Kai-shek and American President Herbert Hoover celebrating the new era of international communications for China. Matsunaga Tadao, chief of the Japanese Government's Telegraph Office in Shanghai, lamented that not a single Japanese was present at the ceremony, nor was the Rising Sun among the foreign flags flown at the site.[44] By then, Japan's hope of cementing its communications influence in China through a wireless monopoly had ended in utter failure. That same year, after well over a decade of service as the advisor to the government, Nakayama Ryūji left China when his contract was discontinued. Within a decade, Japan had lost many of the advantages it once held in telecommunications in its informal empire.

OFFENSIVE IN MANCHURIA

After coming to power in 1927, the Chinese Nationalist government launched the so-called revolutionary diplomacy in order to recover China's sovereign rights lost in the "unequal treaties" with foreign powers. In late 1928, the Chinese Ministries of Communications and

---

43. For detailed descriptions, see George Street and Cecil Bailey, "C. G. R. A.: China Completes the World's Air Circuits," and M. Pavlovsky and H. Sauve, "French Equipped International Wireless Station in Shanghai," both in *FER* 27 (March 1931): 178–80, 181–87. The authors were representatives of the two companies. For an authoritative account by a Chinese official on telecommunications development in China between 1927 and 1937, see Yu Feipeng, "Shinian lai de Zhongguo dianxin shiye," in Zhongguo wenhua jianshehui, comp., *Kangzhan qian shinian zhi Zhongguo*, 365–404; and Youdianbu, *Zhongguo jindai youdian shi*, 174–78.

44. Matsunaga Tadao, "Shinyu mudentai no fukkō to Nisshi kan denki tsūshin no konseki," *TKZ* 370 (June 1939): 39.

Foreign Affairs set up a joint committee to conduct negotiations to restore China's sovereignty over its international telecommunications. It adopted a "Basic Outline for Submarine Cable Negotiations," which stipulated that the Chinese government would assume ownership and operation of all foreign submarine cables in Chinese waters after 1931. Existing foreign monopolies would be abolished, while all land lines leased to foreign cable companies would be recovered. At the minimum, foreign cable companies would have to have a landing permit from the Chinese government and would not be allowed to engage in direct public dealings. The Chinese share of the telegraph tariff was to be increased. In 1929, the Chinese government announced that its existing treaties with foreign cable companies would not be renewed when they expired at the end of 1930.[45] If realized, these stipulations would have had grave consequences for foreign communications in East Asia, with Japan being the most severely affected.

Diplomatic Efforts

Fearing further erosion of its strategic telecommunications interests in China, the Japanese government took the 1930 negotiations seriously and made elaborate preparations. The MOC even made plans for the worst-case scenario: should the Chinese government withhold recognition of landing rights and use of the Japanese cable, the MOC would delegate a private company to operate the Shanghai end of the cable. Fully aware of what expiration of the two foreign submarine cable companies' monopolies would mean, the MOC proposed cooperating with the Great Eastern and the Great Northern, both in order to prolong the right of operating cables in Shanghai and to reduce competition among the three. If the two companies could not reach an agreement with the Chinese government, however, Japan would terminate its agreement with them. Moreover, Japan did not want to fall behind and proposed opening direct wireless communications between major Chinese cities and Tokyo and Osaka. To do so, Japan was even ready to abandon the monopoly claim by Mitsui's Wireless Station and convert it into an outstanding loan. Since further expansion of submarine cable

45. Youdianbu, *Zhongguo jindai youdian shi*, 179–80.

rights in China appeared impossible, Japanese officials thought that wireless was the new area to compete for in China.[46]

In contrast to China proper, however, Japan was not prepared to bargain over its interest in Manchuria. Already, by the late 1920s, Chinese efforts at modernizing communications and transportation began to threaten the status quo in Manchuria.[47] Zhang Zuolin, the Chinese warlord who once worked closely with the Japanese, launched a "Communication Project of the Three Eastern Provinces," which included building one of the most powerful shortwave wireless stations in East Asia. Built with imported German Telefunken equipment in 1927, the station in Mukden (Shenyang) enabled China to communicate directly with Germany for the first time. These programs not only continued after Zhang's assassination by Japanese army officers in 1928, but also expanded to include international wireless communication with the United States and France. In November 1930, the largest Chinese city, Mukden, also opened its automatic telephone service.[48] If many of the modernization schemes threatened the predominant positions of foreign cable companies like the Great Northern, they were particularly ominous for the Japanese.

Keenly aware of its military implications, the Japanese Army requested that the Japanese government do its best to avoid the issue of Japanese communications in Manchuria in its negotiations with China. If necessary, the Army was prepared to sacrifice some of Japan's telecommunications interests elsewhere in China to ensure that Manchuria's communications remained intact.[49] From Tokyo's perspective, Japan's strategic and economic interests in Manchuria would be in jeopardy especially if the control over submarine cables were lost. To counter Chinese demands to regain control over telecommunications in

---

46. "Ni-Shi denshin kōsho ippan hōshin," "Teikoku Nagasaki Shanhai sen min'ei keikaku an gaiyō," "Tai-Daihoku, Daito kaisha koshoan," all in MOC Papers I-75.

47. On the competition between proposed Chinese railway and Japan's SMR, see Matsusaka, *The Making of Japanese Manchuria, 1904–1932*.

48. Mantetsu, Harbin jimusho, *Manshū no densei*, 248–86; "Mukden Automatic Telephone Exchange," *FER* 27 (January 1931): 46–48; Youdianbu, *Zhongguo jindai youdian shi*, 174–76.

49. Rikugunshō, "Tai-Shi tsūshin seisaku ni kansuru gunjijo no iken," transmitted from Sugiyama Hajime (vice army minister) to Imaida Kiyonori (vice minister of communications) October 6, 1930 (MOC Records II, 200).

Manchuria, the Japanese argued that telecommunications were included in the administrative rights associated with the former China Eastern Railway and even appealed to "customary law." A survey of telecommunications in Manchuria published by the SMR in 1930, however, admitted that it was "regrettably difficult" to find a clear legal basis for Japan's telecommunications operations in Manchuria.[50]

In retrospect, the Japanese government negotiators did a splendid job in Shanghai. Led by Shigemitsu Mamoru, a young diplomat newly appointed minister to China whose star would rise after the mid-1930s, they managed to keep the negotiations bogged down over minor concessions. The talks produced little in the way of concrete results, in large part thanks to Japan's delaying tactics. Moreover, China's revolutionary diplomacy was often louder in words than in actions, since China had underestimated the resistance not just from Japan but also from other foreign countries. Indeed, when the Kwantung Army made final preparations for its attack on the Chinese garrison in Manchuria in September 1931—almost exactly a year after the negotiations began—all of Japan's telecommunications facilities in Manchuria and elsewhere in China remained intact.

Military Offensive

The Japanese military takeover of Manchuria had been planned for quite some time. Frustrated by what it perceived to be the steady erosion of Japan's interests under the increasingly independent warlord Zhang Zuolin in Manchuria and by Tokyo's reluctance to make an aggressive response, officers in the Kwantung Army began to take matters into their own hands. In June 1928, Kwantung Army conspirators had staged a successful explosion on the railway, in which they assassinated Zhang and hoped to take advantage of the resulting chaos as an excuse to intervene. Lack of coordination and support from home turned it into a fiasco. The action only served to stiffen the anti-Japanese stance of Zhang's son, Zhang Xueliang, known as the Young Marshall, who declared allegiance to the Nanjing government not long after. By 1929, senior Kwantung Army staff officers had concluded that they had to

---

50. Mantetsu, Harbin jimusho, *Manshū no densei*, 1: 383; Nippon denshin denwa kōsha kaiteisen shisetsu jimusho, *Kaiteisen hyakunen no ayumi*, 205–7.

act quickly or face the imminent re-incorporation of Manchuria into a unified China. Although the Nanjing government's revolutionary diplomacy turned out to be more gesturing, many senior army officers reached the same conclusion. The only remaining difference between staff officers in the Kwantung Army and their sympathizers in Tokyo was timing.[51] Given the larger forces at play, technology alone could not have altered the course of history at this juncture.

On the night of September 18, 1931, Japan's Kwantung Army, which was stationed in Manchuria, launched a military takeover of the three northeastern provinces of China. The rumor that the Kwantung Army was plotting action had been going around Tokyo for some time. Four days earlier, on September 14, after being admonished by the emperor as well as cautioned by the foreign minister, Japan's senior army leaders held an emergency meeting in Tokyo and decided to dispatch Major General Tatekawa Yoshitsugu to Manchuria to caution the restive officers against rash action. Leaving on September 15 and traveling by Japan's rail system through western Japan and the Korean peninsula, Tatekawa was expected to arrive in Mukden on September 18. The night of September 14, however, a staff officer in Tokyo sympathetic to the aspirations of the Kwantung Army sent a secret telegram to the Mukden Special Service Agency, alerting it that the general was coming to prevent the Kwantung Army from taking action. Reaching the recipients on September 15, the telegram was a shock to the Kwantung Army conspirators, who met that evening and, after tense discussion, decided to move the date of the planned incident from September 28 to September 18. Upon arriving in Mukden at 7:00 P.M. on September 18, Tatekawa was promptly taken to a Japanese inn, where he was entertained. About 10:00 P.M. that evening, an explosion was set off on the South Manchuria Railway (SMR) track north of the city, an incident that the Kwantung Army immediately used as a pretext for launching its premeditated attack on Chinese positions. By the next morning,

51. For a recent work discussing these developments, see Matsusaka, *The Making of Japanese Manchuria*, especially Chap. 9. An older Japanese work that examined developments in the Kwantung Army in detail is Shimada Toshihiko, *Kantōgun* (Tokyo: Chūō shinsho, 1965).

Mukden was in Japanese hands.[52] What might have transpired in Manchuria had the Kwantung Army officers not been tipped off, or had Lieutenant General Honjō Shigeru, the Kwantung Army commander, been told of the emperor's admonition by telegram immediately after the September 14 meeting?[53] The plot might have been called off or, more likely, postponed. There is little doubt that the secret telegram from Tokyo gave the Kwantung Army plotters time to hasten their preparations for attack. Thus, speed of communications became a critical factor in shaping the outcome of the Kwantung Army conspiracy.[54]

Japan's military operation in September 1931 was a smashing success. Even with reinforcements dispatched from the Korean Army without orders from Tokyo, the Kwantung Army had slightly over 14,000 men; the much larger Chinese force of some 300,000, however, offered little or no resistance.[55] Moreover, the Japanese military action benefited from vigorous training, superior equipment, and meticulous planning; it also took full advantage of excellent transportation and communications facilities operated by the semi-governmental South Manchuria Railway Company. Japanese employees in the Communications Bureau of the Japanese-administered Kwantung Leased Territory also joined the action the very night of the conflict, cutting off Chinese telegraph and telephone lines and repairing Japanese lines severed by the Chinese forces. They also censored incoming and outgoing Chinese telegrams.[56]

---

52. Seki Hiroharu, "The Manchurian Incident, 1931," in Morley, ed., *Japan Erupts*, 201–6, 222–30. According to Sadako Ogata (*Defiance in Manchuria*, 58–59*n*25), no record of this telegram exists.

53. This suggestion was raised at the meeting but not adopted, for fear that such a telegram might be misinterpreted (Yoshihashi, *Conspiracy at Mukden*, 155).

54. Other means of communications also played roles: after the meeting on September 14, Tatekawa immediately contacted Ōkawa Shūmei, known for his ultranationalist writings and links to right-wing groups. Ōkawa in turn arranged for one of his men to fly to Manchuria to notify Colonel Itagaki Seishirō, one of the Kwantung Army ringleaders. Arriving on September 16, the envoy was able to contact Itagaki on September 17. On the other hand, on the morning of September 18, Japan's consul general in Mukden sent a request to General Honjō to stop any possible provocations from Japanese elements. The request, sent to Liaoyang by express letter (*sokutatsu*), was intercepted by a staff officer and did not reach Honjō; see Seki, "The Manchurian Incident, 1931," in Morley, ed., *Japan Erupts*, 204–5.

55. For a recent work on the subject of Chinese resistance and collaboration, see Mitter, *The Manchurian Myth*.

56. Sakurai Manabu, "Jihen go ni okeru tsūshin jigyō no gaikyō," MOC Papers I-176.

Although the military phase of the takeover proved relatively easy, it was by no means a foregone conclusion that the Japanese military would be able to keep the spoils of war. Many civilian politicians and even some senior military leaders in Tokyo were initially opposed to the Kwantung Army action, but they were presented with a *fait accompli* and could not back off completely. Fanned by the jingoistic mass media, the public backed tough action. In taking over the telecommunications facilities belonging to the Chinese government, the Kwantung Army was particularly aware of the value of the high-power Mukden Radio Station, with its two 20-kW transmitters built by RCA and Telefunken, respectively, and one 10-kW Telefunken transmitter. As a broadcasting station, the Kwantung Army hoped that Mukden Radio would help justify its actions in Manchuria and rally public support in Japan. The station could also bring domestic news and even entertainment programs to the Japanese in Manchuria, which the Army considered "extremely beneficial" to unifying national opinion on Manchuria policy.[57] Less than two months later, when Japanese residents in Manchuria staged a mass assembly to voice their support of the military actions, the Kwantung Army used Mukden Radio to broadcast the event live to audiences in Japan, with the help of the Nagoya Broadcasting Station. This was the first time relay broadcasting was carried out between Japan and Manchuria. On New Year's Day of 1932, Army leaders in Mukden and Tokyo successfully had their speeches broadcast in both Japan and Manchuria, marking another first: a Japan–Manchuria exchange broadcast.[58]

In addition to broadcasting, the Kwantung Army anticipated an increased demand for telegraphic and telephone communications both within Manchuria and between Manchuria and Japan. Once the Kwantung Army broke out of the narrow SMR Zone, new communications facilities became necessary for controlling the vast territory of Manchuria. In January 1932, the Kwantung Army sought to operate telecommunications in Manchuria by setting up its own Special Communications

57. Rikugunshō, "Tōhoku mudentai no shori hōshin" (December 1931), reprinted in Gaichi denki tsūshin shi hensan iinkai, comp., *Gaichi kaigai denki tsūshin shiryō* (hereafter cited as *GKDTS*), 13: 88–89.

58. Shindo Sei'ichi, "Nichi-Man denshin denwa ni tsuite," TKZ 330 (February 1936): 148–49.

Department. Although headed by a major from the Kwantung Army's corps of engineers, the department of 40-odd employees was staffed almost entirely by Japanese civilians from the SMR and by MOC officials dispatched to Manchuria as adjunct members of the military. The department was put in charge of military communications, military propaganda aimed at foreign countries, relay broadcasting and phototelegraphy between Mukden and Tokyo, and local broadcasting. Due to damages to the existing land lines between Mukden and Harbin, for example, more than 200 Japanese-language telegrams a day had to be relayed by wireless from October 1931 onward. Moreover, after the Kwantung Army moved its headquarters from Liaoyang in the south to Changchun, now renamed Shinkyō (Xinjing in Chinese, also spelled Hsinking), the Special Communications Department started wireless telephone between these two cities. It was also prepared to operate international public communications if the necessary arrangements could be made.[59]

Communication Without Recognition

Japanese authorities in Manchuria were eager to reopen international telegraph service partly to gain international recognition for their *fait accompli* in the area.[60] In November and December 1931, the Army Ministry contacted the MOC and the Ministry of Foreign Affairs to seek their support. Nakatani Hikota, a former SMR employee now with the Kwantung Army's Special Communications Department in Manchuria, came to Tokyo to discuss the matter with the MOC. Both ministries were concerned that it might appear improper to the international community that the formerly Chinese Mukden Radio Station was now being operated by the Japanese Army. The Army Ministry also agreed that direct military management of the station was unnecessary. Instead, it proposed placing the station under nominal joint operation by the

---

59. "Kantōgun meirei" (January 12, 1932), reprinted in *GKDTS*, 13: 92–93; *Kantōgun tokushu tsūshinbu gyōmu yōhō* 1: 1–9, diagram; Kantōgun tokushu tsūshinbu, *Kantōgun tokushu tsūshinbu gyōmu yōhō* 2: 26.

60. Kantōgun tokushu tsūshinbu, *Kantōgun tokushu tsūshinbu gyōmu yōhō*. Some Japanese officials in Manchukuo even considered floating foreign loans in building the new capital, Xinjing, in order to show that they had an "Open Door" in Manchuria. The idea was, however, bitterly opposed by the military, and the project had to be scaled down. See Koshizawa, *Manshūkoku shutō keikaku*, 100–108.

Kwantung Army and the new "Manchurian" government then being established. Inexperienced in such matters, the Japanese Army was concerned that the lack of a qualified operating agency might dismay foreign telecommunications companies when it came to business matters such as calculation of the telegram tariff. Nakatani assured officials in Tokyo that because there were existing agreements between Mukden Radio and foreign companies, this would not be a problem.[61] Nakatani's pragmatic prediction turned out to be quite correct.

Due to the opposition from the Army leadership in Tokyo, officers in the Kwantung Army had to abandon their original plan to annex Manchuria as a Japanese colony. In early 1932, they created the state of Manchukuo, which was nominally independent.[62] Despite the fact that Japanese military occupation of Manchuria met with general criticism and disapproval from the international community, the need for international public communications with Manchuria remained strong. Foreign residents in the region were numerous and most stayed. Although many Western governments censured Japan's actions in Manchuria, they nevertheless planned to retain their diplomatic posts, business operations, and other interests there. Communication with these establishments in Manchuria thus became indispensable. This was especially true for Germany and the United States, the two countries that had made prior wireless communications arrangements in the region with the Chinese authorities. The German and American companies that had sold equipment to Zhang were interested in reopening telegraphic service in order to receive debt payments from the revenue. As the German consul in Tianjin informed his Japanese counterpart, the German firm

---

61. Memo G1965 (December 7, 1931) about the visit to the MOC by the chief of the Army Ministry's Defense Section on November 30, and memo G1967 entitled "Hōten musen dentai ni okeru shōgyō tsūshin toriatsukau kaishi ni kansuru ken" (December 7, 1931), MOC Records I, 180. See also "Hōten musen denshinkyoku ni kansuru ken" (December 2, 1931), MOC Records II, 92. Years later, Nakatani was to reminisce that the bureaucrats in the MOC withheld support because the new Manchukuo had not received foreign government recognition or acceptance into the international wireless treaties, complaining that things would work better without the bureaucrats. See Nakatani Hikota, "Kokusai musen denshin no kaishi," in Manshū denden tsuiokuroku kankōkai, *Akai sekiyō*, 42.

62. On this subject, see Yamamuro Shin'ichi, *Kimera: Manshūkoku no shozō*; translated into English as *Manchuria Under Japanese Domination*.

Siemens was owed a total of $32,395.63 as of April 40, 1931. Similarly, the U.S. secretary of state told the Japanese ambassador in Washington that, given RCA's stakes in Manchurian telecommunications, Japanese authorities must not only provide proper protection of the wireless facilities there but also reopen them for public service.[63] Without some *modus vivendi* with the occupying Japanese authorities, however, nothing could happen. The Japanese sought to work with the Americans, who had a greater stake in the wireless loans. When RCA sent its Far Eastern agent from Japan to Manchuria in mid-March 1932, the Army granted him the favor of traveling on a Japanese airplane.[64] As one Japanese newspaper remaindered noted at the time, "In view of the forthcoming visit to Manchuria of the League of Nations Inquiry Commission, those concerned in the venture are eagerly looking forward to the success of the tests [for the resumption of communications]."[65] Here the Americans and the Japanese seemed to have found a common cause. Communication with the United States resumed in April and with Germany in July.

Left out of the picture were the Chinese. Japan's military takeover of Manchuria brought its uneasy relationship with China to the brink of war. In early 1932, as the Kwantung Army was preoccupied with mopping up Chinese resistance in Manchuria and with fighting international censure, Japanese forces provoked a fierce clash with Chinese forces in the international city of Shanghai. The Chinese government appealed to the Western powers and world opinion for assistance in stopping Japan's invasion.

After the establishment of the puppet regime of Manchukuo, China's Nationalist government responded with a policy of nonrecognition and instituted a "communication blockade." It immediately withdrew all government employees at the Northeastern Telegraph Administration from the region to Beijing. When Manchukuo released its own postage stamps, the Chinese Postal Service in the area was also withdrawn. The Chinese Ministry of Communications notified all foreign countries that mail and telegrams bound for China would not travel

63. Kantōgun tokushu tsūshinbu, *Kantōgun tokushu tsūshinbu gyōmu yōhō*, 109–11.

64. Memo G521 (March 17, 1932), MOC Papers I, 180. See also Gaimushō, *Nihon gaikō bunsho*, Series II, Part 1, 1: 29–30, 35–37, 40.

65. "Mukden Radio Station Plans to Resume U.S. Broadcasts," *Japan Advertiser*, March 25, 1932.

through Manchuria and ordered that postal and telegraph offices not handle communications with Manchuria.[66] Such a communications blockade, it was hoped, would deepen Manchukuo's isolation and might even bring about its collapse.

Japanese communications officials had anticipated the Chinese blockade and prepared countermeasures in advance, taking advantage of the apparatus of its informal empire in China to do so. Since Japan and China had not formally broken off diplomatic relations, all existing bilateral treaties between them were still in force. Therefore, Japanese authorities in Manchuria could simply reroute all post and telegraph traffic through the Japanese post and telegraph offices in the Kwantung Territory and SMR Zone, both of which were protected in bilateral treaties. This scheme worked. In this way, mail and telegrams from Europe continued to travel through Siberia and Manchuria.[67] As a result, the Chinese "communication blockade" proved ineffective.

When it learned about the ongoing exchange between Americans and Japanese over international telecommunications in Manchuria, the Chinese government was understandably dismayed. In early April 1932, Chinese diplomats in Washington asked the American government to restrain RCA from entering into a contract with the Japanese in Manchuria. The United States replied that the "American government had no authority by which it could prevent action of this type by an American company," while reassuring China that it was fully aware of the grave implications of such a contract in view of America's nonrecognition policy of Japan's occupation. Two days later, after meeting with the Chinese diplomat, Stanley Hornbeck of the State Department told an RCA representative that the transaction in Manchuria would have to be considered strictly as one that the corporation entered on its own responsibility and at its own risk since the American government could neither approve nor disapprove.[68] Although international communica-

---

66. Okumura Kiwao, "Man-Shi tsūyu mondai no kaiketsu," *TKZ* 318 (February 1938): 6–8.

67. "Tai-Man densen fusa ni taisuru taisaku," MOC Papers I-176. Japan would use the same argument—without success—to try to gain acceptance of the MTT at international conferences, portraying it as a Japanese company headquartered in Dalian, a Japanese leased territory.

68. Memo by Hornbeck of a conversation with Colonel Davis of RCA (April 11, 1932), in *FRUS*, 1932, 3: 685–86.

tions did resume, the Kwantung Army's bid for formal recognition had not worked out as expected.

Worse, the League of Nations dispatched the Lytton Commission to Manchuria in early 1932 to investigate the conflict. The Japanese authorities adopted a variety of tactics to strengthen their position. In late April, the MOC director of telecommunications sent a confidential telegram to the chief of the Kwantung Communications Bureau, instructing the latter to "send by mail copies of all outgoing and incoming telegrams of the League of Nations Commission" to Tokyo. Since the matter was highly secretive, it was to be directly handled by a ranking official. Moreover, all official telegrams sent by the Lytton Commission were to be transmitted via Japan unless a specific route was requested.[69] The MOC attempt at communication espionage did not seem to have helped. Several months later, the commission issued a report that called for return to the status quo before September 1931. This was not acceptable to Japan. By then, the Kwantung Army had created and stood behind its puppet state of Manchukuo. And after leaving the League of Nations in March 1933 to protest adoption of the Lytton Report, the Japanese government would face its own deepening isolation in the world. Ironically, although its international communications had been restored, Manchukuo remained largely a pariah in the international community. What the Chinese communication blockade failed to accomplish, the Japanese government achieved by itself.

The new and rapidly advancing technology of wireless telegraphy and telephony seemed to offer unlimited opportunities for expanding Japan's communication capacity at home and influence abroad. In reality, technological innovations often produce needed institutional adjustment as well as other unforeseen consequences.

As this chapter illustrates, state control over telecommunications in Japan was neither omnipotent nor static.[70] Although telecommunications

---

69. MOC Director of Telecommunications to Chief of Kwantung Communications Bureau, "Renmei chōsain no denpō ni kansuru ken" (April 22, 1932), nos. 1524–25, MOC Papers I-176. It is not clear how and whether the order was carried out.

70. Takahashi Tatsuo, *Nihon shihonshugi to denshin denwa sangyō*. Amazawa Fujirō and others also emphasize the "state character" of telecommunications in prewar Japan; see Amazawa, ed., *Gendai Nihon sangyō hattatsu shi XXI*.

were still considered vital for military operations and government functions, they were increasingly seen as indispensable for economic and business development as well, and their role in cultural affairs and in the formation of public opinion also came to be recognized. At home, problems with the state monopoly became increasingly obvious in the deteriorating telephone and telegraph service. This realization—coupled with the intense international rivalry for wireless frequencies—compelled the communications bureaucracy to seek new institutional arrangements for wireless communications.

Equally important, new wireless technology proved to be even more of a double-edged sword. Although the wireless gave Japan high hopes of gaining more autonomy in its international communications, similar attempts by China as well as intensified international rivalry in wireless communications in that country would undermine Japan's ascendance on the Asian continent. Despite its early success, Japan's expansion in its informal empire suffered serious setbacks in the 1920s. China's own state-building efforts and rising nationalism as well as Japan's failure to provide a technologically competitive alternative or to find a political solution contributed to this decline.[71] The Kwantung Army's takeover of the Chinese Mukden Radio, once the most powerful wireless station in China, epitomized Japan's failure of peaceful expansion.

A brief survey of the history of telecommunications in Japan in the early decades of the submarine cables and the wireless reveals the importance of both institutional control and technological change in the evolution of imperial strategy. Having successfully consolidated its formal colonies, Japan encountered increasing difficulties in the informal empire throughout the 1920s. Domestic and international political crisis, exacerbated by the worldwide economic recession, exerted a powerful influence in shaping Japan's new empire-building agenda on the continent. In the following decade, remarkable progress in fields such as telecommunications as well as institutional innovations continued to play a significant role in Japan's continental expansion. It is the interplay of technological innovation, imperial expansion, and institutional adaptations in the 1930s to which we now turn.

---

71. For a classic discussion of this subject, see Iriye, "The Failure of Economic Expansion, 1918–1931," 237–69.

PART II

Technology, 1931–1940

In practically annihilating space, the telegraph is one of the strongest links between distant countries, and its importance from a sentimental point of view should not be despised. There is no question that direct and unbroken Imperial telegraphy can do much, not only to stimulate commercial activity between the Mother Country and the Colonies, but also to strengthen that sense of unity and that community of feeling and policy on which the cohesion of the Empire under present conditions depends.

—Charles Bright, 1903

If the discovery of gutta-percha can be said to have overcome the oceans of Britain and America, the invention of the Pupin Coil probably drew continental Europe closer [together]. It is not an exaggeration, then, that the invention of the Non-Loaded [Carrier Cable] System created ties that bound together a new East Asia with powerful and secure cables.

—Watanabe Otojirō, 1943

CHAPTER 3

Toward a New Order on the Continent

In the center of Shinkyō, renamed from its Chinese name Changchun when it was made the capital of Manchukuo, lies Datong Square, a large rotary with a wide, tree-lined boulevard unknown in Japan proper. Under the Japanese, the city was transformed from a medium-sized railway town of 130,000 into a modern metropolis whose population would reach one million by the end of the war. Numerous Japanese architects, city-planning officials, and construction companies participated in the unprecedented project, making use of cheap and abundant Chinese labor.[1] Occupying an entire block at Datong Square, between the Manchuria Central Bank and the Capital Police Headquarters was a massive modernist building: the headquarters of the Manchurian Telegraph and Telephone Company (MTT). Two large stone *qilin* (in Japanese, *kirin*), a mythical Chinese creature known throughout East Asia, adorned its entrance.[2] Both the MTT's central location and its grand physical appearance reflected its importance to Japan's new empire-building project.

For the fourteen years between the Japanese takeover and the collapse of Japan's empire in 1945, Manchuria would be Japan's new frontier for national defense, industrial development, and mass migration. It would serve as the testing ground for state economic planning and bureaucratic reorganization, as well as for new forms of territorial control. Last but not least, Manchuria would become a launching pad for

1. Koshizawa, *Manshūkoku shuto keikaku.*
2. Miki Shūjō, "Sōgyō tōji no kaiko," *Akai sekiyō*, 19.

Japan's further expansion on the Asian continent. The unfolding events in Manchuria had a far-reaching effect on the country and the rest of East Asia. The process of empire-building in Manchuria would not only redefine the orientation of Japanese telecommunications policy but help reshape the techniques of Japanese imperialism as well.

BETWEEN STATE MONOPOLY AND PRIVATE ENTERPRISE

In early 1932, as it made final preparations to unveil the state of Manchukuo, the Kwantung Army launched a massive, systematic study of economic policies in post-occupation Manchuria. Moreover, at the Kwantung Army's request, the SMR had set up an Economic Research Council, based in its Research Section, to examine economic policy issues. Headed by Sogō Shinji, an SMR director who had supported the military from the beginning, the council consisted of 28 subgroups in six departments covering a wide range of topics.[3]

In a policy memo drafted in March 1932, the Kwantung Army clearly defined its major objectives in the area of communications: (1) to ensure that Japan controlled communications, (2) to satisfy military needs, and (3) to expand communications facilities. It also sought to smooth internal and external relations of the new state. Meeting military needs was of paramount importance, one of the Army's plans stated, but the goal would be difficult to achieve if establishing the communications network in Manchuria was left to the Japanese government, due to its well-known budgetary and other constraints. The plan cited examples in other countries to show that a special chartered company could improve efficiency. It proposed a gradualist approach. Although the ultimate goal was a semi-public special corporation under Japanese management, until foreign countries recognized Manchukuo, the status quo would be maintained. The new Manchukuo government would be in charge, with Japanese placed in key posts so as to exercise de facto control. After foreign recognition of Manchukuo, a special corporation would be established with Japan contributing the majority of the needed capital and personnel. Eventually, control of the corporation would pass from the new

3. On the founding of the council, see Yamada Kōichi, *Mantetsu chōsabu*, 100–110.

government to Japan completely. After two months' deliberation, the Kwantung Army finally adopted the "Communications Policy Toward Manchukuo" in late July. Underscoring the benefit of shared facilities and revenue, it included broadcasting under the mandate of the new telecommunications company. It also emphasized that Japanese military officers would participate in the establishment and operation of the company in order to carry out Japan's political and military demands thoroughly.[4]

Although the Japanese agreed that control over communications in Manchukuo was "absolutely necessary for the implementation of Japan's national policy," not all parties initially accepted the Army plan. In late March 1932, the Economic Research Council of SMR began to conduct its own deliberations on communications policies in Manchuria, with emphasis on the scope of operations, management methods, and institutional forms. It came up with a different proposal. The council concluded that, for the time being, telecommunications should remain a government monopoly divided into two separate spheres: (1) a monopoly by the Japanese government in the Kwantung Territory and the SMR Zone, and (2) a monopoly by the Manchukuo government in the rest of Manchuria. As the next step, it recommended the establishment of a "special company," incorporated under Japanese law, to operate telecommunications in both areas on behalf of both governments. Although its own study showed that the SMR owned only a small fraction of existing long-distance communication lines in Manchuria (5 percent, with the Kwantung Communications Bureau owning another 5 percent), the Economic Research Council saw a major role for the SMR in the creation of the new company to operate telecommunications in the two areas. To help ease the transition to a unified operation in Manchukuo, it suggested that telecommunications and broadcasting in Manchukuo initially be run by the SMR. Later, telecommunications in the Kwantung Territory would also become part of the SMR operation. Finally, all public telecommunications would be separated from SMR operations and become the responsibility of the

4. Kantōgun sanbobu sōmuka, "Man-Mō tsūshin tōsei narabi kanri ni kansuru hōsaku" (March 1932); Kantōgun tokushu tsūshinbu, "Man-Mō tsūshin jigyō tōsei ni kansuru iken" (March 1932); Kantōgun shireibu, "Tai-Manshūkoku tsūshin seisaku" (July 23, 1932), all in Mantetsu keizai chōsakai, *Ritsuan: Manshū tsūshin jigyō hōsaku*, 32–33.

special company. Having operated telecommunications in the railway zone, the SMR felt confident about expanding into all of Manchukuo and naturally preferred to retain its autonomy. Thus it suggested spinning off its telecommunications operations to create a new company for the entire area of Manchukuo.[5]

While the Kwantung Army and SMR were busy drawing up plans for running telecommunications in Manchuria, the Tokyo government was far from standing idle. Some officials there favored leaving matters such as communications to the Manchukuo government, on the grounds that it was important to maintain the appearance of Manchukuo as an independent state. The Ministry of Foreign Affairs, for instance, was concerned that the proposed joint Japan-Manchukuo ownership of telecommunications might be seen as violating the Open Door principle and hence alienating any potential foreign support for Japan's position. It proposed setting up the company as a Manchukuo corporation with capital from both Manchukuo and Japan and leaving the details to a secret bilateral treaty.[6]

Through officials posted in the Communications Bureau of the Kwantung Leased Territory, the Ministry of Communications kept a close watch over these developments. In addition the MOC dispatched several groups of officials immediately after the 1931 Manchurian Incident to investigate telecommunications conditions in the region. Okumura Kiwao, a talented young official in the MOC who toured Manchuria at the time, returned to the ministry to rally support for more MOC involvement.[7] Fujiwara Yasuaki, a section chief in the MOC, traveled to Manchuria with several younger officials. During their six-month stay, they discussed policies for communications in Manchukuo with the Kwantung Army. Fujiwara saw the Army's plan for private operation of telecommunications as a thinly disguised way of placing telecommunications under the Army's control and was con-

---

5. Miyake Mitsuharu to Sogō Shinji, March 22, 1932; Mantetsu keizai chōsakai, "Man-Mō ni okeru tsūshin jigyō no tōsei oyobi keiei hōshin an [kari kettei]" (May 1932), and "Manshū ni okeru denki tsūshin oyobi hōsō jigyō tōsei an" (June 1932), all in Mantetsu keizai chōsakai, *Ritsuan: Manshū tsūshin jigyō hōsaku*, 39–47, 15–19.

6. "Manshū ni okeru tsūshin jigyō ni kansuru ken" (August 9, 1932), *Shōwa zaisei shiryō*, microfilm.

7. Kubo Shigeru, "Nichi-Man yūbin jōyaku teiketsu," in Teishin gaishi kankōkai, *Teishin shiwa*, 2: 206–7.

cerned that this would damage Manchukuo's image as a sovereign state. Fujiwara's unease with the Army's high-handed approach is evident in a confidential letter he sent to Tokyo shortly after his arrival in Manchuria. Fujiwara candidly described the stark consequence of Japan's total domination:

> The angry voice against Japan can be heard everywhere, and people are nostalgic for the Zhang Xueliang era. Inside the government, it is obvious that [Manchurian] officials both high and low are suppressed by the Japanese, and they harbor discontent. If both official posts and profits are taken by Japan, all of Manchukuo will consider Japan an enemy. If that is the case, no matter how strong the Japanese troops are, can they confront 30 million people?[8]

Moreover, Fujiwara also felt that, given the underdeveloped infrastructure in Manchuria and the need to build many military communications facilities, it was financially impractical to leave telecommunications to a private company. Even with the future incorporation of telecommunications operations in the Japan-administered Kwantung Territory and the SMR Zone, a private telecommunications company almost certainly could not survive on its revenues alone.[9]

That such an argument came from MOC officials is hardly surprising. After all, it was the government that had operated telecommunications in Japan and its colonies since their inception. Although MOC skepticism toward a private telecommunications enterprise was not unexpected, its opposition to the Army plan caused considerable delay in setting up communications operations in Manchuria. This situation began to worry the Kwantung Army, which mounted an aggressive lobbying effort in Tokyo. Nakatani Hikota, the SMR employee who had been working for the Kwantung Army, waged an all-out campaign in Tokyo to overcome opposition to the Army plan in various government bureaucracies.[10] Surprisingly, he found some invaluable allies in the MOC itself. One supporter was an engineer, Kajii Takeshi. After making an extensive tour of Manchuria in the wake of the incident, Kajii became convinced that although a state monopoly was correct in theory, the

8. Fujiwara Yasuaki, "Manshūkoku yūsei sesshu shori keika" (August 1932), MOC Records I, 177.

9. Ibid.

10. Nakatani, "Kokusai musen denshin," in Manshū denden tsuiokuroku kankōkai, *Akai sekiyō*, 7–9.

Kwantung Army's plan for private management was more realistic for Manchuria. He tried to persuade his colleagues. With the help of Tanabe Harumichi, an influential ex-MOC bureaucrat, Kajii won over MOC Administrative Vice Minister Ōhashi Hachirō, who reopened the discussion within the MOC itself.[11]

In reality, the MOC position on state monopoly was far more complicated. By the late 1920s it had become aware of the financial problems of government-run telecommunications in Japan only too well. Faced with the chronic budgetary constraints in the past, the MOC had already considered various modifications to the government monopoly. For instance, in order to meet rising demand in Japan, MOC Minister Gotō Shinpei had once proposed a semi-private telegraph and telephone construction company modeled after the SMR, before the powerful Finance Ministry killed the plan. In 1927, then-MOC Minister Kuhara Fusanosuke, a businessman turned politician, had ordered a comprehensive investigation of how best to manage telecommunications in Japan. The results gave little cause for optimism for private management, however. MOC bureaucrats concluded that, if privately managed, telecommunications in Japan would not yield a profit of even 1 percent. The idea of private management remained largely theoretical within the MOC until 1929, when Finance Minister Inoue Jūnnosuke in the Minseitō Cabinet adopted a fiscal retrenchment policy aimed at balancing the budget without floating further loans. No longer able to receive enough funds for telecommunications expansion, the MOC faced the biggest crisis in its history. In 1929, it set up a Provisional Telegraph and Telephone Research Group to explore alternatives. According to its recommendation, all government telecommunications facilities were valued at 410 million yen, and could join with private capital to form a semi-public company capitalized at 600 million yen. In the end, the MOC opted for a compromise solution by accepting a Special

11. Kajii Takeshi, "Manshū denshin denwa kaisha setsuritsu no keii," in Teishin gaishi kankōkai, *Teishin shiwa*, 2: 225–31, and "Manshū denshin denwa kaisha setsuritsu tōji no omoide," in Manshū denden tsuiokuroku kankōkai, *Akai sekiyō*, 10–13. In these postwar reminiscences, Kajii explained the Army proposal solely in terms of assisting Manchukuo to establish telecommunications without hurting its appearance as an independent government. See also Tanabe Harumichi denki hensankai, *Tanabe Harumichi*, 186.

Account for Communications Service (SACS), separate from the General Budget, so as to free telecommunications from the government's financial constraints.[12]

Even though the MOC's own "privatization" plan did not work out, a number of younger officials became familiar with the flaws of rigid state budgeting and convinced of the need to reform the existing state monopoly over telecommunications. Although some MOC officials still considered a state monopoly not only superior in theory but even "the trend of the world," most realized that "external circumstances" made it impossible for the Japanese government to run communications in Manchukuo immediately. They finally agreed that a Manchukuo-Japan joint-venture company was preferable to state management by the Manchukuo government.[13] After the MOC finally overcame its internal disagreement and withdrew its opposition to the Army plan in mid-1932, the various parties agreed on private management as the best solution to future telecommunications operations in Manchuria.

In March 1933, Japan's ambassador to Manchukuo and the Manchukuo minister of foreign affairs signed the "Agreement Concerning Joint Management of a Communications Company." Two months later, the cabinet in Tokyo approved the plan and ordered a committee formed to prepare for establishment of the company. Headed by Lieutenant General Yamanouchi Shizuo, a Kwantung Army officer with an engineering background, the committee consisted of fifteen members from Manchuria and fifteen from Japan proper. Its task was to navigate the complex legal and administrative regulations involved in creating a corporation with dual nationality while retaining control in Japanese hands. Okumura Kiwao, the young MOC official who had toured Manchuria and argued in favor of MOC involvement, applied his formidable knowledge to clearing legal hurdles for the new communications enterprise in Manchukuo. One precedent he considered was the Oriental Development Company, which Japan had set up in Korea in 1908 on the basis of laws promulgated in both countries. However, Okumura found this example unsatisfactory because it had applied to Japan's protectorate, Korea, on

---

12. For a detailed discussion, see Naikai Asajirō, *Tsūshin tokubetsu kaikei no umareru made*, 1–140.

13. "Manshūkoku tsūshin jigyō tōsei shian," MOC Records I, 176; "Manshū ni okeru denshin denwa jigyō no keiei keitai ni kansuru rikai," MOC Records I, 177.

the eve of its annexation. This would subject the company law to approval by the Imperial Diet, a procedure that would inevitably cause delays. Eventually, the committee decided to base the new company not on any existing legislation but on an administrative agreement signed between the two governments. By adopting this highly unusual route to minimize possible political interference from Tokyo over the use of government property, the Japanese finally cleared the way for what one leading Japanese authority on international law described as "a rare species unheard of in other countries."[14]

In August 1933, the Manchurian Telegraph and Telephone Company (MTT) was established with the broad mission to operate telegraph, telephone, and broadcasting facilities in Manchukuo. Headquartered initially in the city of Dalian, its operations were to extend throughout Manchuria, including the Kwantung Leased Territory and the SMR Zone. Branch offices were set up in Japan to procure equipment and materials and to recruit and train new Japanese employees. The MTT was placed under the joint supervision of the Manchukuo minister of communications and the Japanese governor of the Kwantung Leased Territory (after 1934, the Japanese ambassador to Manchukuo). The newly created Manchurian Affairs Bureau under the prime minister would also exercise supervision over the company. As an agent of the "national policies" of both Japan and Manchukuo, the MTT enjoyed a range of special privileges, including exemption from various taxation and other government levies as well as a status equal to that of a government agency in terms of land use, construction of lines, use of transportation, and charging of tariffs. Existing telecommunications and broadcasting facilities owned by both governments made up 450,000 of the total 1 million shares. The Japanese government's contribution consisted of telecommunications facilities in the Jiandao area on the Korean boarder, submarine cables between Dalian and Chefoo (Yantai), and submarine cables between Dalian and Sasebo. Estimated to be worth 15 to 18 million yen, it was about three times the amount of the Manchukuo government's contribution. The remaining 550,000 shares were to be subscribed by private interests in both countries.

---

14. Okumura Kiwao, "Manshū denshin denwa kabushiki kaisha no setsuritsu," 149–53.

Both governments guaranteed that private shares would be compensated first but set a limit on returns.[15]

Although a self-styled semi-private, joint-venture enterprise, the company roster showed that the Kwantung Army was firmly in charge, if indirectly. Two senior officers from the Kwantung Army, Lt. Gen. Yamanouchi Shizuo and Maj. Gen. Inoue Otsuhiko, became president and chief of the General Affairs Department, respectively. The officer who had headed the Kwantung Army Special Communications Department was appointed a section chief. Two MOC officials were tapped to lead the business and technical departments. The job of directing the Finance Department went to Nishida Inosuke, an executive from the SMR widely regarded as a business genius. For window dressing, a Mongolian prince named San Duo was named the vice president. The Japanese were predominant at all management levels: there were only seven Chinese among 154 senior executives in 1939; two years later, all but 10 of the 223 high-ranking executives were Japanese. Overall, Japanese employees outnumbered Chinese by almost four to one by the end of 1941.[16]

In this regard MTT was not exceptional. Even those who had once expressed concern about overwhelming the Manchukuo government with Japanese officials had to come to terms with the new reality of Japanese dominance. For instance, MOC official Fujiwara Yasuaki apparently had dropped his initial strong reservations and accepted the appointment as the head of the Postal Division in the Manchukuo Ministry of Communications. If anything, the trend toward Japanese dominance in the Manchukuo government accelerated. By 1935, the proportion of Japanese in the Manchukuo government exceeded even the Kwantung Army's original quota by several times. In key government depart-ments, such as the all-important General Affairs, more than 80 percent of personnel were Japanese.[17] The high-sounding rhetoric of racial harmony and Manchurian independence notwithstanding, the

---

15. Documents related to the MTT establishment are included in Mantetsu keizai chōsakai, *Ritsuan: Manshū tsūshin jigyō hōsaku*, 133–60.

16. MTT, *Manshū denden tōkei nenpō* 9 (1942): 14; *DDJS*, 6: 385. According to the company's own personnel ranking system, a senior executive refers to someone at either the *sanji* (company director) or the *fuku-sanji* (associate director) level.

17. Yamamuro, *Kimera*, 170–73.

structure of Japanese dominance was firmly in place in almost all government and private institutions in Manchukuo. Manchukuo was not a puppet state in the usual sense, for the puppet-masters appeared on center stage themselves.[18]

A "NATIONAL POLICY COMPANY" IN ACTION

Shortly after its founding, the MTT moved from Dalian to the city center of Shinkyō. A MTT recruitment brochure described the company as "occupying an important post for the three National Policies of Manchukuo [i.e., the Five-Year Industrial Plan, Colonialization, and Development of the Northern Frontier] and being on the cutting edge of all aspects of modern civilization."[19] Like other Japanese "national policy companies," the MTT's activities both as a business enterprise and as a military subcontractor constituted a major component of Japan's new order on the continent.

From the beginning, all parties in Manchukuo accepted the importance of telecommunications to the military. As one top MTT executive reminded readers of a Japanese trade journal, it was impossible to understand telecommunications in Manchuria if one neglected its military significance in terms of the geographical environment and internal conditions.[20] Even before founding the MTT, the Kwantung Army had spelled out its detailed expectations for the new corporation. For instance, the company was to set up special telephone circuits between all major military garrisons, on which the military would install equipment to ensure the secrecy of conversations. In addition, the MTT would maintain long-distance facilities that the military could use in time of need.

For the first few years, the MTT concentrated on strengthening the trunk lines linking major cities and on converting bare wires to cables. Additional repairs were required as the destruction of many overhead

---

18. For an in-depth study of the Manchukuo ideology, see Duara, *Sovereignty and Authenticity*.

19. Manshū denshin denwa kabushiki kaisha, *Denden*.

20. Nakada Suehiro, "Manshū ni okeru denki tsūshin jigyō no tokuisei," *Denki tsūshin* (hereafter *DT*) 3.8 (1940): 18–20.

wires by anti-Japanese guerrillas after the Japanese invasion caused interruption to Japanese communications.[21] The company gave higher priority to the "nutrition lines" (*eiyōsen*)—highly utilized commercial lines such as the Dalian–Mukden–Shinkyō line—as well as to circuits with Korea and North China that generated most of its revenues.[22] This strategy worked to the satisfaction of both the company and the military.

MTT also built new facilities to meet the new demands of communication and propaganda. Shortly before the MTT was founded, the Japanese decided to build a high-power wireless station in the new capital of Shinkyō that would supersede Mukden Radio. The 100-kW Hsinking Radio, completed in June 1934 at a cost of 3,000,000 yen, was the largest in East Asia. It was capable of communicating with Japan and North China, as well as with San Francisco and Berlin. All the equipment, it was reported, "from the gigantic antennas to delicate tools," had been made in Japan.[23] Before the outbreak of the Pacific War, Hsinking Radio handled more than 50,000 international telegrams annually, primarily with the United States and Germany.[24] Working there, as one Japanese employee recalled years later, was the "pride of communications men" in Manchuria:

> In the middle of Manchukuo, which is frowned upon by other countries, communication remained a necessity since mankind cannot stop it. As a result, we exchanged wireless waves day and night with San Francisco, Berlin, Paris, and so on; those countries, which did not recognize Manchukuo nonetheless paid full attention to our every move. It was an incredible joy.[25]

Expansion of the Japanese-language telegram service area was the key to overall Japanese activities as well as to the company's business success. One of the MTT's first priorities was extending this service

---

21. For example, telegraph and telephone lines between Harbin and Changchun were frequently interrupted due to Chinese sabotage; see Director, Telecommunications Bureau to Director, Osaka Communications Bureau, July 23, 1932, in MOC Records I, 175.

22. *DDJS*, 6: 399.

23. "Hsinking Has Biggest Radio Station," *FER* 31 (June 1935): 234; *DDJS*, 6: 404–5.

24. MTT, *Manshū denden tōkei nenpō* 9 (1941): 329.

25. Inomomo Tetsuo, "Shinkyō musen," in Manshū denden tsuiokuroku kankōkai, *Akai sekiyō*, 237–41.

from the Kwantung Leased Territory and SMR Zone to other areas under Japanese control. The MTT aspired "not to leave a single Japanese resident outside the Japanese-language telegram zone."[26] Initially, the Japanese Communications Bureau in the Kwantung Territory had concerns about possible protests by foreign cable companies against such expansion. As this proved unfounded, the bureau introduced low rates for Japanese-language telegrams between areas previously under Chinese administration and the Kwantung Territory.

Japanese in northern Manchuria had to wait longer than other areas, because sending Japanese-language telegrams required that local Chinese telegraph operators receive special training. Japan's purchase of the remaining portion of the China Eastern Railway in 1935 opened up a vast area for MTT operations in northern Manchuria. As part of Stalin's policy to avoid confrontation with Japan, the Soviet Union indicated in 1933 that it was willing to sell all of the China Eastern Railway that was under Soviet administration. According to the agreement signed in 1935, Manchukuo purchased the entire 1,700-km railway, as well as other properties, for 170 million yen. The Railway Bureau of Manchukuo was to operate the rail service while the MTT took over China Eastern Railway's telecommunications facilities and telegraph service. The Japanese population in Harbin jumped from a mere 4,000 at the time of the Manchurian Incident to 20,000 after the railway takeover. Three years after the Japanese began telecommunications operations in Harbin, telegrams had increased eightfold, from 19,567 a month in 1932 to 164,513 a month in 1935. By the end of 1937, Japanese-language telegram service was available in all 691 telegraph offices throughout Manchukuo. One former Japanese employee of MTT later recalled with pride the popular saying, "Wherever the Japanese went [in Manchukuo], you can find sashimi and Japanese-language telegram service."[27]

---

26. "Man-Mō kakuchi wabun denpō toriatsukau kaishi ni tsuite" (n.d.), MOC Records I, 175; Maeda Naozō, "Manshū ni okeru denshin denwa jigyō no keiei," *TKZ* 330 (February 1936): 155–56.

27. Machida Itsuyoshi, "Denshin oboegaki," in Manshū denden tsuiokuroku kankōkai, *Akai sekiyō*, 35.

TENTACLES INTO NORTH CHINA

Understandably, bilateral relations between Japan and China remained deeply strained after 1932. Internally divided, the Chinese government had to opt for a pragmatic policy of buying time for self-strengthening.[28] Emphasizing the "special relationship" between the two countries, the Japanese government pressed for closer "economic cooperation" and warned that China's dependence on Anglo-American assistance was tantamount to "playing barbarians against barbarians." Better communications between the two countries was regarded as not just a practical necessity but also emblematic of the nature of the relationship. For instance, given the large volume of telegraphic traffic between Japan and China, Japan considered it abnormal that no direct wireless connections existed between the two countries. In 1936, under repeated Japanese pressure, China not only opened wireless telegraphy but inaugurated its first international radiophone service, between Shanghai and Tokyo.[29]

Restoring and expanding transportation and communications routes to North China became the top priority for the Japanese. Following the signing of the Tanggu truce between Japan and China in 1933, the Kwantung Army temporarily halted its southward advance. Japanese expansion into North China did not come to a stop, however. Fujiwara Yasuaki and Okumura Kiwao, both sent to investigate conditions in Manchuria immediately after the incident, secretly traveled to Beijing to make preliminary arrangements for the reopening of communications.[30] As part of the agreement, the Japanese demanded reopening of communications between China proper and Manchuria. Under the premiership of Wang Jingwei, the Chinese government in Nanjing adopted a conciliatory position toward Japan. In September 1934, Fujiwara Yasuaki and an MTT official arrived in Beijing for negotiations with representatives of the Nationalist government. As a result, the Chinese

---

28. For a discussion of domestic Chinese politics, see Parks Coble, *Facing Japan: Chinese Politics and Japanese Imperialism* (Cambridge, Mass.: Harvard University Council on East Asian Studies, 1988); in Japanese, see Lu Xijun, *Chūgoku kokumin seifu no tai-Nichi seisaku, 1931–1933* (Tokyo: Tōkyō daigaku shuppankai, 2001).

29. Chu Chia-hua, *China's Postal and Other Communications Services*, 156. Chu served as China's Minister of Communications (Jiaotong bu).

30. Okumura Kiwao, *Teishin ronsō*, 421–46; Fujiwara Fumiaki, "Teishinshō denmukyoku jidai," in Okumura Katsuko, ed., *Tsuioku Okumura Kiwao*, 33–34.

postal blockade was lifted at the beginning of 1935, after two years and seven months.[31]

Although postal communication was restored, telegraph and telephone arrangements were far from satisfactory to Japan. The Japanese demanded more than resumption of telecommunications links with locations inside China proper. The Kwantung Army, in particular, was keen on setting up telegraph and telephone links with North China. Not long after the Manchurian Incident, the Kwantung Army and civilian planners had come to the conclusion that Manchuria alone would not suffice for Japan's projected autonomy in strategic resources. North China, rich in coal and other materials, also had to be incorporated into an economic union with Japan.[32]

At the Kwantung Army's insistence, negotiations over telecommunications began in August 1935 between Chinese officials and Kwantung Army representatives, since the Chinese government still refused to recognize Manchukuo. The Japanese negotiators—MTT executives, MOC officials, and Kwantung Army colonels—met with Chinese government representatives over a period of several months. Telephone connection was discussed first and agreed upon without much difficulty. A 24-hour telephone connection was established between Tianjin and Beijing in North China and Mukden, Dalian, Andong (present-day Dandong), Yingkou, and Jingxian in Manchukuo. Inclusion of Shinkyō and Harbin was to be discussed later. Both sides agreed to connect telephone lines at Shanhaiguan on the border, but each would collect its own terminal fees (considered an "abnormal method" by Japan).

Both sides wrangled over proportions of telegram revenue, rates, and service areas of Japanese-language telegrams. The Japanese demanded that Japanese-language telegram rates, which were at the same level as European-language rates, be reduced to the same level as Chinese-language telegrams, on the grounds that "Japan finds it unacceptable

---

31. For a Japanese newspaper account of the negotiation, see *Hōchi shinbun* (Manchuria-Korea edition), January 15, 1935; Youdianbu, *Zhongguo jindai youdian shi*, 184.

32. Peattie, *Ishiwara Kanji and Japan's Confrontation with the West*; Barnhart, *Japan Prepares for Total War*. For a more recent work, see Marjorie Dryburgh, *North China and Japanese Expansion 1933–1937: Regional Power and the National Interest* (London: Curzon Press, 2000). For a Japanese study of Japanese economic activities in North China, see Hagiwara Mitsuru, *Chūgoku no keizai kensetsu to Nitchū kankei: Tai-Nichi kōsen e no jōkyoku, 1931–1937* (Kyoto: Mineruba shobō, 2000).

that China should treat its neighbor of same race the same way it treats Whites." They suggested that China "should first clarify the racial significance (*jinshūteki igi*) and make use of bilateral friendship" with Japan.[33] They rejected earlier MOC agreements that gave China a greater share of revenues and instead demanded an equal division.

The Japanese also wanted to expand the service area of Japanese-language telegrams beyond Tianjin to include Beijing and Jinan. In addition, Japan proposed including the Dalian–Yantai submarine cable to handle telegrams between China proper and Manchukuo. Clearly, Japan was interested in expanding links with China proper. The Japanese prevailed in most of their demands, although they had to accept Japanese-language telegram rates slightly higher than Chinese-language rates, and no Japanese-language service to Jinan for the time being.[34] In October 1935, agreements dealing with telegram rates and service areas were signed.

Even before these hurdles were cleared, Japan's military was extending its encroachment on North China. The Army authorities in the field, both the Kwantung Army and the China Garrison Army based in Tianjin, were less patient than leaders in Tokyo with such peaceful means. Directed by Colonel Doihara Kenji, the North China Autonomous Movement was in full swing, aimed at reducing the influence of the central government in the area, with the ultimate goal of making it into another Manchukuo. In 1935, the semi-independent Hebei-Chahar Political Council had been founded in Beijing under General Song Zheyuan. In the demilitarized area bordering Manchukuo, the Japanese helped set up the East Hebei Anti-Communist Autonomous Government in November 1935. Headed by a Japanese-educated politician, Ying Rugen, it immediately declared independence from the Nanjing government.[35]

The MTT's role as a subcontractor to the military thus went beyond Manchukuo. As a private company, the MTT could conveniently skirt the thorny question of Chinese nonrecognition of the Manchukuo regime. The MTT was to provide a springboard for further extension of

---

33. "Tsūden ni kansuru Kantōgun gawa an no kosshi ni taisuru setsumei fui" (September 7, 1935), MOC Records I, 187.

34. "Man-Ka kan denshin oyobi denwa renraku ni kansuru ken," MOC Records I-187.

35. Kahn, "Doihara Kenji and the North China Autonomy Movement," 177–210.

Japan's telecommunications influence into China proper, a task it undertook with great enthusiasm. At the request of the army, MTT set up a small office in Tianjin in August 1935. The following year, MTT dispatched a number of employees to join a technical committee, established within the China Garrison Army to survey telecommunication facilities in North China and Inner Mongolia.[36]

To be sure, MTT was not the only player from Japan. Despite occasional differences with the military over means, the MOC saw North China as part of its larger operations in China proper. Responding to a request from the China Garrison Army, the MOC dispatched several officials to serve as advisors to the East Hebei regime and to monitor the situation in North China. In his report to the MOC, Nagaoka Shinsho stressed the need to unify all the administrations involved and to establish a close relationship with the Japanese forces in the region. To establish a strong North China regime independent of China's central government in Nanjing, Nagaoka recommended that special attention be paid to communications facilities.[37]

The preferred techniques of control were familiar: by extending various loans, the MTT and MOC were able to place Japanese advisors in the Communications Committee of the Hebei-Chahar regime in Beijing. All these activities were coordinated by the Army's Matsumuro Special Service Agency. The Army calculated that, through these Japanese advisors, Japan would better understand the inner workings of the North China regime and thus gain the confidence of its leaders, since "the Chinese national character places emphasis on interpersonal relationships."[38]

MTT's presence in North China soon made a difference. After the MTT set up a telegraph office in Shanhaiguan, strategically located on the North China–Manchukuo border, Japanese-language telegrams

---

36. *GKDTS*, 8: 128–39.

37. Nagaoka Shinsho, "Kōtsū iinkai to teishin jigyō" (January 5, 1936), MOC Records I, 243.

38. Not all MOC officials easily capitulated to the military, however. At a dinner banquet held at the residence of the commander of the China Garrison Army, MOC advisor Satani remarked sarcastically that "all these years of civilian service have made me ignorant of international or military affairs, so please forgive me if I should say something disrespectful." This episode was reported in Tominaga, "Hokushi kinkyō," MOC Records I, 250.

could then be relayed as far as Tianjin after an eight-hour railway journey. Though still time-consuming, this was a significant improvement over the previous time required between Dalian and Tianjin, which had been two to five days. Beginning in early 1935, the Japanese Resident Association in Beijing began accepting telegram exchanges through Shanhaiguan.[39] In June 1936, the East Hebei regime obtained a loan of 1.5 million yen from the MTT and contracted with the company to upgrade telecommunications facilities in the area. The major part of the expansion was construction of a telephone line from Tianjin to Shanhaiguan. The MTT also repaired the existing line linking the city of Tongzhou—seat of the East Hebei Autonomous Government—with other cities in the region. Municipal telephone lines in Tongzhou and 21 other towns were also started or expanded. Clearly, the emphasis was on strengthening the regime and its ties with Manchukuo.[40]

What actually happened went even beyond the original plan, however. Of the four telephone lines linking Tongzhou and the Japanese Infantry Regiment barracks in Beijing, two were installed without notifying the Chinese authorities. Telephone lines in Beijing were then extended to Ying Rugen's private residence in the city. In December 1936, after protest by the Chinese, a compromise was reached to withdraw the lines between the infantry barracks and Ying's residence. Another case was discovered by accident. Again without notifying the Chinese, the MTT extended the telegraphic line between Shanhaiguan and the headquarters of the China Garrison Army to the Japanese Resident Association in Tianjin. Upon completion of the construction, the MTT sent telegrams celebrating the beginning of Japanese-language service. By mistake, some of these telegrams were sent through the Chinese Telegraph Office and became incriminating evidence, to the embarrassment of the Japanese. Even the MOC engineer admitted that incidents like these served to poison the atmosphere and complicate further negotiations. After the Chinese protested to the Japanese consulate, the Kwantung Army and MTT were consulted and a face-saving measure was concocted: the Japanese agreed to halt Japanese-language telegram service at the Resident Association, on the condition that the

---

39. "Shina ni okeru denshin kankei zakken," Japanese Ministry of Foreign Affairs Archives (hereafter JFMA).

40. Hokudenkai, *Kahoku denden jigyō shi*, 43–44.

Chinese strive to reduce the wireless radiogram rates between Japan and China and to begin accepting greeting telegrams.[41] Although these conflicts were diffused, the tempo of further expansion into North China certainly was not.

An incident like this by no means deterred Japan's attempts to consolidate its influence in East Hebei. In mid-1937, the MTT concluded a series of formal agreements with the East Hebei Anti-Communist Autonomous Government, upgrading telegraph and telephone communications between Manchukuo and areas under the control of the East Hebei regime. Two Japanese army officers, representing the Kwantung Army and the China Garrison Army, respectively, signed the document, in addition to Nishida Inosuke, the department head of MTT, and Ying Rugen, head of the East Hebei government. Even Tokyo could now enter into direct communication with Tongzhou, the seat of the East Hebei government, via Manchukuo.[42] The agreement was but one of the many blatant encroachments on North China—and efforts to weaken Nanjing's influence—that paved the way for the clash that broke out near the Marco Polo Bridge barely two weeks later.

OCCUPIED CHINA AND THE MTT MODEL

On the night of July 7, 1937, a relatively minor skirmish occurred between the troops belonging to the Chinese 29th Army and the Japanese China Garrison Army on the outskirts of Beijing. That clash, as well as the subsequent escalation, was not entirely unexpected. Already, Japan's media had turned their attention to North China as tension between China and Japan mounted. *Nichinichi shinbun* became the first Japanese newspaper to report the outbreak, thanks largely to measures taken in anticipation of a conflict. Since news telegrams sent from Beijing or Tianjin in North China were subject to military censorship, *Nichinichi* had devised "special measures" by setting up its own wireless network.

41. Murata to Arakawa (January 31, 1937), 12th report, MOC Records I, 250; Kishi to Foreign Minister (received on January 16, 1937), MOC Records I, 183.

42. For the texts of the agreements, see "Manshū Kitō kan tsūshin renraku kyōtei teiketsu no ken" (June 26, 1937), in Imperial Japanese Army and Navy Archives Microfilms, Reel 108.

Its operators in North China would send news stories to a transmitter located in Mukden, which forwarded them to Japan by regular public channels. In this way *Nichinichi*'s branch office in Tianjin beat other news agencies in Japan by almost two hours in reporting the news.[43] *Nichinichi*'s swift reporting of the Marco Polo Bridge was but one example how prior developments in North China had already set the stage for a showdown.

Internally divided, the Japanese military was initially prepared to localize the conflict, but only after teaching the Chinese a severe lesson and obtaining concessions. The determined efforts of the Nationalist government, facing strong anti-Japanese popular opinion at home and hopeful of international support to resist Japanese aggression, made Japan's immediate goal increasingly impossible. By mid-August 1937, both sides were preparing for a final showdown as China opened a second front in Shanghai. With reinforcements arriving in both North and Central China, Japan was drawn into a prolonged military campaign on the continent.[44]

Controlling telecommunications facilities was an urgent task from the beginning, yet the Japanese lacked the infrastructure similar to that in Manchuria. Unable to handle these mounting tasks while fighting a war, the China Garrison Army (now the North China Expeditionary Army) asked the MTT through the Kwantung Army for immediate technical assistance in operating telecommunications facilities. On July 8, the day after the initial clash took place, MTT set up a special committee to cope with the crisis in North China. On the same day, a group of MTT employees departed for North China to help repair and operate the crucial telecommunications lines between Tianjin and Shanhaiguan.[45] Censorship of telegraph and toll telephone service, which yielded valuable intelligence, received high priority in affected areas as well. As the fighting escalated and spread, restoring damaged telecommunications facilities in North China became a top priority for the Japanese, although the need varied according to local conditions. Whereas Beijing

43. *Mainichi shinbun 70-nen shi*, 320. The question as to who fired the first shot remains the subject of endless discussion and controversy. For a recent comprehensive study in Japanese, see Hata Ikuhiko, *Rokkōkyō jiken* (Tokyo: Tōkyō daigaku shuppankai, 1997).

44. For an English-language study on the problem of establishing Japanese control in North China, see Lincoln Li, *The Japanese Army in North China, 1937–1941.*

45. *GKDTS*, 7: 139–41.

was handed over to Japanese troops peacefully, Tianjin, the largest commercial city in North China, was hard hit in the crossfire between Chinese and Japanese forces. To prevent Chinese snipers from taking advantage of tall buildings, Japanese troops detonated several such structures near the large Japanese settlement, destroying a major telephone switching station in the process. In early August 1937, only days after the city had fallen into Japanese hands, a second group of MTT executives, headed by MTT Director of Business Operations Nishida Inosuke, flew to Tianjin and set up the headquarters of MTT's North China Special Operation. Within a month, the company dispatched 300 of its Japanese employees to the area, including some twenty Japanese women chosen from cities in Manchukuo to operate the newly installed telephone switching in Tianjin. By the end of the year, MTT personnel in North China increased to 655.[46]

The Japanese, who had already made extensive plans for expansion into North China, also wanted to establish de facto control of telecommunications in the area in order to gain the advantage at any armistice.[47] As it became clear that the conflict was to be a protracted affair, Japan began to consider measures for long-term occupation. Initially interested in creating a second Manchukuo in North China, the North China Expeditionary Army, headed by Field Marshall Terauchi Juichi, was fully aware of the importance of the MTT's expertise and wanted to retain all the dispatched MTT personnel. Seeing this as a good opportunity for the company to boost and consolidate its influence in China proper, MTT President Hirose Hisasuke demanded that the company be given complete control over operations in North China in return. Otherwise, he threatened to call every one of his people back. As a result, the relationship between Japanese authorities in North China and Manchukuo began to show signs of serious strain.

Enter the MOC. Already actively involved in the region even before the conflict, the ministry responded to the conflict by first dispatching several officials to North China in early August 1937. Ostensibly visiting

46. Suzue Shizuo, "Jihen chū no Hokushi (Tianjin, Peiping) densei gaikyō" (August 23, 1937), MOC Records I, 255; Hokudenkai, *Kahoku denden jigyō shi*, 49; Manshū Denshin denwa kabushiki kaisha, *Sōgyō 5-shūnen*, 138–73.

47. Technician Nakamura, "Hokushi shisatsu hōkoku" (September 18, 1937), MOC Records I, 255.

Japanese troops on the front, their real mission was to persuade the Japanese military authorities there to accept Tokyo's proposal of a telecommunications enterprise for North China separate from the MTT. Reporting from North China in September 1937, MOC official Nakamura Jun'ichi argued that "implementation of continental communications policy and overseas expansion of Japan's communications strength is by nature a task the MOC should take up." Operation in North China was not only outside the scope of MTT activities, he emphasized, but beyond its capability.[48] By implication, only the MOC was qualified to operate and expand telecommunications in occupied China. It is no surprise that many in the MOC were concerned that the MTT would extend its operations into China proper. By November 1937, the MOC became convinced that a telecommunications enterprise in North China independent of MTT was desirable to reassert Tokyo's control over telecommunications in that area. Any formal ties with the MTT, whether a subsidiary or an MTT-run operation, would harm the new enterprise, the MOC insisted, since it would obstruct the flow of capital, technology, and personnel from Japan into North China.[49] In December 1937, the MOC set up a Committee on Administration of Communications in China to coordinate its activities in China and to take advantage of new developments. Although the MOC played only a supplementary role in the establishment of a new communications enterprise in Manchuria, it apparently entertained great hopes of influencing future telecommunications policy vis-à-vis the Asian continent. Predictably, the MOC's entry tipped the balance against the MTT. As 1937 drew to a close, the MTT was no longer predominant in North China. Hundreds of MOC employees had arrived in the area to serve in the Military Postal Services or to assist in restoration of telecommunications facilities.

The MOC team was headed by Watanabe Otojirō, a young, energetic official then in charge of the Investigation Section of the Telecommunications Bureau at the ministry. Unlike most government bureaucrats with degrees from prestigious imperial universities, Watanabe had grad-

---

48. Ibid.; Nakamura Jun'ichi, "'Daitōa tsūshin seisaku' jidai kara denden kōsha danjō zen'ya made," in Teishin gaishi kankōkai, *Teishin shiwa*, 3: 444–46.

49. "Manshū denshin denwa kabushiki kaisha no Hokushi shinshutsu ni taisuru iken" (November 1937), MOC Records I, 252.

uated from the MOC's own training school and had passed the Higher Civil Servant Exam in 1923. Already known for his expertise and prolific writing, he had served in the Government-General of Korea as a secretary to Superintendent General Imaida Kiyonori. While in Korea, Watanabe had published a widely read work on telecommunications management, the first systematic work on the subject in Japanese. Returning to Tokyo in 1934 at the age of 32, he occupied key posts in the Telecommunications Bureau and was responsible for a number of telecommunications expansion plans, earning a reputation as "the genius of the MOC."[50] En route to a study tour of Europe in 1937, Watanabe was summoned back to Tokyo as soon as the fighting broke out in China. His arrival represented the ministry's determination to boost its presence in North China.

As expected, the MOC sided with the Army against the MTT in North China. In early December 1937, tension between the MTT and the North China Expeditionary Army seemed to be reaching crisis proportions. An apparently agitated Watanabe telegraphed the MOC in Tokyo that "the president of MTT was opposed to the Army plan" and that the "negotiation was breaking down completely, with MTT expected to withdraw its personnel from North China." Watanabe urged that immediate secret preparations be made to dispatch MOC technical personnel to North China in the event of MTT's sudden withdrawal. Embroiled in the rivalry between the Army and MTT, Watanabe was asked by the Army to organize the new effort to establish a separate telecommunications enterprise. Pressured by the MOC, he accepted the offer and would remain in China for the next eight years, cutting short an illustrious career in the ministry.[51]

It soon became clear that support for extending MTT operations into North China was far from unanimous even inside Manchukuo. Fearing a drain on its resources, the Manchukuo government turned against the idea. Perhaps most important was the fact that the Kwantung Army also gradually came to oppose the MTT's continued involvement in North China, on the grounds that the company should be primarily devoted to strengthening telecommunications in Manchukuo, especially

50. "'Kirinji Watanabe Jirō' no eikō to aikan," in Teishin dōsōkai, comp., *Senpai ni kiku*, 655–71.

51. Watanabe to Utsubo, December 6, 1937, MOC Records I, 204.

for its own operations in northern Manchuria. One colonel of the Kwantung Army's First Section was particularly adamant, demanding that if it was impossible to recall MTT personnel from North China all at once, they should be replaced by less capable employees so that expertise in Manchuria would not be depleted. Another officer objected to continued MTT involvement unless appropriate replacements could be found. Interestingly, both officers agreed that some consideration be given to employing Manchukuo's experience in "continental management" when developing North China.[52]

After several days of negotiations, representatives of the North China Expeditionary Army, Kwantung Army, Army Ministry, and MTT finally reached an agreement on December 10, 1937. A separate North China Telecommunications Administration (NCTA) was to be established in Beijing as an official agency to operate telecommunications in North China for the time being. Two days earlier, the East Hebei regime and the Japanese authorities had reached a separate agreement that the new NCTA would take over all previous MTT operations in that area.[53] Organizationally, the MTT would be out of North China, but MTT Director of General Affairs Inoue Otsuhiko, a former army major general with a background in communications engineering, would head the new entity. Endō Goichi, Watanabe Otojirō, and Handa Mitsuhisa (all MOC officials) were appointed associate directors. In principle, MTT employees in North China would become employees of the NCTA if they wished to remain; for those desiring repatriation to Manchuria, the agreement stipulated that the MTT provide replacements so that the total strength remained the same.[54] Eventually, the NCTA would consist

52. "Uchida jimukan hōkoku (Hokushi kankei)" (received November 29, 1937), MOC Records I, 204.

53. For the text of the agreement, see Hokudenkai, *Kahoku denden jigyō shi*, 45–46.

54. "Hokushi densei shori ni kansuru oboegaki" (December 10, 1937), MOC Records I-243. The semi-official history of the NCTA written by former Japanese employees provided a somewhat different story: the policy of the Tokyo government—the Army Ministry and MOC—was rejected by the staff officers of the North China Army and Kwantung Army, in a typical *gekokujō* (juniors dominating seniors) fashion in which subordinates rebel against their superiors. Since it was difficult to implement national policy this way, the government dispatched staff officers who shared its views to North China. The MTT, under the influence of the Kwantung Army, inevitably experienced friction with MOC. In the end, the headship of NCTA did not go to Nishida, who was considered "a firebrand with strong local color ill-suited for improving MTT-MOC re-

of nearly 500 former MTT employees, plus a smaller group of Japanese from the MOC in Japan. Among the latter was the future technical chief Asami Shin, considered one of the most popular young engineers in the MOC.[55] Because these MOC veterans would become employees of a nongovernment corporation in another country, salary and benefits were important considerations for transferees. As one NCTA executive recalled, balancing between the MTT and the MOC employees became an intricate game for the personnel department of the new company.[56]

Once the status of the NCTA in relationship to the MTT was clarified, its own character became the subject of much discussion for several months. Since the NCTA was set up in a hurry, would telecommunications continue to be run by the new government to be established in North China? Just as various parties finally worked out a deal in the field, the barely two-month-old Cabinet Planning Board in Tokyo drew up a comprehensive Economic Development Plan for North China. Adopted as a cabinet decision on December 21, 1937, the plan clearly rejected the idea of monopoly management throughout Manchukuo and North China by single companies in the area of transportation and communication. However, it laid out the additional understanding that the personnel, technology, and expertise of the SMR and MTT would be utilized in North China.[57] In late December, the Army Ministry, Navy Ministry, and MOC held a joint discussion in Tokyo that resulted in the "Outline of Policies Toward Telecommunications in China." This reflected a consensus that the military conflict in China presented "a perfect opportunity for achieving the goal of securing 'de facto control' (*jisshutsuteki shihai*)" over China's domestic and international tele-

---

lations, but was given to the more amiable Inoue" (Hokudenkai, *Kahoku denden jigyō shi*, 50).

55. Manshū denshin denwa kabushiki kaisha, *Sōgyō 5-shūnen*, 141–42; Hokudenkai, *Kahoku denden jigyō shi*, 64, 114; "Zadankai: Densei sōkyoku zengo no koro," in ibid., 304–14; reminiscence by personnel chief Okui, in ibid., 229–31.

56. Stress and friction took its toll when the head of technical affairs, Handa Mitsuhisa, died a sudden and mysterious death in Beijing shortly after arriving in China (Hokudenkai, *Kahoku denden jigyō shi*, 310–11).

57. The SMR did not give up its effort to bring about "unified management of Manchurian-Chinese transportation," however. In mid-1938, the SMR considered methods of exerting control over the North China Transportation Company, a subsidiary of the North China Development Company, but did not succeed (Nakamura Takafusa, "Japan's Economic Thrust into North China, 1933–1938," 249–50).

communications, including its cable, wireless, and broadcasting facilities. The outline also envisioned that a new telecommunications company created in North China would later move into Central and South China.[58]

This time, the MOC readily embraced the idea of a telecommunications company. In a proposal drafted in December 1937, Watanabe Otojirō expressed concern that, although required by its military and public character, "state-owned, state-run" telecommunications would be subject to the financial constraints of the new regime in North China, making it difficult to expand. The form of management must depend on a concrete plan for the next five years, he argued. For a privately owned and run enterprise to attract capital and enjoy flexibility, a large amount of initial capital was needed. Whatever the form of management, Watanabe emphasized, the Japanese must be in firm control as supervisors. Although in principle those Chinese of excellent quality "with an understanding of Japan's national policy" could be employed, he explained, capable Japanese should be placed in key positions.[59] Another "local plan" sent from Beijing by an MOC official in early April 1938 contained perhaps the most elaborate analysis of the merits and demerits of various forms of ownership and management (see Table 2). Although its late arrival probably precluded it from much consideration, it nevertheless reveals the Japanese rationale behind various institutional forms. The proposal considered factors affecting the future tasks facing the new enterprise—finance, consolidation of operations, driving out foreign influence, and, last but not least, ease of asserting Japanese control. In conclusion, it recommended a modified privately owned, privately run monopoly limited to North China.[60]

In early 1938, preparations for the North China Telegraph and Telephone Company (NCTT) proceeded at full steam. A Preparatory Committee was set up in February to discuss the company prospectus and to evaluate existing facilities, business plans, and other issues. The

---

58. "Shina densei taisaku yōkō" (December 21, 1937), MOC Records I, 253.

59. Watanabe, "Densei kanri keiei hōshin an" (December 1937), MOC Records I, 245.

60. Kamio to Director, Engineering Bureau, "Chūgoku shinseiken ni okeru denki tsūshin jigyō no kigyō keitai ni kansuru ken" (received April 1, 1938), MOC Records I, 245.

Table 2
Comparing Enterprise Forms in Telecommunications
Under the New Regime in China, 1938

Main issues	Privately owned, privately run	Privately owned, state-run	State-owned, privately run	State-owned, state-run
Purchase of privately run facilities	Easy	Somewhat easy	Not easy	Not easy
Gathering funds	Easy	Somewhat easy	Not easy	Not easy
Contacts with business	Good	Good	Not good	Not good
Japan-Manchukuo-China cooperation	Very easy	Somewhat easy	Not easy	Not easy
Operations management	Economical	Somewhat good	Somewhat good	Uneconomical
Business administration	Efficient	Somewhat good	Somewhat good	Easy, inefficient
Relation to state budget	Not related	Somewhat related	Closely related	Easy, closely related
Relation to political change	Not related	Related	Related	Closely related
Driving out foreign influence	Suitable	Unsuitable	Unsuitable	Unsuitable
Supervision	Need to strengthen	Somewhat little	Somewhat little	Somewhat little
Use of government guaranteed facilities	Difficult	Somewhat difficult	Somewhat difficult	Easy
Speedy and accurate communication	Good	Somewhat difficult	Somewhat difficult	Easy, not good

SOURCE: Kamio to Director, Engineering Bureau, "Chūgoku shinseiken ni okeru denki tsūshin jigyō no kigyō keitai ni kansuru ken" (received April 1, 1938), MOC Records I, 245.

Chinese side was represented by Wang Keming and three ministers in the puppet regime in Beijing. In contrast to Manchukuo, the new Chinese regime in North China strove to maintain a semblance of autonomy and expressed concern that MTT would be written in as a shareholder, although it had yet to be recognized by the Chinese. The Chinese representative also demanded Chinese parity in top executive positions as well as among senior employees. An evaluation of Chinese government telecommunications facilities also proved difficult because the Chinese side asked for a higher estimate. In the end, however, all these attempts proved futile.[61]

As the armed conflict escalated, Japan's political objective in China also evolved. A number of regional governments were set up by the field armies. The Inner Mongolia Autonomous Government debuted in October 1937. In December, Japan's North China Expeditionary Army set up a Provisional Government in Beijing, and the Central China Expeditionary Army established a Restoration Government in Nanjing the following March. By October 1938, both Canton and Wuhan had fallen into Japanese hands. The war would enter a stalemate.

Immediately north of the newly designated North China was Inner Mongolia (Mongolian Borderland; Mōkyō), an area located between Manchukuo, North China, and the Mongolian People's Republic. With roughly 500,000 sq km but a population of only 6 million, the area had been the Kwantung Army's turf for some time, thanks to its strategic location bordering Soviet-influenced Outer Mongolia. After an invasion in August 1937, the Kwantung Army set up a string of small "autonomous" regimes that were eventually amalgamated into the Inner Mongolia Autonomous Government in the city of Zhangjiakou in September 1939.

As in North China, the Kwantung Army immediately called on the MTT to restore telecommunications facilities in Inner Mongolia. From late August 1937, the MTT dispatched over 300 employees to that area. To meet military needs, the Army set up the Inner Mongolia Telecommunication Facility Company (IMTFC) in March 1938, capitalized at 12 million yen (see Table 3). The company featured a Mongolian as its

61. Shirozaki Fumio, "Kahoku densei sōkyoku ni haken sarete," *TKZ* 362 (October 1938): 67–69.

Table 3
Capital of Major Japanese-Controlled
Telecommunications Companies in Occupied China, 1938
(million yen)

Source	NCTT	CCTC	IMTFC
Local Chinese government	10	5	2
Special regional corporations	12[a]	6[b]	4[c]
Japanese companies	12	4	6
Other	1[d]		
TOTAL	35	15	12

[a]North China Development Company.
[b]Central China Revitalization Company.
[c]Bank of Mongolia.
[d]Public issue.
SOURCE: DDJS, 6: 418, 448, 455.

president but was run by the Japanese. In addition to some 300 Japanese sent from Japan, 400 local residents were employed. Actual operation was placed under the Governmental Bureau of Postal and Telecommunications, which was headed by a Japanese official from Manchukuo. Japanese military authorities concluded a secret agreement with the local regime that ensured Japan's "guiding authority" through the now-popular formula of "comprehensive and strong internal guidance."[62]

If North China and Inner Mongolia showed the preponderant influence of the field armies, Central China was a different matter. On August 13, 1937, conflict broke out between Chinese and Japanese forces in Shanghai, the hub of China's commerce as well as of international interests. The MOC's involvement in Central China had traditionally been strong, due to the presence of the Japanese Government Telegraph Office in Shanghai since the 1910s. Through its officials posted in Shanghai, the MOC enjoyed a steady flow of information, even in the months of fiercest fighting between Chinese and Japanese forces. Despite these differences, the Japanese goal of gaining effective control of communications in the area was similar, as was the language used to describe it. In November 1937, before the Japanese military launched its

62. "Mōkō rengo chiku ni okeru shori hōshin" (January 20, 1938), MOC Records I, 237; Manshū denshin denwa kabushiki kaisha, *Sōgyō 5-shūnen*, 150–52.

assault on the Chinese capital of Nanjing, Matsunaga Hangorō, head of the Japanese Telegraph Office in Shanghai, drafted an outline for operating communications in the Lower Yangtze area. According to his plan, Chinese government telecommunications should be kept out of the Shanghai area, where a joint Sino-Japanese company would be set up.[63] In late November, local representatives of the Army, Navy, Communications, and Foreign Affairs ministries agreed to take over Chinese communications facilities in Shanghai by force. Beginning on November 27, 1937, the Shanghai Telegraph Office and domestic wireless facilities were placed under Japanese control. Chinese employees were encouraged to stay on while the Japanese installed censorship officers at the office.[64]

One of the most urgent Japanese objectives in Central China was the control of information through censorship of international communications, given Shanghai's position as the news center of East Asia. Such measures were particularly crucial in view of the negative foreign press Japan's military operations in the Shanghai area had received. Their efforts soon met strong protest from the foreign communities. In January 1938, Japanese censors at the telegraph office prevented Harold Timperley of the *Manchester Guardian* from sending news dispatches alleging widespread Japanese military atrocities in the Lower Yangtze area. Timperley protested and enlisted British consular officials to his support. In the meantime, Japanese authorities in Shanghai forwarded Timperley's dispatch to Tokyo in an effort to prepare for the potential international fallout from the fiasco.[65]

Almost paralleling the process in North China, the forms of Japanese control of telecommunications in Central China underwent extensive discussion. Higashi Hirohito, director of the Nagoya Communications Bureau, was dispatched to Central China following Japanese

---

63. Matsunaga Hangorō, "Shanhai chiiki ni okeru tsushin keikaku yōkō" (November 1937), MOC Records I, 205.

64. Kōain kachū renrakubu, *Jihengo ni okeru kyū Kōtsūbu densei kikan sesshū keii narabini misesshū bubun no sesshū hōsaku ni kansuru chōsa kenkyū*, 2.

65. A copy of Timperley's original dispatch, hand-copied on Japanese military stationery, can be found in the GNTC archives in Shanghai. Ironically, when the Japanese Foreign Ministry forwarded the telegram to its embassy in Washington, it was intercepted and deciphered by U.S. intelligence. See Daqing Yang, "Convergence or Divergence: Recent Historical Writings on the Rape of Nanjing," *American Historical Review* 104, no. 3 (June 1999), 851*n*47.

reinforcements. Shortly after his arrival in Shanghai, Higashi was quick to point out that, although economic considerations were important, the foremost criterion in deciding the ownership form was ease of Japanese control, which in turn would depend on the nature of the new Chinese regime to be set up after the ceasefire. If a Manchukuo-type regime were set up in which Japanese could easily participate in decision making, he reasoned, then a state-run telecommunications company was preferable, despite possible difficulties in raising capital. Otherwise, a privately run company would allow Japanese control through infusion of capital, personnel, and technology. Higashi also pointed out that a private company would make it easier to arrange communications with Japan and other parts of the occupied areas should the new regime not receive immediate international recognition.[66]

In early 1938 the Cabinet Planning Board discussed the question whether to place telecommunications operations under an umbrella regional development company or to exert greater control directly from Japan. "Since telecommunications rights are one of the sovereign rights of a state," some officials argued, "criticism would be unavoidable if a Japanese 'special company' were to operate telecommunications in Central China." But others noted that "it is difficult for a company primarily dedicated to economic development to fulfill the missions in military, diplomatic, and other national policy matters." The Planning Board decided that a special company would give the appearance that "the communications sovereignty of the [Chinese] regime is respected, although in essence it is easy for Japanese capital and technical personnel to enter and take control of the substance of operations."[67] Finally, there appeared to be an emerging consensus that a semi-private company was the best option for ensuring Japanese control.

But the battle was far from over, since there was no agreement as to which Japanese would control it directly. Even when faced with their common adversaries, bureaucratic rivalries among the Japanese were not uncommon. The problems of North China were replicated in Central China, albeit with different configurations. In January 1938, Kajii

66. Higashi Hirohito, "Chūshi denki tsūshin jigyō no keiei keitai ni tsuite" (January 1938), MOC Records I, 265.

67. "Kikakuin dai-3-iinkai no naka Shina toshi kaisha setsuritsu yōkōan ni taisuru iken" (February 18, 1938), MOC Records I, 200.

Takeshi, MOC director of engineering, went to Shanghai for nearly two weeks to discuss plans with the Japanese military authorities as well as with the MOC's official-in-residence. They reached a basic agreement about how to guarantee Japan's "autonomous communications rights," a euphemism for control over Chinese telecommunications.[68] In reality, however, one of Kajii's missions was to mediate between the Army and Navy over the choice of the head of the future company. The Japanese Army's influence was much weaker in Central China than in North China, whereas the Navy, through its Navy Landing Party permanently stationed in Shanghai, was able to exert more influence over a wide range of matters, including telecommunications. Kajii, who happened to be a high-school acquaintance of the North China Army representative and also knew the navy officer in charge, broke the stalemate. In the end, Fukuda Kon, a former personal secretary to ex–Prime Minister Admiral Okada Kensuke, and hence considered a "Navy man," was picked to head the new Central China Telecommunications Company (CCTC). Knowledgeable about communications through his earlier experience in the Japan Wireless Telegraph Company, Fukuda was considered to be well prepared to deal with complicated foreign telecommunications interests in the Shanghai area. Katayama Katsuzō, chief of the Engineering Section in the Government-General of Taiwan (GGT), was appointed the general manager.[69]

From the start of the hostilities, broadcasting also topped the Japanese agenda in both North and Central China. Within days of the outbreak of fighting in North China, Japanese broadcasting from Dalian was relayed in Tianjin and broadcast to Japanese residents in the area. Meanwhile, a 100-kW transmitter that had been taken over was used to jam radio broadcasting from Nanjing by broadcasting Chinese music at the same frequency. This broadcasting was also carried out by smaller transmitters in Dalian, Mukden, Harbin, and Chengde. Radio

---

68. "Teishinken no kakuritsu e Chūshi ni tokushū kaisha—Kajii kakuchō genchi chōsa kanryō," *Chūgai nippō*, January 12, 1938.

69. Kajii Takeshi, "Kahoku Kachū Mōkyō ni okeru denshin denwa kaisha no setsuritsu," in Teishin gaishi kankōkai, *Teishin shiwa*, 2: 455–59. Matsuo Matsutarō, a Japanese involved in telecommunications in Manchukuo, would later blame Fukuda for his tendency to bypass regulations and institutions, as he was not a proper "communications man" from MOC; see Matsuo, "Tairiku ni katsuyakushita teishnjin (3)-Kachū, Kanan," in Teishin gaishi kankōkai, *Teishin shiwa*, 2: 471.

broadcasting was considered so urgent that, later in October 1937, the Cabinet decided to establish a high-power radio station in North China to counter the influence of the Chinese broadcast from Nanjing. Since it would begin broadcasting on January 1, 1938, the North China Expeditionary Army asked the NHK to construct and operate broadcasting facilities in Beijing and other major cities in North China.[70] In Central China, the Army News Department set up broadcasting in December 1937. As in North China, the NHK provided technical personnel to build and operate broadcasting facilities.[71] The government in Tokyo considered broadcasting in China to be a crucial component in diplomacy and set up an inter-ministerial committee to supervise the program. In contrast to Manchukuo, where broadcasting had developed as part of the MTT and had a highly successful advertisement program, broadcasting remained separated from telecommunications companies in occupied areas in China—a set-up along the lines of the Japan Broadcasting Corporation in Japan and semi-public broadcasting corporations in its colonies.

By the early fall of 1938, the dust finally settled over the control of telecommunications operations in occupied China. The greater infusion of MOC personnel reflected the greater influence of Tokyo in communications matters in China than had been the case in Manchukuo. In a June 1938 internal memo, MOC had specifically addressed the control of the new telecommunications companies that were to be established in North and Central China. Since the MOC was to be in charge of unified control over telecommunications in East Asia, it argued, MOC must adopt specific measures to establish effective control (*jisshitsuteki shihai*) over the two companies: (1) a certain number of company employees should come from the MOC so as to reflect MOC policies in their business activities; (2) the MOC could also influence the new companies through investment by telecommunications companies under its direct supervision; and (3) the MOC would either directly supply staff or make recommendations on appointments in new government

70. "Hokushi hōsokyoku setchi ni kasuru kakugi" (August 28, 1937); Kita-Shina hōmengun, "Hokushi hōsō zantei shori yōkō no seitei" (November 25, 1937), in NHK, *Hōsō 50-nen shi*, 67–68.

71. Chūshi hakengun hōdōbu, "Kōha musenden kantokushu no setchi" (April 1, 1938), in NHK, *Hōsō 50-nen shi*, 67–68.

agencies concerned with communications in China.[72] As a whole MOC seemed to have accomplished these objectives.

It was ironic that, although the MTT itself was kept out of China proper, a modified MTT model was finally adapted and transplanted to North China and Central China. Only in South China, where the Japanese occupation began later and was limited to several smaller geographical areas, did the model of a single telecommunications company fail to gain acceptance. The region generally lacked the extensive telecommunications infrastructure already in place in North and Central China, although the Pearl River Delta around Canton and Hong Kong enjoyed relatively good telecommunications service provided by foreign companies. Japanese operations in various occupied cities such as Canton (Guangzhou) and Swatow (Shantou), therefore, took on a more local character. Given its proximity and prior activities, the Government-General of Taiwan played an active role in supplying technical personnel and building facilities for the Japanese military. Finally in late 1940 a small Xiamen (Amoy) Telecommunications Company was established along the MTT lines to operate the local facilities as well as the few submarine cables.[73]

Between 1931 and 1938, Japan's informal empire in China underwent drastic transformations. Its political history is well known. Manchuria, where Japan had enjoyed extensive treaty rights, was now under Japanese control through the creation of a puppet regime. Japan's continued military encroachment on North China further escalated tensions with the Chinese Nationalist regime in Nanjing and hastened the outbreak of the war with China in 1937. As Japan occupied major urban centers and many adjacent areas in eastern and central China and prepared for full exploitation of their natural resources, telecommunications infrastructure not only ensured the success of military operations but also assumed additional importance and became an objective on its own merits.

In this context, the establishment and the operation of the MTT—Japan's first full-service quasi-"private" telecommunications company—

72. "Hokushi oyobi chūshi ni okeru tsūshin kaisha ni taisuru tōsei yōryō" (draft; June 1, 1938), MOC Records I, 264.

73. *DDJS*, 6: 474–78.

is instructive for several reasons. Considered a special "national policy company," the MTT experiment, like Japan-sponsored state-building in Manchuria, represented a major departure from established practices in telecommunications in modern Japan.[74] The Manchurian experiment would not only influence reorganization of the Japanese economy at home but be applied in other parts of the new empire as well. As Japan began extending its military and political influence south of the Great Wall in the late 1930s, the MTT also became part of Japanese strategies in controlling telecommunications in China and elsewhere in Asia. To ensure adequate funding for building new facilities to meet the increased demand and to maintain effective Japanese control under the facade of political independence of China, there was finally a consensus that semi-private telecommunications companies were best suited to Japan's strategic needs on the continent. The MTT was further demonstration of the adaptability of Japan's imperial expansion.[75]

Of course this was not completely new. Although the military was undoubtedly the main driving force, civilian technical experts—in both government bureaucracies and semi-private companies—were no less enthusiastic and proved indispensable. The military in Manchuria in particular might be considered the midwife of the MTT—which was the progenitor of such companies—but those Japanese long involved in "national policy companies" such as the SMR provided the indispensable expertise. As this chapter has shown, the creation and the transplantation of the MTT model were not the work of a single mastermind but the result of extensive political negotiation. The process of setting up these companies also generated considerable tension, not just between the field and the center, but also between armies based in different regions. Even the matter of choosing a company head in Central China touched off interservice rivalries. Local initiatives outside Tokyo played a large role in the creation of these new public policy companies. None of this took place in isolation from Japan, however. Even many "communications men" from the MOC in Tokyo, while frustrated at losing

74. Noda keizai kenkyūjo, *Senjika no kokusaku kaisha*, 380–87.

75. One earlier example of such remarkable adaptability in the continental expansion is the South Manchuria Railway Company; see Matsusaka, *The Making of Japanese Manchuria*.

the initiative to men in the field, came to give their support. Without these supporters, the military would not have been able to pull it off. Just as the MOC and other bureaucracies sought to extend their influence on the continent, the drastic events on the continent would impact technological as well as political developments at home.

CHAPTER 4

Inventing Japanese Technology

In March 1932, the same month the Kwantung Army established Manchukuo, the *Journal of the Institute of Telegraph and Telephone Engineers* in Tokyo published an article entitled "Proposal to Use Non-Loaded Cable as Long-Distance Telephone Line." The creation of a puppet state on the continent was a watershed event for Japan, making irreversible its military invasion of Manchuria and inviting further international isolation.[1] In contrast, few people outside electrical engineering circles noticed the article on telecommunications technology co-authored by three young MOC engineers. Certainly no one at the time anticipated a direct connection between these two events in the near future. Yet within the space of a few years, the non-loaded cable (NLC) proposed in the article would be celebrated as the vital technology in Japan's effort to build a strategic communications link between the home islands and Manchukuo.

The importance of this technological invention was not limited to Manchuria. As the Japanese telecommunications expert Watanabe Otojirō claimed during the Pacific War, NLC was the technological equivalent in Japan's new empire-building endeavor to the gutta-percha submarine cable in the creation of the British empire. In the meantime, NLC would be heralded as a quintessential "Japanese-style technology" and a milestone in modern Japan's quest for technological autonomy. Even decades later, many in Japan were still convinced that "consis-

---

1. Nish, *Japan's Struggle with Internationalism.* On Japan's multifaceted mobilization for Manchukuo, see Young, *Japan's Total Empire.*

tently in every step from invention to application, it was literally a domestically produced technology (*kokusan gijutsu*), worthy of international pride"[2] and the development of NLC was "clearly the starting point of the leap forward of our telecommunications technology to the world's top level."[3] Such lavish praise seems puzzling given the fact that NLC has hardly been known outside Japan, then or now.[4]

It is necessary, then, to examine the development of this new long-distance telecommunications technology in the context of Japan's political and economic development during the first half of the 1930s. Japan's decades-long drive for technological autonomy and its new search for a technological solution to its rapid territorial expansion were both crucial. Then we can assess how electrical engineers, generally portrayed as faceless and passive, became advocates of technological autonomy at home and Japan's leadership role in Asia, and thus helped formulate a new ideology of Japan's empire. Seen in this light, the debut of the new technology and the rise of the technology bureaucracy in Japanese politics in the second half of the 1930s were not accidents; rather, they were closely linked to Japan's new empire-building enterprise.

THE LONG ROAD TO TECHNOLOGICAL AUTONOMY

To understand how NLC was first created and later became an embodiment of "Japanese-style technology," it is necessary to examine briefly the history of technological development since the late nineteenth century. Telecommunications, then on the cutting edge of technological innovation, offers a good case. Although a handful of Japanese knew of the electric telegraph from Dutch works, it was the American Matthew Perry who brought the first telegraph set to Japan and demonstrated it in 1854. It was not until after the Meiji Restoration

---

2. *Denki tsūshin jishu gijutsu kaihatsu-shi: Hansō denwa-hen* (hereafter *DTJGKS*), 4.

3. Musōka hōshiki kankōkai, *Gijutsu kaihatsu e no michi*, 214.

4. Indeed, the fate of NLC stands in sharp contrast to a better-known technological innovation that originated in Japan in the early twentieth century, the wireless antenna of Yagi Hidetsugu. Whereas Yagi's antenna technology was first turned into practical use overseas and then re-imported to Japan later, NLC was almost immediately moved on to manufacturing and application in Japan even though it remained virtually unknown elsewhere.

of 1868, however, that the Japanese government began building nationwide telecommunications facilities with the help of Western technical guidance. Electrical engineers were prominent among the "hired foreigners" who oversaw the first wave of infrastructure-building in the 1870s and 1880s. Telegraphy was an important field of technology transfer actively encouraged by the Meiji government. Under British engineer W. E. Ayrton, the government Engineering College, established in 1871, set up one of the first telegraphy departments in the world in 1877 (renamed the Electrical Engineering Department in 1884). In 1888, the Association of Electrical Engineering was founded in Japan. Gradually, Japanese engineers, trained by foreign teachers in Japan or educated abroad, assumed control over most projects.[5]

The state played a leading role in promoting technological development in Meiji Japan. Systematic research and development in telecommunications technology began when the government established a small laboratory within the Ministry of Engineering in 1875, under the supervision of a British engineer. This lab was transferred to the MOC in 1885 and was renamed the Electric Institute.[6] In 1911, three engineers in the MOC invented the TYK wireless telephone (the name derives from the first letters of their last names). In 1917, the MOC founded the Telegraph and Telephone Society to promote internal research. Increasingly, scientists in these government-funded institutions made important contributions in electrical engineering.

Since the end of the nineteenth century, the Japanese military had been a leading force in research and development in telecommunications technology. The stunning speed with which Japan began research in wireless technology is a case in point. In contrast to its attitude toward the telegraph and the telephone at a time when Japan lacked technological and organizational bases for the infusion of the new technology, the Imperial Japanese Navy demonstrated a keen interest in wireless communications and was ready to experiment on its own almost immediately after their invention. The military services

5. The best English introduction is Beauchamp and Iriye, *Foreign Employees in Meiji Japan*. For a Japanese work on Westerners hired in the field of communications—both electrical and postal—see Takahashi Zenshichi, *Oyatoi gaigokujin* (7): *tsūshin*.

6. For English-language accounts, see Odagiri and Goto, *Technology and Industrial Development in Japan*, 155–78.

also established their own integrated research facilities; foremost among them was the Naval Research Institute established in 1923. As Kozo Yamamura pointed out, the military not only benefited from the progress in telecommunications technologies but also stimulated technological development in Japan.[7]

Manufacturing was a different story, however. Since the late Tokugawa period, numerous Japanese had attempted to produce Western technological gadgets. Some were successful. In 1876, only two years after Alexander Graham Bell obtained a patent for his telephone, Tanaka Daikichi and a few other Japanese produced two telephone sets based on an American import. As with many other modern technologies, these individual efforts did not lead to commercial production, however.[8] As the "high tech" industry of the day, Japan's telecommunications industry, which involved manufacturing electronic equipment and transmission wires, began with the infusion of foreign capital and technology: the Nippon Electric Company started as a Western Electric subsidiary, and Furukawa was in a partnership with the German firm Siemens und Halske. Of Japan's leading domestic telecommunications manufacturers, only Fujikura Wire Works, a manufacturer of electric wires and cables, was largely Japanese in origin.[9] Dependent on their foreign partners, these private businesses often served as the conduit of technology transfer from overseas.

World War I provided the first occasion for the Japanese government to emphasize the importance of indigenous manufacturing (*kokusanka*), as many foreign imports became temporarily unavailable to Japan.[10] Financial concerns soon became another major factor. In the wake of the Great Earthquake of 1923, the government began to promote indigenous production aggressively to reduce demand on its limited foreign reserves. The Ministry of Commerce and Industry took the

---

7. For a stimulating essay on the relationship between technological development and overseas expansion, see Yamamura, "Success Ill-gotten?" 113–35.

8. Nihon kagakushi gakkai, comp., *Nihon kagaku gijutsu shi taikei*, 65. As one Japanese historian noted, those who attempted to reproduce Western products at the end of the Tokugawa period were still under the influence of "seclusion," since they were hoping to stem foreign imports by producing domestically. In this sense, they were pioneer "techno-nationalists" (ibid., 28–29).

9. Kudo Akira, *German-Japanese Business Relations*.

10. Nihon keieishi kenkyūjo, *Oki denki 100-nen no ayumi*, 103.

lead in 1926, setting up a Committee for Promoting Domestic Production, followed by the announcement of an "Outline for Domestic Production of Telecommunications Equipment" in 1929. The MOC, which held the monopoly on telecommunications service in Japan, gave preference in procurement to companies that were at least 51 percent Japanese-owned. By the late 1920s, there had been considerable progress through a combination of indigenization, reverse engineering, and licensing.[11] Partly as a result, Japan's imports of telecommunications equipment decreased drastically. In 1932, the MOC imported only 700,000 yen of equipment, mostly for experimental purposes, a sharp decline from 16 million yen in 1925.[12] Foreign shareholdings in Japanese manufacturers were reduced in the meantime. Japan's Sumitomo group, for instance, would become the largest shareholder in NEC.[13]

As MOC was aware, domestic manufacturing often came at a higher cost. The production of gutta-percha (GP) submarine telegraph cable in Japan reveals that the government had to balance domestic capability against financial considerations. Gutta-percha—a rubber-like substance grown only in Southeast Asia—was used to insulate underwater telegraphy cables, and Britain dominated its worldwide production. As gutta-percha refining and coating facilities were unique and required a large initial investment, at least 1,000 to 1,500 km of GP cables had to be produced annually to be profitable, exceeding Japan's need of 500 km each year without new expansion. Japan imported all its GP cables until World War I, which interrupted this import, as with many other industrial goods from Europe. In 1917, to replace the old cable between Shimonoseki and Pusan, the MOC had to use two 220-km inferior rubber-coated cables produced by two Japanese manufacturers, Yokohama Wires Company (established in 1896) and Sumitomo Electric Wires (established in 1920 as a joint venture between Sumitomo and NEC). Due to stiff competition from foreign imports, which resumed after the war and were being sold at low prices, cable production in Japan came to a stop after 1920. In the same year, however, the Furukawa Company won a Chinese government order for 535 nautical miles (1,000 km) of GP cable between Shanghai and Chefoo (Yantai). Financed by the

---

11. Yamamura, "Japan's Deus ex Machina," 65–95.

12. Takahashi Tatsuo, *Nihon shihonshugi to denshin denwa sangyō*, 313.

13. Fransman, *Japan's Computer and Communications Industry*, 31–34.

Japanese government loan to China, the advance of 1.5 million yen enabled the company to invest in new equipment and open a new factory. GP cable was thus produced in Japan for the first time. The 1923 earthquake, however, destroyed it.[14]

Despite some progress in its domestic manufacturing industry, Japan remained largely dependent on American and German firms for patents of advanced telecommunications technology. In July 1931, for instance, Sumitomo Electric Wires obtained a license from a subsidiary of the Western Electric Company in the United States in order to produce the new para-gutta submarine telephone cables.[15] As late as 1933, when Sumitomo participated in the MOC bid to manufacture submarine telephone cable to link Hokkaido and Karafuto, the company manager had to appeal repeatedly to the American side to lower its royalties on the para-gutta cable so that Sumitomo could reduce costs and win the bid against other domestic competitors.[16]

Although financial concerns continued to loom large as Japan's sources of foreign currency shrank, the cost of technological dependency began to take on an ideological dimension. The Great Depression that began in 1929 seemed to confirm Japan's worst fear of being shut out by increasing protectionism abroad. The early 1930s were quite different from the late nineteenth century, when advanced technology had been readily available to a Japan devoted to modernization. After all, in an increasingly hostile world in the wake of the Great Depression and the Manchurian Incident, many Japanese feared, open access to advanced technology from the West could no longer be guaranteed. As a result, Japan's bureaucrats and industrialists alike considered it a high priority to move from what they called "inauthentic indigenization" to independent technological development. Building on the momentum of the Domestic Production Movement that began in the 1920s, the gov-

---

14. Nihon taiyō kaitei densen kabushiki kaisha, *Nihon taiyō kaitei densen kabushiki kaisha shashi*, 3–11; Nippon denshin denwa kōsha kaiteisen shisetsu jimusho, *Kaiteisen hyakunen no ayumi*, 207–10.

15. Sumitomo denki kōgyō kabushiki kaisha, *Shashi Sumitomo denki kōgyō kabushiki kaisha*, 620–23; Nihon taiyō kaitei densen kabushiki kaisha, *Nihon taiyō kaitei densen kabushiki kaisha shashi*, 13–15.

16. Correspondence between Sumitomo Wire's general manager Akiyama Takesaburō and Western Electric executives is from the AT&T Archives. I thank Andrew Robertson for sharing these documents with me.

ernment sought to bolster Japan's technological independence by emphasizing a distinctively Japanese path of technological innovation. Only by developing innovative technologies on its own, it argued, could Japan eventually reduce its vulnerability to other industrialized countries for raw materials and find alternatives that would utilize resources available within Japan's new sphere of influence in Asia.[17] Under the new circumstances, technological independence assumed unprecedented importance for Japan's own techno-nationalists.

New developments in long-distance telephone transmission technology illustrate this newfound urgency. Germany, which had to maintain communication with Eastern Prussia via long-distance cable, was a leader in this field. In the 1920s, the German firm of Siemens demonstrated the possibility of using submarine telegraphy cable for telephone transmission.[18] For an island-nation with overseas colonial possessions, such technology was of great strategic value. By the beginning of the 1920s, however, almost all submarine cables in Japan were used exclusively for telegraphy. Only 2 percent of all submarine cables (1.5 percent in terms of length) were used for telephony. Japan expressed much interest in obtaining such technology but was politely turned down. Partly as a result of this refusal, the MOC mobilized its technical strength and enlisted the cooperation of the private firm NEC in an all-out effort to develop long-distance telephone transmission technology on its own.

In May 1931, a group of Japanese engineers arrived in Pusan on the southern tip of the Korean peninsula, across the strait from Fukuoka. Their goal, as defined by the MOC, was to make telephone communication possible via cable between Japan and its colony of Korea, which were separated by the Korea Strait. They would do so by opening part of the existing telegraph cable and converting it for voice transmission. The experiment also involved construction of relay stations, reconstruction of land cables, and the laying of new cables. Four months into the experiment, in September, the Kwantung Army launched the all-

17. *DTJGKS*, 82–83; Hoshimi Uchida, "Western Big Business and the Adoption of New Technology in Japan: The Electrical Equipment and Chemical Industries, 1890–1920," in Okochi and Uchida, eds., *Development and Diffusion of Technology*, 145–62; Morris-Suzuki, *The Technological Transformation of Japan*, 136–42.

18. The technical background of carrier telegraphy and of using telegraphy cable for telephony (named Krarup Cable) is discussed in Siemens, *History of the House of Siemens*, 2: 155–62.

out invasion of Manchuria with the assistance of the Japanese garrison troops in Korea. Since the cables linking Korea and Japan were fully utilized during the day for telegram transmissions, these engineers had to conduct their experiments at night. Despite the interruptions, they made progress by installing a carrier telephone system on existing telegraph cables. In early 1933, telephone communication via submarine cables between Japan and the Asian continent became possible for the first time. The well-publicized fact that almost all equipment used in this experiment was made in Japan boosted confidence in domestic technology.[19]

ANATOMY OF AN INVENTION

Among the MOC technicians sent to Korea was a 30-year-old electrical engineer, Matsumae Shigeyoshi. A native of Kumamoto prefecture in Kyushu, Matsumae had studied at the Electrical Engineering Department of Tōhoku Imperial University, one of Japan's leading centers for research in electrical engineering. Among his teachers was Dr. Yagi Hidetsugu, a world-renowned authority on antennae, whose design quickly became adopted worldwide. While a student, Matsumae began research on the vacuum tube, a relatively novel electronic device vital to signal transmission. Upon graduation, he could have stayed at the university as a researcher or worked for a leading private business firm, but decided to enter the MOC instead. In 1925, Matsumae became a junior engineer in the MOC and was soon assigned to the Nagasaki Communications Bureau as a telephone engineer. A year and half later, he returned to Tokyo. Yagi's warning that "only blockheads go to work in that uninspiring place, and the work they do is equally uninspiring," turned out to be true for a few years for young government engineers like Matsumae. He had nothing to do. Somewhat disillusioned, Matsumae became involved in the Christian movement under the influence of Uchimura Kanzō. He also began to conduct his own research and participated in the project in Korea with much enthusiasm.[20]

In his experiment, Matsumae applied the amplifier—an apparatus developed by Nukiyama Heiichi of Tōhoku Imperial University—as a

---

19. On this experiment, see *DTJGKS*, 11–12, 73–82.

20. For a somewhat self-serving autobiographical account, see Sakamoto, *A Lion Aroused: Conversations with Shigeyoshi Matsumae*, especially 82–101.

frequency equalizer and achieved a surprisingly clear quality of voice transmission. During his assignment in Korea, Matsumae received a telegram from his wife in Tokyo, informing him that their baby son was suffering from children's dysentery and that his condition was serious. Since the disease was considered fatal at the time, Matsumae was eager to learn more than a tersely phrased telegram could convey. One of his colleagues suggested using the telephone then under experimentation. Matsumae immediately made the necessary arrangements and called his home in Tokyo. The quality of the voice was so good that his wife initially thought he had returned to Tokyo.[21] About the same time that Matsumae's group succeeded in applying carrier telephony on submarine telegraph cables across the Korea Strait, a gutta-percha loaded cable produced by Sumitomo Electric Wires was used to connect Tsushima and Pusan, making telephony by submarine cable possible between Japan and Korea.[22]

The success of the cross-strait experiment had a far more important impact on Matsumae and his colleagues. For some time, Matsumae had been experimenting with a new method of applying carrier waves in long-distance telephone transmissions. The standard technology of the day was the loaded cable system, first devised by Dr. Michael Pupin of Columbia University at the beginning of the century. The diminution effect on electric signals transmitted over distances posed an obstacle to scientists in the nineteenth century. To prevent the electric current from being exhausted over a long distance, Pupin proposed placing so-called loaded coils along the transmitting cable at predetermined intervals. These coils, later named Pupin Coils, reinforced the electric current as it passed through. By using this technology, American Telephone and Telegraph (AT&T) extended the New York–Denver line to San Francisco, thus achieving the first transcontinental telephone communication in the United States in 1915.

In the mid-1920s, loaded cable was introduced to Japan. Under the direction of AT&T engineers, an 800-km loaded cable was constructed

21. The story goes that after finding out more about the symptoms, Matsumae told his wife to see a different doctor, who made a correct diagnosis and thus saved their son's life (ibid., 121–22).

22. Nippon denshin denwa kōsha kaiteisen shisetsu jimusho, *Kaiteisen no hyakunen no ayumi*, 239. In late 1929, Sumitomo had already connected Kyushu and Tsushima but with a similar cable made in Germany.

between Tokyo and Okayama in central-western Japan. The loading coils used on this line were imported from the Western Electric Company, the production wing of AT&T. There were some problems with this method, however. The loaded cable system would not work properly beyond a certain frequency, thus severely limiting its capacity in transmitting human voices. An even more serious defect was the limitation of distance. The time-lag in communication over long distance, as well as the varying speed of the electric waves in the cable due to different frequencies within the human voice range, led to a loss of natural voice quality and made it impossible for a telephone network to span a great distance.[23] As a result, scientists in several industrialized countries tried to improve the features of the loaded cable but could not totally break away from the Pupin method.

Matsumae was, in his own words, "not enslaved by the Pupin theory." The experiment in the Korea Strait had shown him that, without using loading coils, the attenuation loss could be compensated by installing amplifiers and attenuation equalizers. This led him to a reexamination of the fundamental properties of the cable. He suggested removing the loading coils and, to prevent the consequent drop in the electric current, introducing an amplification filter wave-detection device with the aim of achieving a multiplex carrier telegraphy system. For amplification, he suggested using the vacuum tube, which had been invented not long before in America and improved in Japan.[24]

Matsumae's unconventional ideas, still somewhat half-baked, met with considerable skepticism and resistance in electrical engineering circles in Japan. He continued with his research, however, with the assistance of several young engineers in the MOC, such as Shinohara Noboru, a recent graduate of the Electrical Engineering Department of Tokyo Imperial University. Initial field experiments on a new 33-km cable between Oyama and Utsunomiya convinced Matsumae of its technical feasibility.[25] In a second article co-authored with two MOC engi-

23. Sakamoto, *A Lion Aroused*, 116. On technical aspects of wire transmission development in the United States, see Fagen, *A History of Engineering and Science in the Bell System*, vol. 1.

24. Matsumae, "Experimental Study on the Non-Loaded Cable Used as a Long-Distance Telephone Line," in idem, *Hatsumei e no chōsen*, 1073; Sakamoto, *A Lion Aroused*, 116–17.

25. On this experiment, see *DTJGKS*, 160–62.

neers, Matsumae argued that "it is absolutely necessary to adopt the non-loaded cable for important long-distance circuits," since "only the non-loaded cable will have the distinctive feature for future developments from the technical point of view." He even suggested that it "can also be applied to picture transmission, trunk line broadcasting, etc."[26]

Not everyone was persuaded. Only after some initial difficulties did they get their path-breaking article on the special features of low-frequency telephone circuits and relay coils accepted for publication in 1932.[27] Opposition came from most domestic telecommunications manufacturers, who had invested heavily in importing the loaded cable technology from abroad—an investment that had yet to show any returns. Even within the MOC, opinions were sharply divided, as some of Matsumae's colleagues questioned the feasibility of the new technology. Those opposed to Matsumae's proposal favored the established practice of loaded cable and engaged him in a few rounds of open discussions in engineering journals. They pointed to various defects of non-loaded cable as well as to its higher cost.[28] An article published in an engineering journal in late 1932, for instance, argued that with proper additional equipment, current light-loading circuits would be capable of carrying conversations over 5,000 km. Although NLC could be used as a special solution in exceptional circumstances, the author pointed out, there was no need to make it a new standard.[29] Matsumae and his associates mounted a series of spirited responses. Matsumae stressed that NLC was particularly suited for long-distance communication but conceded that loaded cable could remain in use for shorter distances.[30]

---

26. Matsumae, "Experimental Study," in idem, *Hatsumei e no chōsen*, 1047–49. "Trunk line" refers to the main line in a communications network.

27. For more intimate accounts by some of the participants, see also Musōka hōshiki kankōkai, *Gijutsu kaihatsu e no michi: musōka hōshiki no kaihatsu to shido seishin*; and Sakamoto, *A Lion Aroused.* For excerpts of exchanges between Matsumae and his opponents, see *DTJGKS*, 111–18.

28. According to Matsumae, he had to persuade some of his colleagues to challenge his proposal openly so that he could further explain the new technology. Iwai Takanobu, who was considered an authority on loaded cable, may have outdone himself so as to "hurt the feelings" of Matsumae. See Iwai Takanobu, "Sōka to musōka," in Teishin gaishi kankōkai, *Teishin shiwa*, 2: 250–51.

29. The gist of Ōhashi Kan'ichi's article can be found in *DTJGKS*, 114–15.

30. Matsumae, *Hatsumei e no chōsen*, 189–95. He also noted that, as an indigenous technology, NLC enabled "we Japanese to have far more authority."

At this critical juncture, Kajii Takeshi, chief of the Telephone Section of the Engineering Bureau, arranged for Matsumae to spend a year studying in Germany, perhaps in an effort to mitigate the controversy within the ministry.[31] Had Matsumae been working alone, his absence from Japan would have spelled the end of development of the new technology. Fortunately, the research project was already under way. During Matsumae's absence, his colleagues led by Shinohara continued the testing, making use of data Matsumae sent back from Germany. In late 1933, an experiment with relays on loaded cables between Osaka and Kobe was completed successfully. To make non-loaded cable truly operational, a high-performance relay amplifier and cable with a low level of crosstalk were necessary.[32] Soon afterward, MOC engineers conducted the first experiment on a 7.5-km non-loaded cable between Omichi and Miko in western Japan. It was a success.

At this stage, cable-manufacturing technology proved to be the key. Each of the three major electric wire and cable manufacturers in Japan—Sumitomo Electric Wires (now part of the NEC group), Furukawa Electric, and Fujikura Wire Works—supplied cables of equal length so as to compare different properties.[33] Although the non-loaded cable method had its own defects, such as the increase of crosstalk, Shinohara and his colleagues were confident about improving the quality of the cable. In the meantime, in Germany, Matsumae engaged in heated discussions with Dr. H. F. Mayer of the Central Laboratory at Siemens und Halske. A strong proponent of light-loading as the best solution to long-distance telephone communication, Mayer was vehemently opposed to Matsumae's NLC method and presented Matsumae with new data. The debate ended without a decisive conclusion, which Matsumae attributed to his inadequate knowledge of German. But

31. Sakamoto, *A Lion Aroused*, 125–27. Kajii also entrusted him with the mission of negotiating the opening of an international telephone link between Japan and Germany.

32. Kobayashi Kōji, *C & C modan komyunikeeshon*, 43. As a young engineer at the Nippon Electric Company, Kobayashi participated in the development and testing of the NLC.

33, For this experiment, see *DTJGKS*, 29–31, 148, 165–69; Musōka hōshiki kankōkai, *Gijutsu kaihatsu e no michi*, 30–32. Such cooperation between the bureaucracy and several manufacturers is similar to what Richard Samuels has described between the military and the aircraft industry in roughly the same period; see Samuels, *"Rich Country, Strong Army,"* 116.

Matsumae was able to send the important data he had obtained from Mayer back to Japan.[34]

Such experiments might have continued for quite some time had it not been for Japan's pressing strategic needs in the 1930s. After the successful telephone transmission on submarine cables in the Korea Strait, the MOC set up a Committee on Basic Planning for Long-Distance Communications, with Kajii Takeshi as its chairman. What was not immediately apparent from its name was that MOC was drawing up plans for a new long-distance telegraph link between Tokyo and Mukden in the puppet state of Manchukuo. Since the cable was considered a military link and a matter of national priority, there would be no lack of funds, a problem in other domestic projects in the past.[35] Considerable resistance still remained in the MOC to this untried invention, however. In the early summer of 1935, just as Matsumae returned from his year-long overseas tour, the MOC was deliberating on cable technology for the Japan–Manchukuo project. Although before his departure for Europe Matsumae had indicated that he intended to resign and devote himself to education, he apparently changed his mind. Throwing himself into the deliberations, Matsumae made an emotional plea at the MOC meeting:

> Nowhere outside Japan has the non-loaded cable been put to use. If we do it, it is going to be Japan alone. However, the non-loaded cable technology is by no means an adventurous technology. I believe it is Japan's mission to adopt it. For instance, when telephoning from Tokyo to Beijing, one absolutely can't have a satisfactory conversation using a loaded cable. Non-loaded cable is the only method to achieve this objective. Japan's long-distance cable network is at the point of expansion right now, and turning down this technology will forever cause Japan regret.[36]

Matsumae correctly sensed that this was the critical moment for the new non-loaded cable technology. If the widely accepted loading method, or even light-loading, were adopted for the unprecedented proj-

---

34. See chap. 27, "Long Distance Cable and Telegraphy," in Siemens, *History of the House of Siemens*, 2: 155–74. F. H. Mayer proposed lighter loading coupled with amplifiers and succeeded in adding a second speech channel by using a carrier-frequency system. However, since it required closer spacing of amplifiers, it was used only in cases of real necessity and for very important lines of great length in Europe.

35. Sakamoto, *A Lion Aroused*, 135.

36. Musōka hōshiki kankōkai, *Gijutsu kaihatsu e no michi*, 32–33.

ect, it would probably also be used for other parts of the telecommunications network in Japan, and non-loaded cable would remain a theoretical idea. Matsumae had greater hopes for the non-loaded cable than just one project, albeit an important one. It was no coincidence that he shrewdly mentioned Beijing, the political center of North China. At this juncture, the Beijing area was the hotbed of the North China Autonomy Movement, a scheme by the Japanese Kwantung Army to separate an area adjacent to Manchukuo from the control of the Chinese Nationalist government in Nanjing. Since the Manchurian Telegraph and Telephone Company (MTT) was busy extending telegraph and telephone lines into areas east of Beijing, the promise of high-quality telephone connections with North China had important political implications for Japan's continental policy.[37] In this way, Matsumae clearly saw the nexus between technology and empire and consciously linked his innovation to Japan's new expansion on the continent.

Yet more than Matsumae's stirring speech was needed to break the stalemate. In 1935, the new MTT was looking for an appropriate way to strengthen its communication links with Japan by upgrading the Mukden–Andong telegraph line, which would be the first installment of the Japan–Manchukuo cable. Nakada Suehiro, head of MTT's Engineering Department and an old acquaintance of Kajii Takeshi, arrived in Tokyo to investigate technological standards. Although initially inclined toward a light-loaded cable as used by Germany, he became attracted by the merits of the non-loaded method. Shioda Shinji, an MTT employee who accompanied Nakada to Japan, was to comment later that the people inside a government bureaucracy like the MOC often lacked the courage to put new ideas to practical use, even when the ideas themselves were excellent ones. Sensing some reluctance on the part of MOC, Nakada and Shioda decided that the MTT could offer a testing ground for the new technology. This was no means an easy decision for them. Nishida Inosuke, head of the MTT's Business Department, was strongly opposed to a more expensive technology and thought that using Manchukuo as a testing ground was too risky. But by joining forces with younger MOC engineers and obtaining the blessing of Kajii himself,

37. See Shimada Toshihiko, "Designs on North China, 1933–1937," in Morley, ed., *The China Quagmire: Road to Pacific War*, 135–74.

Nakada prevailed. The MTT decided to install non-loaded cable in Manchukuo.[38]

This decision had a far-reaching effect, for it provided the much-needed push to put non-loaded cable across the threshold to the stage of actual application. The MTT's important role in initiating this grand enterprise once again demonstrated that engineers on the "new frontier" tended to be more innovative than their counterparts at the imperial metropole. Matsumae himself was greatly impressed with what he described as the "spirit of the technicians from Manchuria in nurturing something not yet cultivated in the Home Islands."[39] After much effort was devoted to demonstrating the economic advantages of NLC, the MOC concluded that the trunk cable linking Japan and Manchukuo via Korea would use the new technology.

Unlike the newly founded MTT, most major electronic equipment manufacturers in Japan still had reservations about the NLC technology, even though they had participated in the experiment. At the same time, the domestic market was still in a slump, so these companies were eager not to antagonize their biggest customer—the MOC. Personal ties may also have played a role in winning them over. Kobayashi Kōji of NEC, for example, had been a classmate of Shinohara Noboru in the Electrical Engineering Department at Tokyo Imperial University. He would emerge as one of the first strong supporters of the NLC technology. Shinohara would also play a key role in repairing matters whenever the strong-minded Matsumae alienated someone in the manufacturing industry. In the meantime, the MOC also engaged in a publicity campaign. One of their tactics was to have Victor Records make a demonstration audio record highlighting the difference in sound quality between loaded and non-loaded cables, together with a full explanation of the merits of the latter.[40] These efforts seemed to be successful.

---

38. Toya Tokujun, "Chōkyori tsūshin gijutsu," in Manshū denden tsuiokuroku kankōkai, *Akai sekiyō*, 83–84; Kajii Takeshi, *Waga hansei*, 112; *DTJGKS*, 203, 226.

39. Quoted in *DTJGKS*, 226.

40. Yoshida Gorō, "Musōka hōshiki rikai e no doryoku-PR," in *DTJGKS*, 120.

A TRIUMPH OF JAPANESE TECHNOLOGY

In 1936, the Japanese government adopted non-loaded cable for the new Japan–Manchukuo cable network as well as for the long-distance communications networks in Japan, thus establishing the supremacy of the new technology in Japan. In the same year, Matsumae was awarded the Asano Prize by Japan's Association of Electrical Engineering for his ground-breaking contribution to the development of telecommunications technology. Named after one of Japan's first electrical engineers, who oversaw the laying of the submarine cable to Taiwan, the prize of 1,000 yen further consolidated the reputation of NLC as well as that of its chief inventor. Later that year, Matsumae received his doctoral degree from Tōhoku Imperial University. As Kajii Takeshi would describe it a few years later, the NLC technology was "the greatest invention in Japan's telecommunications industry."[41] Now recognized as Japan's unique contribution to the field of telephone transmission, NLC would be celebrated as a major step forward in Japan's quest for technological independence from the West.

Although the adoption of NLC in the Japan–Manchukuo long-distance route was cause for great pride in Japan, it was relatively unrecognized outside the country. To make up for the lack of attention to NLC outside Japan, in 1938 Matsumae managed to publish, in English, a work entitled *Study on the Long Distance Communication System by Non-Loaded Cable*. In it, Matsumae proposed "an ideal communication network of the world" based on NLC. He further suggested that "If the League of Nation[s] existed for the peace of humanity, then the best way to demonstrate its worth of existence is to work for the completion of such a communication network that has the greatest mission of exchanging the civilization of mankind."[42] Toward the end of the book, he seemed to lapse into a religious mood by asking

> how to complete the communication networks on this narrow earth? We communication engineers must consider what type of engineering should be

---

41. "Kajii Takeshi hakasei ni kiku," *TKZ* 379 (March 1940): 62.

42. Matsumae, *Study on the Long Distance Communication System by Non-Loaded Cable*, 1180. Although the original text is in English, I have paraphrased the quote.

> adopted and do not neglect the effort for these objectives. It is for the peace of mankind, for the exchange between civilizations, and for the solemn purpose of creation of universe by God that these engineering skills should be used. And for this purpose, the uncultivated continent of Orient is given to our country, which has long since grown in its history and now has the honor to accomplish the great mission [that] is given to us. We also awake to our position and undertake the independence of our industry and now stand up to exercise freely the great mission of our peaceful reclamation of this given Orient.[43]

Matsumae was not alone in seeing such potential for the new technology. In a speech in 1936 marking the sixtieth anniversary of the invention of the telephone, Kajii Takeshi similarly predicted that NLC technology could be used in a Eurasian telephone cable or transatlantic cable. Together with the wireless telephone, "conversation will become freely possible hardly with any idea of distance."[44]

As Matsumae pointed out in his book on the NLC system, by the 1930s, both the United States and Germany had already constructed their domestic toll lines using the loaded system. Since Germany had indicated its intention to switch from a light-loaded to a non-loaded system, Matsumae speculated, its "future industry may be wholly aimed at exporting to other countries."[45] Shinohara Noboru, who played a leading role in the development, brought with him a four-part documentary film on the NLC when he attended an international meeting on telecommunications in Oslo in 1937. When Shinohara visited Mayer in Berlin after the conference, the German engineer who had once debated with Matsumae allegedly reversed his position and accepted NLC as the best solution.[46]

In one sense, NLC seemed truly "Japanese," since one finds almost no reference to this technological breakthrough in Western literature. That Japan's enthusiasm for NLC was not matched in the West is less of a mystery when we consider the larger context. In hindsight, there are many striking parallels between developments in Japan and overseas, even though they may not have been obvious to those involved

43. Paraphrased from Matsumae, *Study on the Long Distance Communication System*, 121.

44. An excerpt of Kajii Takeshi's speech is reprinted in *DTJGKS*, 169–70.

45. Matsumae, *Study on the Long Distance Communication System*, 1176.

46. Shinohara Noboru, "Musōka keburu hōshiki kaihatsu tōjō no wadai"; and Yoshida Gorō, "Musōka hōshiki rikai e no doryoku—PR," in *DTJGKS*, 100–101, 120.

at the time. Matsumae and his colleagues themselves made references to overseas research on similar ideas by A. B. Clark and H. S. Osborne in the United States. Matsumae even acknowledged that Bell Laboratories had conducted experiments on the non-loaded systems that were "similar to ours," but noted that since none of the details had been published, he was unable to know exactly how similar. On his return journey, Matsumae had visited telecommunications makers in Britain, but seemed to arouse little interest. In the United States, he met with engineers of AT&T but found them, like the Englishmen he had met, unwilling to engage in in-depth discussions about the new technology. Having heard of AT&T's experiment in Morristown, New Jersey, on long-distance telephone cables, Matsumae requested permission to visit but was turned down by the company.[47]

Matsumae's suspicion was not unfounded. As early as May 1928, the Bell Laboratories—the research wing of AT&T—had issued a report on the prospect for carrier telephony on non-loaded or lightly loaded cable pairs. The same month, Bell engineers discussed the properties of the newly improved amplifier with the objective of applying it to the cable-carrier system. Thus, the basic technological elements of the non-loaded cable were already in place. In the beginning of 1930, researchers experimented on a 40-km section of the New York–Chicago telephone cable at Morristown, New Jersey. They concluded that the cable-carrier system would eventually replace the loaded cable on the trunk routes of a long-distance communications network. However, further research in manufacturing technology as well as cost reduction was required to make it feasible.[48] The onset of the worst depression in U.S. history

---

47. *DTJGKS*, 14–16, 119. On Matsumae's experience in Germany, see also Sakamoto, *A Lion Aroused*, 127–34.

48. Quoted in *DTJGKS*, 69. Evidence suggests that removal of loading was carried out much earlier. As noted in a Bell Labs publication during World War II, shortly after the plan for a transcontinental telephone network was approved in 1915, engineers discovered "more attractive possibilities in the use of non-loaded 165 mile lines having additional repeaters to make up for the increased line losses." According to its author, "the non-loaded lines also had important possibilities in the application of carrier telephone systems, the commercial development of which got well started during the 1915–1920 period." Subsequent studies and experiments in the United States showed non-loaded circuits had noticeably better quality of speech transmission and a wider transmission band. On the 104-mile circuits, however, loaded mileage actually increased due to production limitations on repeaters, reaching a peak in 1923. Use of open-wire load-

slowed plant growth abruptly. As Bell Labs' official history notes, the slowdown "effectively ended the speculation about novel cable designs and larger-capacity systems."[49] The depressed economy halted further development in the United States. Although the experiment in Morristown had been eminently successful, the report concluded that "under the present economic conditions there is no immediate demand for the installation of systems of this type." Commercial exploitation of the cable-carrier system known as Type K started during the mid-1930s, followed by a more economical K2 cable-carrier system. Moreover, American engineers were already studying an entirely different and revolutionary technology—the coaxial cable, which promised to carry hundreds of channels.[50]

Regardless of whether it was Japanese or American engineers who *first* invented the NLC,[51] the crucial difference between Japan and other countries was the fact that for Japan the NLC technology was considered "an answer to the demand of the time" (*jidai no yōsei ni kotaeta*).[52] Uncertain business prospects as well as vast existing facilities made AT&T executives reluctant to adopt the new non-loaded carrier technology. Kobayashi Kōji, the NEC engineer and participant in the NLC project, reported that during his visit to the United States in 1938, an American engineer expressed regret that there were so many old cables in the United States, so that "even [if] there is research on the non-loaded cable in the Lab, there is no opportunity to use it in reality."[53] Matsumae had already noted, in an article published in early 1934, that

---

ing in the United States did not cease completely until 1934. See Thomas Shaw, "The Conquest of Distance by Wire Telephony," *Bell System Technical Journal* 23, no. 4 (October 1944): 393–98.

49. Fagen, *A History of Engineering and Science in the Bell System*, 73–75.

50. Thomas Shaw, "Evolution of Inductive Loading," *Bell System Technical Journal* 30, no. 4 (October 1951): 1239–40. Shaw also noted that "for voice-frequency transmission, the use of repeaters [on cable] without loading would have been unduly expensive, due to the high cost of repeaters and the much more expensive distortion-correcting networks and regulating networks that would have been required."

51. According to the official history of the Siemens company, Great Britain was the first to discard the Pupin Coil altogether in 1935, and "it was found possible to transmit twelve conversations simultaneously per pair of 'unloaded' cores" (Siemens, *History of the House of Siemens*, 2: 165).

52. *DTJGKS*, 4.

53. Kobayashi Kōji, "Ō-Bei ni okeru hansō tsūshin gijutsu," *DT* 2.4 (June 1939): 7.

it would be too costly for Germany to convert all its existing underground loaded cable into the non-loaded method.[54] The national agenda—"the demand of the time"—in Japan of the early 1930s was quite different from the business calculations of AT&T and the research priorities of its scientists. Development of the non-loaded cable in Japan highlighted the state's role in promoting new technology at a time when private firms were reluctant to take the risk. Moreover, it was the empire-building agenda that tipped the balance in favor of Japan's own technology. This explains the different paths of NLC development in Japan and the United States.

Ideology clearly played an important role as well. The times demanded not only suitable technologies to fulfill Japan's technical requirements but also a new role of technological leadership for Japan. While Matsumae and his associates sought world recognition for Japan's new technological achievement, there was already a trend toward rejecting foreign technology and using only Japanese technology. Kimura Suketsugu, head of the Engineering Division at the Fujikura Wire Works, was among the first to stress, in an open discussion in 1935, that NLC "stems the unhealthy trend of reliance on the West in our engineering circles."[55] Many without a background in engineering readily shared their views. Imaida Kiyonori, a onetime vice minister of MOC who also served as superintendent general in Korea under General Ugaki Kazushige, underscored the importance of Japan's own technology. At a Diet meeting in 1939, he noted:

> As a result of the invention of the Japanese technology known as the NLC carrier communications method, such long-distance communication has become extremely good [in quality], and plans [for further expansion], too, have become possible. . . . Especially nowadays, [when] international situations have become very seclusionist (*sakoku teki*), and it is difficult for enough foreign technology to come in, research in technology is all the more necessary. This applies not only to the technology of communication but to technology in all areas.[56]

54. The gist of Matsumae's article is in *DTJGKS*, 118.

55. Kimura Suketsugu, in Matsumae Shigeyoshi ronbunshū kankōkai, *Hatsumei e no chōsen*, 597.

56. *Teikoku gikai kizokuin gijiroku* (March 18, 1939), 18.

Just as indigenous technology was vital to Japan's self-sufficiency, many leading Japanese engineers had come to expect telecommunications technology developed by Japan to play an increasingly prominent role in Japan's new overseas expansion. Part of the reason was the need to improve Japan's trade position. As the MOC's brochure on the Japan–Manchukuo cable noted:

> The transition from the loaded cable developed abroad to non-loaded cable promoted (*teishō*) in our country is simultaneously a transition from reliance on foreign technology in our telecommunications equipment industry toward technological independence. This is the beginning, where we add intelligence-based manufactured goods in communications equipment to exports hitherto limited to light-industry goods. This change will contribute greatly to the improvement of our trade balance.[57]

Although NLC stood out as a crowning achievement of Japan's telecommunications technology, there were other important innovations in telecommunications manufacturing. One area was phototelegraphy (*shashin denshin, shashin denpō, shashin densō*) technology. Although sending photographs over telegraph lines had been tried at the beginning of the twentieth century, it was the imperial coronation of Hirohito in 1928 that marked the beginning of its commercial use by Japanese news agencies. Four months before the event, the MOC granted newspapers and news agencies permission to set up private phototelegraphy circuits. They immediately made arrangements to import phototelegraphy equipment from Europe: *Asahi* and a major news agency deployed imported German equipment, and *Nichinichi* arranged to import equipment from France. Due to the poor quality of the French equipment, however, *Nichinichi* had to abandon it and, at the last minute, had no choice but to use Japanese phototelegraphy equipment developed and produced by NEC. The successful transmission of some 253 photographs—half of which subsequently appeared in the *Nichinichi* newspaper—marked another turning point in Japan's own research and development in the new

57. Teishinshō, Kōmukyoku, *Nichi-Man renraku musōka keburu no kansei ni saishite* (September 1939), 24. It admitted that three important foreign patents—all American—had been used in the NLC and other carrier communications. It went on to note that a certain Japanese company had succeeded in producing a new type of carrier equipment that "not only used no foreign patents but also avoided imported raw material as much as possible." As a result, "splendid, full domestic production" (*migoto zen kokusanka*) could be expected soon.

communications technology.[58] Japanese engineers were already making some progress in television technology to transmit live pictures. Improvements continued on the existing communications technology, including the famed non-loaded cable technology. The standard number of telephone circuits on such cables was increased from the initial three to six. The MOC, in particular, issued repeated calls for improving technical levels and indigenous production of telecommunications equipment. Although the public could easily appreciate the benefits of phototelegraphy and facsimile, it felt that this was not the case with other technological progress.

A TECHNOLOGICAL HEGEMONY IN ASIA

The development of NLC technology had important implications for Japan's relations with Asia, especially China. In the 1930s, the ongoing civil war notwithstanding, economic reconstruction and state-building in China gained momentum. Early in that decade, the League of Nations sent many technical experts to China to work on various development projects. Britain and the United States were converting the remainder of their Boxer indemnity payments to funding various cultural and scientific enterprises in China, often with explicit connections to the promotion of their own exports. Apart from the international wireless stations, the Nationalist government's most ambitious scheme in the area of communications was perhaps the domestic long-distance telephone network, which would connect nine provinces in Central and North China. The entire network was bare-wire, except in rivers, where underwater loaded cable was used. AT&T standards were deployed throughout. Due to the distances involved, vacuum tube repeaters were used only on lines linking the capital of Nanjing with major cities such as Hankow and Tianjin. According to the plan submitted by the gov-

58. Niwa Fumijirō and Kobayashi Masaji, "Shashin densō no ichi hōshiki," originally published in *Denki gakkai zasshi* 487 (February 1929), excerpts reprinted in Nihon kagakushi gakkai, comp., *Nihon kagaku gijutsu shi daikei* (19), 226–28. On the efforts of Dentsū and other news agencies to use phototelegraphy during the ceremonies in Kyoto, see Dentsū tsūshin shi kankōkai, *Dentsū tsūshin shi*, 178–85; and Tsūshinsha shi kankōkai, *Tsūshinsha shi*, 945–49.

ernment's Division of Telecommunications in early 1934, the total cost would amount to 3,598,927 yuan. More than 2 million yuan of copper wires and other equipment would be purchased from Britain with the Boxer indemnity earmarked for river management. The remaining amount would be spent locally on wood and labor.[59]

In ways reminiscent of the wireless rivalry in the 1920s, Japan considered China's reliance on Western assistance to be against Japanese interests. The Japanese government objected to this development, seeing it as another manifestation of the age-old practice of China "playing off barbarians against barbarians." In early 1934, Foreign Ministry spokesman Amau Eiji discussed the issue of Western countries extending financial and technical assistance to Nationalist China. The Japanese government opposed this, Amau said, because it fanned anti-Japanese sentiment in China. The so-called Amau Statement stirred up considerable controversy abroad, although it was supposed to be merely an elaboration of Foreign Minister Hirota's statements in the Diet two months earlier to the effect that "Japan, serving as the only cornerstone for the edifice of peace in East Asia, bears the entire burden of its responsibilities." But Amau was more specific:

> The assistance rendered to China by the Powers, even if it is financial, technical, or by some other name, is bound to have political significance, and thus leads to establishment of spheres of influence in China. It is the beginning of international management or division of China. As such, it not only brings great misfortune for China, but also [has] grave consequences for the security of East Asia and consequently for Japan. Therefore, Japan must object in principle.[60]

The association of financial and technical aid with political implications is significant here. Amau's statement was a defensive attack on what had been a trend in China ever since the late 1920s, when Japan's financial and technological influence had begun to decline there. Yet if

---

59. Plan submitted by Director of Telecommunications (Yan Renguang) to Minister of Communications (Zhu Jiahua), January 23, 1934, Chinese Communications Ministry Records I-20(21)-979. Almost immediately, Chiang Kai-shek, based at headquarters in Nanchang, ordered additional lines in areas of anti-Communist campaigns. See Memo by Director of Telecommunications, July 1934, in the same volume. See also W. H. Tan, "Telephonic Communications in China," *FER* 32 (November 1936): 509.

60. The text can be found in Gaimushō, *Nihon gaikō nenpyō oyobi shuyō monjo*, 284–86; see also Crowley, *Japan's Quest for Autonomy*, 196–97.

Japan were to be considered the stabilizing force in Asia, it would have to demonstrate its ability to do so *technologically*. Thus the new sense of mission had to be matched with ability on the part of Japan. For Matsumae, at least, this amounted to Japan's "peaceful reclamation of this Orient" it had been given. And maintaining a lead in technology was increasingly seen as essential to shoring up Japanese influence in China at a time of growing Western presence there. In this sense, the Amau Statement was probably Japan's last opportunity to consolidate its informal empire in China by means of technological superiority.

In late 1934, as part of a larger effort to "normalize" the troubled relationship, the Foreign Ministry entered into negotiations with the Chinese government over outstanding Japanese loans. The Chinese wanted to repay the loans, but the Foreign Ministry expressed the strong desire that about half the funds be reinvested in enterprises that were "financially sound and necessary to the [Japan's] national policy." Telecommunications were particularly favored. Tsujino Sakutarō, an MOC engineer who was serving as a communications advisor to the Chinese government, drew up a "Ten-Year Telephone Expansion Plan for North China," in which 15 million dollars would be spent toward telephone expansion in North China over a ten-year period. By way of reinvestment short of demanding direct control, the Foreign Ministry reasoned, it could save "face" for the Nanjing regime and avoid foreign accusations of instituting a monopoly in China.[61] The MOC worked closely with the Foreign Ministry and drew up a new proposal for telecommunications expansion in North China along similar lines. Using the monthly Chinese loan payment, Japan would embark on a five-year plan at a cost of over 9 million yen, which included automation of telephone exchanges in Beijing and Tianjin, expansion of communications facilities in three other provinces, and wireless communications between North China and Japan and Manchukuo.

It was no coincidence that those involved in non-loaded cable development became leading advocates of an aggressive new technology policy for Japan in East Asia. Telecommunications were far from being well developed in China.[62] As far as the MOC was concerned, Japan's

61. Hikita Yasuyuki, "Nihon no tai-Chūgoku denki tsūshin jigyō tōshi ni tsuite," 44.

62. By 1937, there were only 210,000 telephones in all of China (the United States had 27 million); see Forman, *Changing China*, 282.

technological strength promised a preferable course. Because China was to be the first target for overseas expansion, the MOC set up a Committee for Investigating Telecommunications in China in early 1935. Headed by the MOC vice minister, the committee was charged with investigating Chinese telecommunications policies and practices, the presence of foreign interests, the status of Japanese interests, and previous loans. A detailed list of telecommunications matters in China was to be studied to determine Japan's course of action. In 1935, the committee dispatched Kajii Takeshi, the newly appointed director of the Engineering Bureau, on a tour of China. In Shanghai, Kajii took part in the East Asia Industrial Exposition, along with a delegation of Japanese businessmen and engineers. After meeting with many Chinese officials and discussing the advantages of the new NLC technology, Kajii left to survey telecommunications in North China.[63]

Returning to Japan, Kajii submitted a 71-page confidential report "concerning communications relations between Japan and China." In the report, Kajii stressed that the national character of China was different from that of Japan, and recommended a more pragmatic and flexible approach to solving the problems between the two countries. As Kajii saw it, the idea of "serving the powerful [party]" had always been strong in China. In modern mechanical civilization, the Chinese considered the West to be advanced. Although Japan was superior to China in material and institutional development, the Chinese believed that Japan was still far behind Western countries in aspects of civilization that originated in the West (*taisei bunka*). Kajii noted that since World War I, Japan had not only caught up with the West in industrial technology but had surpassed it in some areas, such as textiles, synthetic fabrics, ship-building, and telecommunications. The solution to the problem with China, Kajii felt, was to convince the Chinese of Japan's newly developed industrial and technological superiority so that the Chinese could be brought to rely on Japan instead of blindly following Anglo-American interests. If China adopted Japan's telecommunications technology, this would expand Japan's communications rights

---

63. Kajii, *Waga hansei*, 143–45. The delegation was headed by Inoue Tadashirō, a former railway minister who later became the first president of the Technology Board.

(*tsūshinken*), strengthen communications ties between the two countries, and promote Japanese exports.[64]

Kajii shared his impressions of China with members of the Electrical Engineering Association in Tokyo the following June. Declaring China to be a country lacking the power to absorb modern aspects of civilization, Kajii concluded that electric communication was not suited to China's national character. In his view, the communications enterprise in China had proceeded from an economic, profit-making motive, not as the nerve system of the state. The new telecommunications projects in China, such as the Nine-Province Long-Distance Telephone Network, he noted with dismay, were based almost entirely on imports from Britain bought with Boxer indemnity funds. After showcasing non-loaded cable to the Chinese as representative of Japan's progress in telecommunications technology, Kajii came away with the impression that "if Japan were to advance into China in the future, unless we can present them with Japan's unique technology, the Chinese will not entirely agree [with us]." He felt this was the case because in the past Japanese exports to China had consisted mostly of products identical to those of Britain and America. As a result, the Chinese had concluded that, rather than purchasing them from a late-developing country (*kōshinkoku*) like Japan, it was better to buy directly from advanced countries like Britain and America that had more reliable technologies. Kajii specifically opposed the idea of selling Japan's surplus old-style telephone sets to China, not only because China was upgrading to automatic telephones but because doing so would only strengthen the impression in China of Japan's technological backwardness.[65]

Kajii recommended that, as the closest neighbor to China, Japan should provide economic assistance ahead of Western countries. Shifting his interest from Manchuria to China proper, Kajii concluded that by cooperating with China, Japan could confront Europe and America and establish real peace in East Asia. According to Kajii, Japan's lifeline was neither Manchuria nor North China. In a rhetorical question, Kajii

64. Kajii Takeshi, "Nisshi tsūshin kankei ni tsuite," October 1935, NCTT Records 2028/55(2).

65. Kajii Takeshi, "Shina ni okeru denki tsūshin jigyō," *Denshin denwa gakkai zasshi* 161 (August 1936): 14.

asked his audience, "Isn't China our real lifeline?" More specifically, he suggested

> developing China by Sino-Japanese cooperation and increasing China's purchasing power so that it becomes the outlet for Japan industrial products. By thorough cooperation between Japan and China, we can maintain peace in East Asia and eliminate Western oppression of East Asia, can't we? I firmly believe that we have to promote Sino-Japanese cooperation by improving China's communications facilities with the hands of Japanese technicians. This is our mission; this is our ideal.[66]

Kajii's vision of China-Japan cooperation was remarkable in two ways. First, he subtly took issue with the military emphasis on Manchuria and instead stressed the market in China Proper. Second, he saw in Japanese technology the possibility of competing with Western influence in China. People like Kajii represented the last remaining advocates of economic expansion into China. In some ways, Kajii merely provided additional support for the notion of "industrial Japan, agricultural China," although he reformulated the rationale for economic expansion on the basis of Japan's new confidence in its technology. It was no coincidence that many others in the MOC also seemed to place a high priority on expanding Japan's telecommunications influence overseas. Renewed confidence in Japanese technology offered added hope that Japan was again capable of competing in foreign markets. They noted Japan was still doing poorly in comparison with other industrialized countries in exports of telecommunications equipment. According to a Japanese report, in 1935 Japan's shares of total world exports of wireless and cable products were a meager 4.0 and 0.7 percent, respectively.[67] Kajii recommended a more expansionist approach that would strengthen the domestic telecommunications industry and establish Japan's telecommunications hegemony in East Asia by "our unique technology and superior indigenous products." There was great opportunity for export expansion in East Asia, he noted, because telecommunications facilities were extremely underdeveloped in the region—22 percent of the world's population had only 0.7 percent of the world's telephone sets.[68] In early

66. Ibid., 1–22.

67. Naikaku, Jōhōbu, *Tōa ni okeru denki tsūshin seisaku* (1939), reprinted in *KSS*, 139–46.

68. Kajii, "Shina ni okeru denki tsūshin jigyō," 1–22.

1936, a confident Kajii Takeshi even proposed a conference on communications technology in East Asia:

> As Japan has become an important industrial nation not only in the East but also in the world, Japan has come to possess unique, superior domestic technology independent of the West. This promises the future of Japan as the leader of a communications network in the Orient. Just as conferences on communications technology exist in Europe, [so] Japan should convene such a conference in the Orient, both to establish [its] telecommunications influence in the Orient and to export Japan's communications technology.[69]

Kajii's vision was soon put to the test. In March 1937, the Chinese Ministry of Communications announced an open bid for an underground telecommunications cable between Nanjing and Shanghai. The significance of this cable linking China's capital with its most important economic center was obvious. The project, based on light loading, would include a 300-km-long duplex steel-shield cable as well as loading coils, repeater stations, and toll switches. The contenders included several major foreign cable manufacturers. Apart from the German firm Siemens and the China Electric Company (CEC), a subsidiary the International Standard Electric Company had established in partnership with Sumitomo Electric Wires, there were two leading Japanese firms—Furukawa Electric and Fujikura Wire Works (represented by Mitsui Bussan). Busy promoting the new non-loaded cable technology, the MOC saw in this project a golden opportunity for Japan to break into the increasingly competitive China market. To coordinate strategy, the MOC gathered major Japanese participants in Tokyo and reached the agreement that (1) as a national policy, Japan would present a project estimate based on NLC; (2) Furukawa and Sumitomo-NEC would present estimates based on loaded cables through Siemens and CEC, respectively, although NLC would remain the ultimate goal; and (3) if any of the Japanese bidders won, the project would be shared among all three Japanese manufacturers. To coordinate the effort, MOC officials were dispatched to Shanghai, where they joined Nakayama Ryūji, the veteran Japanese expert on telecommunications expansion in China, and another MOC official in residence in Shanghai.

---

69. Kajii Takeshi, "Tōyō denki tsūshinmō yori mitaru Manshū no chii," *TKZ* 330 (February 1936): 175.

Kajii was realistic about the prospect of Japan winning the bid in China. Fully aware that "the Nanjing government was not favorably disposed toward Japan," Kajii reasoned that if China were to adopt such a "uniquely Japanese method as NLC, it will be under the control of Japanese technology in the future." Believing that China would not adopt NLC, Kajii considered it best for Japan to first accept the bid in loaded cable, as the Chinese government desired. This would allow Japan to promote the NLC method by first building ties with the Chinese Ministry of Communications.[70] The Japanese side thus fully appreciated the larger implications of technological diffusion as well as technological dependence.

In early July, the Japanese seemed to be gaining ground. They had even ascertained that, through the CEC, Sumitomo and the newly established Japan Telegraph and Telephone Construction Company (JTTCC) would be given the contract. On the night of July 7, Chinese and Japanese forces clashed outside Beijing. Earlier that same day, the small group of Japanese engineers from MOC had gathered, anxiously awaiting the final results of their bid for the Chinese government telecommunications project in Shanghai, more than 1,000 km to the south. But just as the final decision was about to be made, the confrontation in North China began to look increasingly ominous, with reinforcements pouring in from both sides. On July 18, the Japanese group abandoned the project and left Shanghai.[71] Many of them would return to China on the heels of the Japanese Army several months later, however, with the new mission of consolidating and expanding control of telecommunications in areas that came under Japan's occupation.

Japan's attempt at economic expansion into China's telecommunications market thus became one of the first casualties of the war, as China and Japan embarked on a prolonged, bloody conflict that was to last for eight years. The outbreak of war in the summer of 1937 ended some of the uncertainties in Japan's policy deliberations, at least for the time being. It by no means brought an end to conflicting goals and interests, however.

---

70. "Shanhai-Nankin kan denwa keburu kōji ukeoi nyūsatsu ni kansuru ken" (June 5, 1937), MOC Records I, 250.

71. Inoue Fumisaemon, "Maboroshi no Nankin-Shanhai kan keburu," *DTJGKS*, 263–64.

If the outbreak of war spelled the end of a peaceful expansion in China based on technological hegemony, enthusiasm for technology did not wane. If anything, government engineers like Matsumae now emphasized the greater urgency for technological mobilization in Japan. Matsumae published another essay in *The Journal of the Electrical Society* at the beginning of 1938. Emphasizing the technological inferiority of both Manchukuo and China, he noted that there was nothing more than open wire in their telecommunications facilities. As he wrote, "It is impossible to build an important long-distance circuit in East Asia on the basis of such primitive open-wire technology. To fully implement our policy encompassing politics, diplomacy, and industry in East Asia, we must plan the expansion of an East Asian long-distance communications network using cable, which can ensure secrecy and technical stability as soon as possible." As he was eager to point out, the cable link soon to be completed by Japan was more than simply a communications trunk line between Japan and Manchukuo. It would facilitate Japan's economic expansion not only into China, but also to all of East Asia and even Europe. It would hold profound significance as a vehicle for "Japan's great mission to become the foremost leader in guiding the culture and opinions in East Asia."[72]

Matsumae was not the only one to emphasize the central importance of technology in Japan's new empire-building venture after the outbreak of the China War, nor was the interest in Japan's technological superiority vis-à-vis Asia limited to the MOC. A growing sense of confidence in Japan's technology was shared by many Japanese engineers and bureaucrats, especially after they saw for themselves conditions in neighboring China. Perhaps the clearest explication of the importance of technological capability in Japan's new empire-building in China was provided by Miyamoto Takenosuke, a Home Ministry bureaucrat trained in civil engineering at Tokyo Imperial University. Having visited Manchuria several times and attended the 1936 East Asian Industrial Exposition in Shanghai, Miyamoto was familiar with the level of technology in China. Writing in 1940 as head of the Technology Department of the Asia Development Board, Miyamoto noted that technology,

72. Matsumae Shigeyoshi, "Nichi-Man musōka keburu ni tsuite," *Denki kyōkai zasshi* (January 1938), reprinted in Matsumae Shigeyoshi ronbunshū kankōkai, *Hatsumei e no chōsen*, 825–49.

like science, was losing its universal character in favor of monopolies by specific countries. To carry out the "great enterprise of Asian development," he opined, Japanese technology must therefore have its own special characteristics. Miyamoto considered there to be three basic requirements: Japanese technology must be (1) advanced, (2) comprehensive, and (3) adaptable to local conditions. In his view, the foremost element was the "absolute necessity for Japan to maintain its technological lead over China and the rest of Asia":

> If China's technology progresses by one step, Japan's technology must progress by two. In this way, the half-century technological gap between the two countries will not be reduced but will continue forever. In other words, it is absolutely necessary that Japanese technologies brought to the continent must be superior not only vis-à-vis China, but also vis-à-vis the entire world.[73]

If, one day, China's technology developed to such a degree that it could exploit its rich resources without Japan's support, Miyamoto warned, East Asian economic cooperation would collapse from within.[74] Thus, creating technological dependence on Japan in the rest of Asia was essential to the success of Japan's new empire. Here technology became synonymous with power, which was, after all, one of most basic building blocks of imperialism. These were not just abstract ideas with little relevance to policy. Just as technology was facilitating expansion in Asia, so these engineer-turned-ideologues of techno-imperialism were gaining influence within the government. In fact, imperial engineering projects in the empire and the rising fortune of engineers at home became linked.

THE RISE OF A TECHNOLOGY BUREAUCRACY

The question of originality aside, there is little doubt that non-loaded cable was a great achievement and deserves to be considered the "most systematic technology" created by the Japanese in the prewar period.[75]

---

73. Miyamoto Takenosuke, "Kōa gijutsu no mittsu no seikaku," *TKZ* 379 (March 1940): 14. For commentary, see Kawahara Hiroshi, *Shōwa senji shisōshi kenkyū*, 200; Mimura, "Technocratic Visions of Empire," 97–118; Mizuno, *Science for Empire*, esp. Chap. 2.

74. Miyamoto, "Kōa gijutsu no mittsu no seikaku," 14.

75. Nojima Susumu, "Introduction," in Nihon kagakushi gakkai, comp., *Nihon kagaku gijutsu shi daikei* (19), 16, 18.

From research to testing to implementation, a large number of agencies and personnel were involved. In terms of geographical scope and impact, the Japan–Manchukuo long-distance cable was quite possibly the largest engineering project undertaken in the new empire.

The launching of the Japan–Manchukuo Long-Distance Telecommunications Cable also had a profound impact on the government engineers themselves. For one thing, the technological breakthrough greatly stimulated systematic research into advanced communications technology. When Matsumae Shigeyoshi first joined the MOC in the mid-1920s as a young engineer just graduated from university, the ministry had only recently established its Engineering Bureau, based on the Engineering Section in the Communications Bureau and in the Bureau of Provisional Telegraph and Telephone Construction (set up in 1920 as part of the Third Telephone Expansion Plan). Matsumae soon found the atmosphere in the Engineering Bureau "dreary and dispiriting, as if being in a desert." Given the prevailing discrimination in Japanese government bureaucracies against technical personnel in favor of those trained in law or economics, Matsumae had good reason to be discouraged about his career prospects. Scientific research and Christian activities became his only escapes.[76]

The early 1930s saw a marked change in this situation as technological innovation began to have a profound impact on bureaucratic organization within the MOC. Beginning in 1935, the year NLC began to gain recognition in Japan, the MOC began to discuss how to integrate all its technical strength into a single unit that would function as what Matsumae would later call a "technological general staff" (*gijutsu sanbō honbu*). The Investigation Section was added to the Engineering Bureau in 1937, with Matsumae appointed its chief. The reconstituted bureau was specifically devoted to turning new research in technologies into applications. Projected expansion of the domestic telephone network was one factor behind the reorganization. Technicians associated with NLC development—notably Shinohara Noboru and many new graduates from electrical engineering departments—congregated in the section's Transmission Division. This was no coincidence. Transmission over distance had special significance in an expanding empire.

---

76. Matsumae Shigeyoshi, "Teishinshō o chūshin toshite gijutsusha undō," in Teishin gaishi kankōkai, *Teishin shiwa*, 2: 436–39; Sakamoto, *A Lion Aroused*, 93–95.

Table 4
Organizational Changes in the MOC Engineering Bureau, 1932–37

Year	Number of sections	Sections
1932	3	General Affairs, Telegraph, Telephone
1934	4	General Affairs, Wires, Mechanics, Wireless
1936	5	General Affairs, Wires, Mechanics, Wireless Japan–Manchukuo Telephone Construction
1937	8 plus 2 offices	General Affairs, Wires, Mechanics, Wireless, Japan–Manchukuo Telephone Construction (until 1940), Investigation, Experiment; and two offices in Korea: Metropolitan Service and Long-Distance Service

SOURCE: *DTJGKS*, 266.

The Japan–Manchukuo long-distance cable project also had wide-ranging implications beyond technology and communications alone. It greatly boosted the prestige of the technical wing of the MOC. It was, above all, a victory of engineers. Within a few years, several new sections were added to the Engineering Bureau, doubling their total number (see Table 4). The previous convention that section chiefs were selected only from engineers with imperial appointee (*chokunin*) status was abandoned.[77]

This time, the change was not just organizational. Imbued with a newfound vitality as well as a sense of national mission, the MOC became a hotbed of political activism for engineers and technicians demanding greater status and influence. Starting from a fraternity of technicians within the MOC known as the Society of Communications Technicians, the movement evolved to include similar associations in other ministries. Their influence would permeate the Japanese bureaucracy during the early 1940s, when Matsumae and his colleagues organized the Technicians' Movement that led to the establishment of the Seven-Ministry Council of Technicians.[78]

---

77. Musōka hōshiki kankōkai, *Gijutsu kaihatsu e no michi*, 42–46; Kajii, *Waga hansei*, 165–66.

78. *DTJGKS*, 34–36; Matsumae, "Teishinshō o chūshin toshite gijutsusha undō," in Teishin gaishi kankōkai, *Teishin shiwa*, 2: 439–42.

Engineers and technicians were generally viewed as nothing more than robotic tools who faithfully executed orders from military men, politicians, or their administrative counterparts in the civil service. The 1930s saw a marked increase in political activism on the part of technological elites like Matsumae. This was a new phenomenon in the Japanese bureaucracy, never seen since its inception in the early Meiji era. In most scholarly literature, the typical prewar Japanese bureaucrat is portrayed as what historian Robert Spaulding has called the "examination man." Such a bureaucrat had won appointment through the higher civil service examinations. Most likely having studied law, he was given administrative duties in the government.[79] Unlike law graduates, "technical officials" with engineering backgrounds were selected not through examination but through screening (*senkō*). By some counts, *gikan* (technical officials) actually outnumbered *jimukan* (administrative officials) in the prewar Japanese higher civil service (53 vs. 47 percent).[80] These "state-employed technicians and engineers" or "technicians in the government" (*seifu no naka no gijutsusha*) were known as "technology bureaucrats" (*gijutsu kanryō*).[81] The majority of them were relegated to purely technical matters and had little or no say in policymaking.[82] Lack of participation in policymaking as well as lack of promotion became increasingly pronounced in the early decades of the twentieth century, even though the number of graduates in the natural sciences continued to rise, increasing fourfold between 1926 and 1945. Many such graduates were absorbed into private industry.

Dissatisfaction among government engineers also grew, to the extent that a number of them resigned and joined the private sector. Even elite technology bureaucrats were affected, since for a long time they could not expect to rise above the level of section chief. This was almost the case with Kajii Takeshi, who studied electrical engineering at Tokyo

---

79. See, e.g., Spaulding, "The Bureaucracy as a Political Force, 1920–1945," 33–80; see also idem, *Imperial Japan's Higher Civil Service Examination*, 182–83.

80. Koh, *Japan's Administrative Elites*, 27.

81. I have avoided the more popular term "technocrat"—which would also include economists—to highlight the distinctive background of these engineers.

82. Bartholomew, *The Formation of Science in Japan*, 269–70. Such a tradition continued until well after the war; see Okita Saburō, *Nihon kanryō jijō*, 142–44. A postwar foreign minister, Okita was trained as an engineer and worked in North China during the war.

Imperial University and joined the MOC in 1920 as an engineer. Disappointed with his career prospects, he nearly followed several of his colleagues in quitting the MOC to join the Sumitomo Electric Wires Company.[83] In the spring of 1935, when Matsumae returned from his year-long, government-sponsored study abroad, he was still thinking of resigning from the MOC. After the start of the Japan–Manchukuo cable project, however, Matsumae was fired with enthusiasm for a greater role for engineers and technology in national politics. Although they were fighting an uphill battle, engineers like Matsumae found new impetus in Japan's empire-building projects and new allies in other ministries.

Matsumae's career offers some parallels with that of Miyamoto Takenosuke. Like Matsumae, Miyamoto was a leading technology bureaucrat trying to break away from the bureaucratic structure established in the Meiji era. A graduate of the Civil Engineering Department at Tokyo Imperial University, Miyamoto also entered the government in the mid-1920s as an engineer, but in the Home Ministry, where he worked on several flood control projects. Although he never thought of quitting, like Matsumae he was unhappy with his career. Besides the low status accorded to engineers in the Home Ministry, by the early 1930s civil engineers in the government faced the unpleasant prospect of downsizing. In Manchukuo, which Miyamoto visited first before the Manchurian Incident and many times thereafter, he found "the perfect zone for rescuing technicians thanks to abundant opportunities." Although Miyamoto did not produce a landmark project such as Matsumae's work on non-loaded cables, he struggled to elevate technology policy and administration to the level of national policy through "the hegemony of technicians," as his biographer Ōyodo Shōichi put it.[84]

Thus empire-building on the continent in the 1930s gave technology bureaucrats unprecedented opportunities to use their talents. Matsumae and his colleagues in the MOC were perhaps an exemplary group in their active involvement in the Japan–Manchukuo Long-Distance Tele-

---

83. Kajii, *Waga hansei*, 48–51. In his memoir, Kajii recorded the following unpleasant episode: when he was awarded a doctorate in engineering, Okumura Kiwao expressed surprise at the celebration that it was not a doctorate of law (ibid., 173–74).

84. Much of the information on Miyamoto is drawn from the excellent study by Ōyodo Shōichi, *Miyamoto Takenosuke to kagaku gijutsu gyōsei*; see also Ōyodo, *Gijutsu kanryō no seiji sanka*, 118–23.

communications Network. Technicians and engineers in other government ministries, who had also suffered from dominance by law faculty graduates, thus found opportunities in Manchukuo and China attractive.[85] Since the early 1930s, many Japanese engineers had been active in various big projects on the continent, ranging from gigantic hydraulic dams in northern Korea to large-scale urban planning in Manchukuo. Although such large development projects in Manchuria provided the initial push, science and technology also enjoyed a boom in Japan as the country moved toward a war economy. The late 1930s saw new universities established to improve scientific and technological research, and the number of publications on such subjects reached new heights.[86] The heyday of technology bureaucrats had finally arrived.

Such new activism on the part of technology bureaucrats was not a matter of self-interest alone. Engineers like Matsumae and Miyamoto were also aiming at something higher. Welding technology with public policy became their obsession, as they tirelessly called for "harnessing technology for the state." Matsumae proved to be an indefatigable advocate for a "technological New Order" under which all technology would be harnessed for the purposes of the state. The ultimate objective was to bring technology and politics together. He would find fertile ground for this idea after 1937, when Japan and China entered into full-scale conflict. By September of that year, under direction from the Army, the Cabinet Resource Bureau began to deliberate on a National General Mobilization Law. In February 1938, an Association of Technology Policy Concerning China was established to deal with many new issues of Japanese development activities in North China. Kajii, Matsumae, and Miyamoto were key figures in the organization. Starting in May, various technical groups petitioned the Konoe Cabinet, calling for establishment of an agency of "technological guidance inside the central body of China affairs." In September, the Association of Industrial Technology was created, with Miyamoto as director general and Matsumae as executive director. In November 1938, the Asia Development Board established its own Technology Department, headed by Miya-

---

85. Koshizawa Akira, *Zhongguo dongbei chengshi jihua shi*, 289; Ōyodo, *Gijutsu kanryō*, 117–40.

86. Cusumano, "'Scientific Industry': Strategy, Technology, and Management in the Riken Industrial Group, 1917–1945"; Bartholomew, *Formation of Science in Japan*, 276–77.

moto Takenosuke. A number of engineers from the MOC were transferred to the new department. As an advisory committee, the Asia Development Board set up an Asian Development Technology Committee less than a year later.[87]

As historian Carolyn Marvin has reminded us, "electricians were as deeply involved in the field of cultural production as in the field of technical production."[88] The invention of the NLC in Japan shows that a technological breakthrough has ideological implications. By turning attention from artifacts to human actors—the Japanese technicians and engineers who engaged in research and implementation—we can recognize their role as visionaries of empire as well as political actors at home. Belief in Japan's technological advance had an ideological necessity—namely, justifying Japan's self-ascribed role as the leader of an Asia that would be independent of Western interests. The construction of an ideology of Japan's technological superiority, I argue, was central to Japan's project of reordering East Asia into a new "co-prosperity sphere" under its leadership and driving out Western influence. As historian Kawahara Hiroshi points out, "technology" became a symbol of Japan's domination in Asia.[89]

A close investigation of telecommunications technology in the turbulent era of the 1930s sheds light on the construction of "Japanese-style technology." Historians have often emphasized that indigenous production in the 1930s resulted from Japan's isolation in the world, but the story of NLC suggests a more proactive role by Japanese engineers based on a variety of needs: material, ideological, as well as personal. Technological development of NLC in Japan not only contributed significantly to the formation of an ideology of Japan's technological hegemony in Asia, it was closely intertwined with Japan's geopolitical agenda in the 1930s. NLC became the best example of a powerful Japanese technology that enabled Japan's ambitious expansion of the telecommunications network on the continent. In other words, Japan's continental expansion not only created the need for such enabling

---

87. Ōyodo, *Miyamoto Takenosuke*, 236–46.

88. Marvin, *When Old Technologies Were New*, 7.

89. Kawahara, *Shōwa seiji shisōshi kenkyū*, 201.

technologies, it was a direct beneficiary of it. The debut of NLC gave Japan's empire-building effort a much-needed boost, almost an aura of "technological inevitability." Having examined the development of a crucial telecommunications technology in Japan against the backdrop of the country's new expansion on the continent, we shall now explore how technological developments affected the shaping of blueprints for empire in the late 1930s.

CHAPTER 5

Envisioning Imperial Integration

On September 30, 1939, several hundred people gathered simultaneously in Tokyo, Keijō, and Mukden to celebrate the opening of the Japan–Manchukuo long-distance telephone cable. Shortly before noon, Minister of Communications Nagai Ryūtarō took the microphone at the Tokyo Central Telegraph Office. Nagai described the newly completed cable as a "revolutionary invention" and attributed its successful construction to earnest cooperation between the government and the private technicians who had worked with the Engineering Bureau of the MOC. Nagai went on to note:

> As the longest cable in the world, completion of the Japan–Manchukuo connection telephone cable has become the focus of attention of all countries. This cable is not an imitation of the West but was completed with the unique technology on the basis of the Ministry of Communications' research and invention; it is significant as the pride of a scientific Japan. In today's world, full of uncertainties, I believe that as the only leading country of colored people firmly established in the corner of East Asia, Japan has a cultural mission that is both real and grave. Considering the great mission of building the New East Asia that has now fallen on the shoulders of the Japanese people, we are more acutely aware of the responsibility of constructing an East Asian telecommunications network as the first step![1]

To demonstrate the effect of the newly completed facilities, Nagai's speech was broadcast simultaneously through loudspeakers to all main-

1. "Nichi-Man renraku denwa kōji shunkōsu," *TKZ* 375 (November 1939): 120–26.

tenance technicians at the several dozen relay stations along the entire 2,700-km cable, as well as over the radio in Japan. Shortly afterward, as if to test the soundness of Nagai's vision, officials and business leaders in Tokyo and Mukden exchanged greetings over the telephone, followed by businessmen in Tokyo and Keijō, located almost midway on the route. In addition, Nagai and the minister of communications of Manchukuo exchanged written congratulatory messages through the phototelegraphy equipment newly installed on the long-distance cable. Nagai's carefully chosen phrase—"Same Virtue, Same Mind"—was apropos for an occasion when instant communications seemed to have annihilated physical distance altogether; given this technological blessing, it was only to be expected that any existing psychological distance would be eliminated as well.[2]

The four years that the Japan–Manchukuo long-distance cable took to complete witnessed profound changes for Japan and East Asia. After the outbreak of war with China in 1937, Japanese forces occupied much of North and Central China, greatly expanding their political and economic presence on the continent. While consolidating its alliance with the Axis powers in Europe, Japan embarked on building its own New Order in East Asia, culminating in what was known as the Greater East Asia Co-Prosperity Sphere. In the meantime, these four crucial years brought official recognition of the critical importance of telecommunications for Japan's new geostrategy in Asia, as Japan designed ambitious imperial telecommunications networks and formulated a comprehensive regional telecommunications policy.

FROM LINK TO NETWORK

Uniting Japan and Manchukuo

In the wake of the Manchurian Incident, Japan's foremost strategic objective in Asia was to consolidate its vital relationship with the newly created Manchukuo. Given Japan's lack of resources at home, its strategists were acutely aware of the country's dependence on raw materials

---

2. Matsumae et al., eds., *Gijutsu kaihatsu e no michi*, 36–38; *DTJGKS*, 28. Shinohara Noboru (*Hitori no kokoro*, 76) similarly recalled that a person in Harbin calling Tokyo was often asked, "When did you come back to Tokyo?"

from Manchuria and saw a "Japan-Manchukuo economic bloc" as an absolute necessity for Japan's survival in the age of "total war." If the rich resources of Manchukuo promised a new industrial base, its vast landmass offered an outlet for Japan's growing popula-tion. Moreover, the establishment of Manchukuo and Japan's purchase of the Soviet-controlled China Eastern Railway extended Japan's line of frontier defense to the Manchukuo-Soviet border.[3] If constructing the economic bloc required overcoming the distance between Japan and its continental puppet state, then better communications links between Manchukuo and Japan were indispensable.

In an essay entitled "Communications Between Japan and Manchukuo," published in early 1934 in the monthly policy journal *Revue diplomatique*, MOC official Okumura Kiwao described the two countries as "fused in a union inseparable in terms of military, politics, economy, and society." Okumura documented the drastic increase in communication traffic between Japan and the newly established Manchukuo. Regular mail increased 48 percent and small parcels 63 percent, thanks to the military postal service introduced two months after the Manchurian Incident. Volumes of telegrams exchanged between Japan and Manchuria shot up even more, from a daily average of 5,077 before the Incident to 8,799 in March 1933, a 70 percent increase. This was helped by the drastic expansion in the number of telegraph offices in Manchuria handling Japanese-language telegrams after February 1932. Despite the sharp increase in traffic between Japan and Manchuria, Okumura pointed out, there were still only a total of five telegraphic circuits between them—two submarine cables and three land lines via Korea—with wireless connections serving a supplementary role. The result was frequent congestion and delay: occasionally as many as 400 telegrams were held up within the same day; some took as long as seven hours to reach Japan from Manchuria. The close relationship between Japan and Manchukuo notwithstanding, Okumura revealed on another occasion, on average a telegram transmission took 2 hours 32 minutes between Shinkyō and Tokyo, 3 hours 2 minutes between Mukden and Tokyo, and 1 hour 43 minutes between Tokyo and Dalian. In contrast, a telegram between London and Nagasaki took only 19

---

3. See Barnhart, *Japan Prepares for Total War*, esp. chap. 1.

minutes, between Hamburg and Nagasaki 14 minutes, and between New York and Tokyo 25 minutes.[4]

As the integration for a "Japan-Manchukuo bloc" began to take off, expanding the communications capability between the home islands and the continent became a major priority of MOC. The ministry expanded postal service, assigning larger spaces for mail shipment on the Shimonoseki–Pusan ferry and the Pusan–Andong trains and announcing plans to start airmail service between Shinkyō, Mukden, and Japan in November 1933. Earlier in 1933, the MOC had opened a direct wireless circuit between Osaka and Dalian and contemplated building new telegraph lines linking the two areas.[5] The establishment of the MTT and the implementation of its expansion plans, predicted Okumura, would smooth communication not only within Manchuria but also between Manchuria and Japan, thus "realizing the ideal of the Japan-Manchukuo bloc." In particular, Okumura confidently predicted, the completion of the Shinkyō Wireless Station and the start of the Japan–Manchukuo wireless telephone service would completely obliterate the "distance of 1,000 miles between Tokyo and Shinkyō in telephone communication as well!"[6]

As Okumura pointed out, long-distance telephony was the most exciting frontier of communications technology at this time. By the early 1930s, advances in wireless and cable technology had made it possible for telephone service to reach outside Japan's home islands. Telephone service between Korea and Japan via submarine cable began on January 15, 1933. In ceremonies that would be repeated time and again throughout the decade, groups of prominent government, military, and business figures gathered at both ends of the newly established telephone connection. In a short speech read to both crowds by a representative, Governor-General Ugaki Kazushige predicted that the opening of the telephone connection would further strengthen the economic and other

4. Okumura, "Nichi-Man kan no tsūshin kankei," *Tsūshin ronso*, 83–99; first published in *Gaikō jihō* (March 1934); "Manshū denshin denwa kabushiki kaikai no denpō ryōkin," *Tsūshin ronsō*, 160–78.

5. Okumura, "Nichi-Man kan no tsūshin kankei," 96; "Chōsen keiyū nai-Mankan denshin senrō kensetsu hoshuhi nado no futankata ni kansuru ken" (August 22, 1933), MOC Records II, 149.

6. Okumura, "Nichi-Man kan no tsūshin kankei," 98–99.

bonds between the colony and the home islands. Government officials and business leaders in the two cities then took turns exchanging greetings over the newly activated telephone line.[7] "As the long-awaited human voices crossed the turbulent waves of the Korea Strait for the first time," Okumura enthused, "we could not but feel that we had subjugated nature!" The first ten days of service saw a daily average of close to 100 telephone calls, on a single circuit with a maximum capacity of 120. On the basis of the number of telegrams exchanged, Okumura predicted that potential demand would reach the neighborhood of 600 calls per day.[8] The newly founded International Telephone Company opened telephone service by wireless to Taiwan in June 1934, followed by inauguration of similar service to Manchukuo in August. In addition to Japan's colonies, international wireless telephone service from Japan also began and would include the Philippines, the Dutch East Indies, the Americas, and Europe by 1937. The rates were by no means cheap. A three-minute phone call to New York cost as much as 95 yen, more than the monthly income of a government bureaucrat.[9] As an indication of the increased economic activities in the empire, as well as renewed efforts at trade expansion elsewhere in the 1930s, 70 percent of such telephone use was related to business matters.[10]

As Okumura had predicted, the inauguration of wireless telephone service between Japan and Manchukuo in August 1934 generated similar excitement. Initially an average of 30 calls per day were exchanged between Japan and Manchukuo. Despite its prohibitively higher cost—seven yen for every three minutes—the daily average jumped to about 130 in less than two years. Linking some 160 telephone exchanges in Japan and four large cities in Manchukuo, the Manchukuo–Japan wireless telephone connection promised to connect some 400,000 telephone

---

7. "Nai-Sen renraku denwa kaitsū shiki," *Chōsen* 213 (February 1933): 156–58; "Nai-Sen denwa kaitsū shiki," *TKZ* 294 (February 1933): 32.

8. For the daily breakdown and destinations of these calls, see Okumura Kiwao, "Nai-Sen renraku denwa no kaitsū," *TKZ* 294 (February 1933): 42–43.

9. For a rate chart as of September 1937, see *Teishin no chishiki* 1.4 (October 1937): 14–15. For a list of destinations and starting dates of service, see Kokusai denki tsūshin kabushiki kaisha shashi hensan iinkai, *Kokusai denki tsūshin kabushiki kaisha shi*, 433–34.

10. Kokusai denwa kabushiki kaisha, *Kokusai denwa kabushiki kaisha jigyō shi*, 67.

subscribers in Japan and 15,000 in Manchukuo.[11] A 1935 MOC study of projected telephone traffic between major cities in Manchukuo and Japan predicted that major commercial hubs, such as Dalian and Mukden, would generate the largest shares of traffic in the future (see Table 5). The MOC made such predictions on the assumption that "with the formation of the economic 'bloc' between Japan and Manchukuo, their close ties will deepen; so use of telecommunications will be strong." As the demand would soon exhaust the capacity of the single wireless circuit in operation, a long-distance telephone cable would solve the problem.

The planning of the new Japan–Manchukuo telephone cable made perfect sense against such a background. As the MOC later explained its decision to launch the new cable:

> Although basic tasks have been accomplished in Manchukuo in the nearly five years since its founding, it still depends on powerful assistance from Japan to maintain national defense and security, exploit resources, promote culture, and develop fully. To ensure peace in the Orient and to realize co-existence and co-prosperity through the mutual cooperation between Japan and Manchukuo, various facilities and policies are being implemented. Among them, the construction of communications facilities linking Japan and Manchukuo is of utmost importance.[12]

With the progress in NLC technology, telephone cables linking Japan and the continent would bring multiple benefits. Not only would such a link enjoy high public demand, a cable between Japan and Manchukuo was also strategically vital for Japan, since the Kwantung Army was eager to strengthen its military preparations against the Soviet Union and expand communications installations along the northern frontier as well as links with Japan proper. The benefits of a new Japan–Manchukuo telephone cable went beyond expanding communications capacity. Compared with the wireless, the cable link using non-loaded technology promised to improve the quality of transmissions, making a long-distance telephone conversation between Manchuria and Japan

---

11. On the beginning of Japan–Manchukuo telephone service via wireless, see "Nichi-Man musen denwa no kaishi to katsūshiki," *TKZ* 312 (August 1934): 135–39; and Shindo Seiichi, "Nichi-Man denshin denwa ni tsuite," *TKZ* 330 (February 1936): 147.

12. Teishinshō, *Nichi-Man renraku denwa shisetsu sekkei keikaku kōyō*, 1.

Table 5
Projecting Telephone Traffic Between Japan and Manchukuo, 1937

Connections	1936	1937	1938	1943	1948	1953
Tokyo–Shinkyō	45.1	55	68	123	194	292
Tokyo–Mukden	25	31	49	93	150	231
Tokyo–Dalian	45.8	54	114	198	307	457
Osaka–Shinkyō	23.2	28	43	78	123	186
Osaka–Mukden	20.2	25	67	132	214	331
Osaka–Dalian	43.7	52	120	212	333	495
Fukuoka–Shinkyō	0.62	0.76	13	21	31	46
Fukuoka–Mukden	0.56	0.71	12	19	29	43
Fukuoka–Dalian	1.4	1.68	43	74	114	170
TOTAL	205	248	529	950	1,495	2,251
Circuits required besides wireless		3	6	9	16	29

NOTE: Figures indicate average number of calls per day; capacity per toll circuit was set at 90 calls.
SOURCE: Teishinshō, *Nichi-Man renraku denwa shisetsu sekkei keikaku kōyō* (February 1937).

Table 6
Composition of the Japan–Manchukuo–China Cable Network, 1937–42

Section	Type	Length (km)	Date opened (year.month)
Tokyo–Nagoya	NLC	400	1937.8
Nagoya–Fukuoka	loaded cable*	864	
Fukuoka–Pusan	NLC	270	1937.12
Pusan–Andong	NLC	900	1939.9
Andong–Mukden	NLC	260	1937.3
Mukden–Shinkyō	NLC	296	1940.8
Mukden–Tianjin	open wire	704	
Tianjin–Beijing	NLC	130	1938.12
Shinkyō–Harbin	NLC	265	1942.12
NLC SUBTOTAL		2,266	
TOTAL	3,540		

* In early 1939, a 110-km non-loaded carrier cable was installed between Funaki near Nagoya and Fukuoka.
SOURCE: Adapted from *DTJGKS* and Musōka hōshiki kankōkai, *Gijutsu kaihatsu e no michi*, 121.

as clear as a local call. Such an improvement would eliminate the many phone calls that could not be completed due to poor connections. How it would reduce the psychological distance between Japan and Manchukuo could only be surmised.

In mid-1936, the MOC Engineering Bureau set up a new Section for Japan–Manchukuo Telephone Construction devoted to this ambitious project.[13] Construction was to be carried out simultaneously in Manchukuo, Korea, and Japan proper. The first part had begun in the early winter of 1935, in Manchuria, where work crews replaced the previous bare-wire lines between the city of Mukden and Andong. The terrain and weather posed considerable difficulties. Since there was only one railway linking the two cities and it was already busy with regular shipments, a new road had to be made and paved in the mountainous areas along the cable route. Even more serious were the frequent Chinese guerrilla attacks on the Japanese crew. In fact, anti-Japanese guerrilla activities always posed a major threat to the MTT's many facilitates

13. Asami Shin, head of the new section, would later become director of engineering at NCTT.

throughout Manchukuo. At the first meeting of local bureau chiefs in 1935, for instance, MTT executives discussed measures to ensure safe operation by stationing troops at the local relay stations.[14] Several Japanese technicians carried pistols for protection, and the Kwantung Army provided an escort for them. In one particularly tense encounter, thousands of rounds were fired. As a security measure, the construction team built electrified barbed-wire enclosures around the relay stations.[15]

The entire cable from Tokyo to the Manchukuo capital of Shinkyō was some 3,300 km in length, most of which buried a meter underground, with relay stations along the route. Since it would be too costly to replace all the existing loaded-cable trunk lines that had been installed during the 1920s, the MOC actually added only a 400-km NLC line between Tokyo and Nagoya. Still, it was an expensive project. In Korea alone, the cost exceeded the estimated 8 million yen by more than 3 million. According to one estimate, over half a million men were mobilized for the work.[16] The final cost, a staggering 43.66 million yen, was split between the MOC, GGK, and MTT. During the four years of construction, it became apparent that technical standards created in Japan had to be adapted to conditions in the new empire. When the MTT initially designed the cable for the northern route between Mukden and Shinkyō in the spring of 1938, it strictly followed earlier MOC standards

---

14. *Manshū denshin denwa kabushiki kaisha 10-nen shi* (1943), 605.

15. Murano Masahisa, "An-Pō keburu no ki," in Manshū denden tsuiokuroku kankōkai, *Akai sekiyō*, 88–94; Sakamoto, *A Lion Aroused*, 136. Remarkably, even long after the war, Matsumae and other Japanese involved in the project still used the wartime term "bandit" to refer to the armed Chinese attacking Japanese engineers working on the Japan–Manchukuo cable in former Northeast China. See comments by Matsumae, Kobayashi, Shinohara, Inoue, Iijima, in *DTJGKS*, 28. On only one occasion was Matsumae corrected by an interviewer who told him that those "bandits" were actually anti-Japanese guerrillas.

16. On the construction of the Japan–Manchukuo route, see Kuroiwa Kōichi, "Nichi-Man renraku denwa keburu no kensetsu ni tsuite," *DT* 3, no. 11 (1940): 94–95; *DTJGKS*, 209–42. Murakami Motoyuki, a MOC engineer who participated in the project, estimated that including construction and transportation of equipment, the total number of people involved in the project reached 2 million; see Murakami, "Nichi-Man rūto kōji kansei no [hinegai]," *DTJGKS*, 222; idem, *Ichi gijutsusha no shogai*, 369. On the laying of submarine cables linking Korea with Japan, see also Nippon denshin denwa kōsha kaiteisen shisetsu jimusho, *Kaiteisen hyakunen no ayumi*, 258–74.

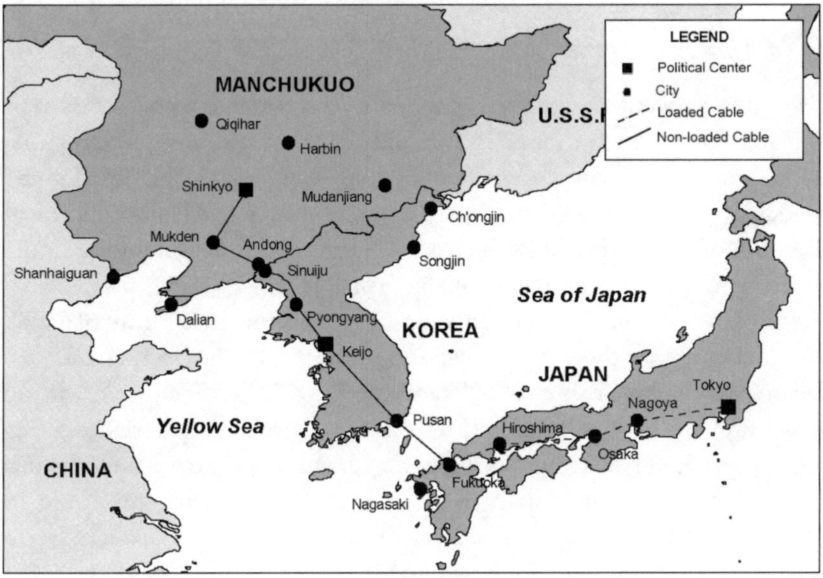

Map 2 The Japan–Manchukuo cable, 1936
(SOURCE: Teishinshō Nichi-Man denwa kensetsuka, *Nichi-Man renraku denwa kaisen nokōsei ni tsuite* [September 19, 1936]).

of setting up relay stations every 50 km, each with six channels. The vast space and small Japanese population in northern Manchukuo differed from Japan proper, however. MTT engineers discovered that the distance between two adjacent Japanese resident communities in Manchukuo varied from 40 to 70 km. To ensure that relay stations were built near Japanese communities so that Japanese technicians could be stationed there for maintenance work, the MTT had to modify MOC standards.[17] Moreover, in anticipation of future use of even higher frequency bands, the MTT decided to build two parallel cables along the busy Mukden–Shinkyō route, a departure from two lines only near relay stations in Japan. As a result, the MTT sent representatives to Tokyo to visit Matsumae and Shinohara at the MOC to seek an understanding and also had to increase the budget by 300,000 yen.[18]

---

17. Toya Noritaka, "MTT no omoide" and "Chōkyori tsūshin keburu gijutsu," in Manshū denden tsuiokuroku kankōkai, *Akai sekiyō*, 247–50, 83–87.

18. Even then, it took considerable effort to reach an agreement with the JTTCC over the price; see Toya Noritaka, "MTT no omoide" and "Chōkyori tsūshin keburu gijutsu," both in Manshū denden tsuiokuroku Kankōkai, *Akai sekiyō*, 247–50, 83–87.

From Link to Network

The enhancement of communications over greater distances was particularly significant for Japan in the mid-1930s. In addition to cementing the special relationship between Japan and Manchukuo, the new Japan–Manchukuo cable would mark the beginning of what Kajii Takeshi of the MOC Engineering Bureau called a "communications network in the Orient." Writing in the pages of the *Journal of the Communications Association* in early 1936, Kajii found an economic bloc consisting only of Japan and Manchukuo "too seclusionist and modest." He noted that the new Japan–Manchukuo cable would facilitate Japan's further expansion on the continent, because it would be less costly to reach the riches along the Yangzi River in Central China via Manchuria than to build a new submarine cable from Japan to Shanghai. Furthermore, this cable to Manchukuo would help expand telephone connection between Japan and Europe, since the Soviet Union was completing its own trans-Siberian cable network.[19]

It was significant that the Japan–Manchukuo cable was already viewed as the first stage of a more ambitious regional telecommunications network in East Asia. As Kajii put it, this cable demonstrated that Japan's continental expansion was a mission defined in geographical, historical, and philosophical terms. Kajii also pondered the historical role of communications across the Korea Strait. "In the past," he wrote:

> Chinese culture came to Japan via the Korean peninsula, with Tsushima as a stepping-stone. This had exactly the same geographical significance as Greek culture using numerous islands in the Mediterranean as stepping-stones to reach the Italian peninsula and producing Roman culture, the source of European culture. Compared with such movements of culture by ships in the past, we in Japan are now taking advantage of these stepping-stones and the peninsula, but using electric waves to transplant culture and develop the economy. Seen with such significance, the Korean peninsula and Tsushima played the role of absorption in the past in Japan's cultural history. Today, they are heavenly blessings for economic development. Today, Japan's expansion to the continent is Japan's mission revealed in the philosophy of historical geography.[20]

19. Kajii Takeshi, "Tōyō denki tsūshinmō yori mitaru Manshū no chii," *TKZ* 330 (February 1936): 168–76.

20. Ibid., 172.

Kajii's vision was not simply the musing of a lone telecommunications engineer. It seemed to be in perfect tandem with the basic tenets of Japan's foreign policy. According to the "Fundamentals of National Policy" adopted at the Five Ministers' Conference held in August 1936, Japan saw itself as the "stabilizing power" in East Asia and would seek close ties among the three countries of Japan, Manchukuo, and China in preparation for future conflicts. Although the policy still emphasized "economic development as the keynote of political policy toward the continent," this was another indication that Japan became increasingly assertive about its own vision of the regional order in East Asia.[21]

It was therefore not a surprise that MOC was already at work on what it termed "an East Asian telecommunications policy." In an internal proposal "concerning the expansion of an East Asian cable communications network," drafted in 1936, MOC declared that in view of the international situation in Asia, it was necessary to maintain extremely close relations among Japan, Manchukuo, and China in military, diplomatic, and economic affairs. As the leader of East Asia, Japan should fulfill the responsibility toward these two countries and bring about the fruits of coexistence and co-prosperity. To cope with international developments and to transmit goodwill among the peoples, Japan must construct a comprehensive East Asian telecommunications network and fully realize its functions. The proposal went on to note that the rapid progress in telecommunications technology in recent years had greatly extended the range of high-quality communications. Wireless telegraphy and telephone could now reach any part of the world, but due to frequency limits, they could not fulfill the need for communications links among Japan, Manchukuo, and China. The centerpiece of the East Asian telecommunications network, therefore, should be the non-loaded cable. Not only was the non-loaded cable the product of Japan's unique research, implemented ahead of the West, the MOC emphasized, but it was also "the most advanced technologically because it is most economical." The proposal also called for "suitable measures for the development of air and sea enterprises" and "preparations for

---

21. For a complete translation, see Lebra, *Japan's Greater East Asia Co-Prosperity Sphere*, 62–64. For a brief discussion, see Beasley, *Japanese Imperialism*, 201–2.

information and propaganda organizations."[22] It is particularly noteworthy that the chief proponents of this policy were those inside the Engineering Bureau who had been involved in construction of the Japan–Korea–Manchukuo long-distance cable. To them, an East Asian telecommunications cable network represented the logical next leap.

The 1936 MOC proposal did not specify exactly how China would be brought into Japan's telecommunications network. This was hardly surprising, since there was no consensus within the Japanese government as to what extent Japan could rely on non-military means to bring China into the "special union" with Japan and Manchukuo. Kajii's preference for expanding Japanese influence in China largely by exporting Japanese technologies contrasted with the military's demand for a more aggressive policy of expanding Japan's physical control of telecommunications facilities on the continent. The Kwantung Army was preeminently concerned with harnessing modern technologies in its preparation for war.[23] At a joint Japan-Korea-Manchukuo meeting on communications matters in late 1936, for instance, Colonel Fukue Shinpei, the head of the Communications Section of the Kwantung Army's General Staff, accused the MOC of being too "old-fashioned" in dealing with the new challenge facing Japan. The MOC was concerned only with matters such as tariffs and maintenance costs, Fukue complained, and lacked the extraordinarily high motivation needed to make communications the true vanguard of Japan's national development. Fukue emphasized that future military conflicts would be nothing like the Russo-Japanese War in terms of communication needs, since military strategies and army organizations had changed considerably, and airplanes as well as poison gas would be deployed. In a nutshell, he argued, when building the great enterprises for the nation, one should not be bound by dogmatic rules.[24]

The military conflict with China that began with the skirmish outside Beijing in July 1937 and escalated into an all-out war soon afterward

---

22. "Tōa yūsen denki tsūshinmō no kakuchō seibi ni kansuru ken," n.d., MOC Records I, Te28n. Although not dated, this document was drafted by the Japan-Manchukuo Telephone Construction Section and used for an internal discussion in October 1936.

23. See Barnhart, *Japan Prepares for Total War*; Coox, *Nomonhan*.

24. Speech by Colonel Fukue Shinpei in "Nichi-Sen-Man tsūshin renraku uchiawase kaigi gijiroku" (October 24, 1936), MOC Records I, o.

removed any lingering differences about how to expand into China proper. As the Japanese forces occupied major political and commercial centers in China, Japan's new continental policy took shape. In early 1938, shortly after the fall of China's capital of Nanjing, Prime Minister Konoe Fumimaro issued the fateful declaration that Japan would no longer deal with Chiang Kai-shek in settling the conflict. Konoe's statement reflected Japan's confidence that it would be able to control China through pro-Japanese regimes.

The war in China also highlighted the urgent question of how to meet the increased demand for communications on the continent. After Japan ordered a general mobilization and sent reinforcements to the continent, military operations in China sharply increased Japan's communications traffic, burdening its existing communications network. Wireless telegraphic communications with Japan's colonies increased 44 percent from June to August 1937. Major gateway stations (*kanmon-kyoku*), such as Tokyo, Osaka, and Nagoya, saw increases ranging from 17 to 34 percent in the number of telegrams handled; moreover, the actual number of words was three times as many as before due to a sharp increase in lengthy telegrams.[25]

The war in China thus created new opportunities for Japan's telecommunications expansion on the continent. As Tamura Kenjirō, the new director of the Telecommunications Bureau, would put it, "The war ushered in a new era of Sino-Japanese cooperation."[26] MOC quickly dispatched officials to the Central and North China fronts to work with the advancing Japanese Army in taking over Chinese telecommunication operations. At the same time, MOC emphasized a telecommunications policy for East Asia as the "foundation of the continental policy, since [telecommunications] would form inseparable ties centered on Japan and would enable complete coordinated defense and economic cooperation [among various nations in East Asia]." Although collapsing time and space to accomplish integration was still considered a crucial function, emphasis was also placed on control. As one of the

25. "Shina jihen to denki tsūshin," *Denmu kenkyū shiryō* (hereafter *DKS*) 16 (October 1937): 150–52. Most of such traffic seemed to have traveled on routes controlled by Japan; the volume through GNTC's Nagasaki station in late August 1937 was 23 percent less than it had been in early July.

26. Tamura Kenjirō, "Denki tsūshin seisaku ni tsuite," *TKZ* 362 (October 1938): 31.

MOC's first policy documents on East Asian telecommunications explained in early 1938, "When a country projects national strength overseas (*kokusei o kaigai ni shinten suru*), it is essential to acquire control over communications in the foreign country."[27] Echoing its earlier positions, the MOC proposed that Japan place all telecommunications operations on the continent under unified control, forming a great "organic communications system" (*tsūshin yūkitai*) by building a Japan-centered network that would extend to Manchukuo and China. To justify unified control, MOC cited the natural monopolistic tendency and public character of telecommunications, which would ensure the "intention for unified control by the state" (*kokka no tan'itsu naru tōsei ishi*) in military, political, and diplomatic affairs. Not surprisingly, the same proposal called for putting such unified control under the MOC.[28] The details of such an "organic communications system," however, had to wait until establishment of a telecommunications advisory committee.

DESIGNING THE EAST ASIAN NETWORK

The Telecommunications Committee

Significantly, MOC's engineers were again among the first to call for a telecommunications advisory committee consisting of influential individuals outside the ministry. Evoking foreign examples such as similar advisory committees in Britain's Post Office, the Engineering Bureau proposed in 1936 abandoning the age-old practice of formulating telecommunications plans only within MOC itself. By "incorporating influential views from business, academia, the military, the diplomatic service, industry, [and] manufacturing," the Engineering Bureau suggested, MOC would be able to come up with the "most appropriate telecommunications policy for the national policy." The proposal made practical sense. In view of MOC's disappointing record in securing funding

---

27. Teishinshō, "Tōa denki tsūshin seisaku kihon taikō" (February 8, 1938), MOC Records II, 697.

28. Ibid.; see also "Tōa ni okeru denki tsūshin seisaku no kakuritsu ni kansuru ken," (February 23, 1938); and "Tōa denki tsūshin seisaku jikko yōryō-an" (February 28, 1938); both in MOC Records I, 245.

for its ambitious expansion plans, such a national communications policy would have a better chance of receiving budget approval in the Imperial Diet or successfully floating a long-term bond issue in the financial markets. The proposal concluded that such a committee would constitute a major contribution to Japan's external communications policy.[29]

The idea of a telecommunications committee initially failed to garner enough support, but the outbreak of war on the continent and Japan's occupation of North and Central China drastically changed the picture. The importance of telecommunications expansion on the continent was now apparent to the military and to other bureaucracies actively involved in the China War. In early 1938, the Cabinet Planning Board, established only a few months before, convened a series of meetings on communications policy for East Asia. Although quite informal, this was the first time officials from the Army and Navy, their respective general staffs, and members of the Foreign Affairs, Colonial Affairs, and Finance ministries sat down with Matsumae Shigeyoshi and other communications officials to discuss the 1936 MOC proposal that called for "establishing a telecommunications policy for East Asia."[30] On the basis of the MOC document, the participants agreed to consider a national telecommunications policy in a systematic manner:

> As the stabilizing force in East Asia, Japan must ensure dominant strength (*shihaiteki jitsuryoku*) in politics, economy, industry, and defense. To accomplish this goal, Japan must assume the unified authority over telecommunications, which is the basic infrastructure of all activities of the state. It is necessary to build a powerful communications network under the principle of unification of communications in Japan, Manchukuo, and China. Such a network will enable Japan to face the communications networks of Europe and America, ensure that Japan becomes the guiding center of communications policy, and establish firm ties between defense and the economy in Asia.[31]

---

29. Kōmukyoku, "Denshin denwa kakuchō seibi ni kansuru chōsho" (May 25, 1936), NCTT Records 2028.

30. "Kikakuin shuzai tsūshin seisaku ni kansuru kyōgikai gaiyō" (February 26, 1938); "Dai-2-kai tsūshin kyōgikai kaigi keika" (March 3, 1938); "Dai-3-kai tsūshin kyōgikai kaigi keika" (March 7, 1938); all in MOC Records I, 200.

31. "Tōa ni okeru denki tsūshin seisaku no kakuritsu ni kansuru ken" (February 16, 1938), MOC Records II, 693.

In the first few months of 1938, issues concerning telecommunications in occupied China took center stage at these meetings. As discussed in the previous chapter, the MOC worked closely with the Cabinet Planning Board and the military to establish several telecommunications enterprises in Inner Mongolia and North and Central China. By August, these new "national policy companies," together with the original MTT, became the building blocks for a larger telecommunications network in East Asia.

In September 1938, the Japanese government established an interministerial Telecommunications Committee as the highest advisory body on telecommunications policy in Japan. This unprecedented cabinet-level committee involved all concerned bureaucracies, including the Cabinet, its Planning Board, the Manchurian Affairs Bureau, and the China Development Board, as well as the Foreign Affairs, Colonial Affairs, Finance, Army, Navy, and Communications ministries. With the prime minister serving as committee chairman, the committee membership consisted of the ranking vice ministers from each of these ministries and agencies (two from Communications). The committee also enlisted over a dozen prominent scholars and experts in telecommunications, including Dr. Yagi Hidetsugu of Tōhoku Imperial University, and Kajii Takeshi, who had recently left the post of MOC director of engineering to head a major manufacturer. Provisional members, directors, and secretaries were appointed from among bureau directors and section chiefs of various ministries. In addition to the minister of communications, who served as the vice committee chairman, a number of current or former MOC officials occupied important posts. Matsumae Shigeyoshi was among those selected as a secretary and charged with the investigation work.[32]

In a speech at the inaugural meeting of the Telecommunications Committee, Prime Minister Konoe evoked Britain's success in "placing the world's submarine cable network under its control a mere 40 years after its invention." Specifically, Konoe mentioned Britain's Imperial Communication Committee, whose guidance had "contributed much to the full realization of the function of telecommunications."[33] As

32. "Denki tsūshin iinkai kansei"; and "Denki tsūshin iinkai meibō" (January 11, 1939), MOC Records II, 693.

33. "Naikaku sōri daijin aisatsu an," MOC Records II, 693.

MOC pointed out in "Explanations Concerning Establishment of the Telecommunications Committee," whereas electric communications had played an important role in the rising "national fortunes" (*koku'un*) of the Western powers, Japan had been woefully late in developing its telecommunications capabilities. It was imperative, therefore, to set up the Telecommunications Committee because a comprehensive telecommunications policy would have a direct impact on Japan's political, military, and economic policies toward the Asian continent. It was necessary to establish the fundamental principles of an East Asian telecommunications policy immediately—principles that Manchukuo and China must adopt as well. Moreover, the satisfactory implementation of such policies must rely on the efforts of both the government and the public, including academic authorities. The committee was

> to cope with the changes in telecommunications in China and to develop a solid East Asian telecommunications bloc centered on our country, because effective, comprehensive unification of separate operations of telecommunications on the continent is a precondition to the full realization of the functions of telecommunications. . . . [In] view of the new developments in East Asia and the telecommunications policies of other countries, it should be the urgent task at this moment to firmly establish a communications policy in our country that unifies East Asia areas, so as to make our country the nexus of telecommunications in East Asia and bring about comprehensive development of facilities in all areas.[34]

Establishment of the Telecommunications Committee thus indicated a basic agreement in the government about the importance of telecommunications to Japan's new geostrategy. As MOC bureaucrats had consistently demanded, a comprehensive telecommunications policy for East Asia was finally becoming a top national priority for Japan. This was no small victory for MOC. To be sure, the Telecommunications Committee did not have the unlimited authority over telecommunications policy that MOC had desired. The Cabinet specified from the outset that "the committee's deliberations should in no way hinder the execution of policies toward Manchukuo and China."[35] In other words,

---

34. "Denki tsūshin iinkai setchi shuishō an" (June 11, 1938), MOC Records II, 693. Related documents can be found in *Kōbun ruiju 62-hen* (1938), vol. 4-3, National Public Records Office (Kokuritsu kobunshokan), Tokyo.

35. "Denki tsūshin iinkai ni kansuru ryōkai jikō" (Cabinet decision of September 6, 1938), MOC Records II, 693.

other bureaucracies and the military actually involved on the continent would enjoy considerable autonomy on telecommunications matters already under their control. Still, although MOC seemed to have been reduced to one of many agencies involved in formulating a comprehensive telecommunications policy, the committee's later recommendations echoed earlier MOC proposals on key issues. In fact, MOC was charged with preparing policy recommendations to be discussed and approved by the entire Telecommunications Committee.

Two policy inquiries were issued at the first general meeting of the Telecommunications Committee in September 1938. The "First Policy Inquiry to the Telecommunications Committee" concerned basic guidelines for telecommunications in East Asia. MOC had noted that generally speaking, telecommunications facilities were enhanced by expanding their geographical scope in a comprehensive and coordinated manner. Three elements—speed, reliability, and economy—thus were essential. Taking note of the situation in East Asia after the outbreak of the China War in 1937, the report claimed that "we aimed at forming a genuinely closely integrated communications bloc by creating these conditions fully. By doing so, we can expand Japan's leadership role in East Asia and at the same time counter the communications policy of the Powers." "In view of the new state of affairs in East Asia, as well as the present conditions of telecommunications facilities there," MOC noted in its inquiry, "it is of paramount urgency to establish a [strong] East Asian telecommunications network centered on our country that brings Japan, Manchukuo, and China into an organic unity."[36]

The new imperial telecommunications policy was to be comprehensive. In addition to ensuring rapid transmission of information, electric means of disseminating information were now considered a powerful tool for mobilization at home and for propaganda warfare abroad. Intrinsically linked to telecommunications and of critical importance to Japan's cultural mobilization and integration was radio broadcasting, considered another potent vehicle for the formation of an East Asian culture defined by Japan. At the first meeting of the Telecommunications Committee, MOC simultaneously made a "Second Policy Inquiry to the Telecommunications Committee" concerning Japan's broadcasting capabilities: "At the present, the national mission of broadcast wireless

36. "Tōa denki tsūshinmō seibi yōkō" (1939), MOC Records II, 693.

telephone is becoming increasingly grave. In view of our domestic and foreign situations as well as the trend of broadcasting in the world, it is an urgent task to expand external broadcasting."[37] "In view of the great power of dissemination and profound thoroughness of broadcasting," the Telecommunications Committee was expected to make recommendations in order "to bolster our guiding spirits of development on the Asian continent and to complete the fusion of Japanese, Manchurian, and Chinese cultures, by coordinating broadcasting policies in East Asia with our national policy."[38]

Blueprints for a Network

Charged with the new mission, MOC immediately set up an internal Research Group on the East Asian Long-Distance Telecommunications Network. Divided into functional areas and led by section chiefs like Matsumae, the group conducted a wide range of detailed studies, including comprehensive estimates of construction and maintenance costs, communications demand, and projected revenues.[39] After months of further deliberation within the ministry and consultation with the military, the MOC submitted the study to the Telecommunications Committee, which issued it as its first policy recommendation in January 1939:

> In view of both the internal and the external state of affairs as well as the present conditions of telecommunications in East Asia, consolidation of an East Asian telecommunications network, which is necessary to ensure the close mutual assistance and linkages in national defense, politics, and cultural affairs among Japan, Manchukuo, and China, is an indispensable component with great urgency for the establishment of the New Order in East Asia.
>
> Therefore, planning is required immediately for a secure and strong communications cable that links political, military, economic, and cultural centers in Japan, Manchukuo, and China, which have the closest relationship to the realization of our continental policy.[40]

---

37. "Teishin daijin Shimon dai-2-gō," MOC Records II, 693.

38. "Tōa ni okeru tsūshin keburu kansen no seibi ni kansuru ken" (1939). I am indebted to Professor Hikita Yasuyuki for a copy of this document during the early stage of my research.

39. "Tōa chōkyori denki tsūshinmō chōsakai no ken" (December 9, 1938), MOC Records II, 362.

40. "Denki tsūshin iinkai shimon tōshin" (January 16, 1939), MOC Records II, 693.

Specifically, the first stage of the grand overseas telecommunications expansion was the construction of a cable network that would function as "the great artery of Japan–Manchukuo–China communications" (see Map 3). The network would consist of two major cable routes with four branches. They would extend first from Tokyo to Fukuoka and Nagasaki in western Japan via Nagoya, Osaka, and Shikoku. From Fukuoka, one would cross the Korea Strait to reach Tianjin and Beijing in North China via Pusan, Keijō, Andong, Mukden, and Shanhaiguan, with a branch that would extend from Keijō to Nanam via Wonson and Ch'ongjin in northeast Korea. The second cable would extend from Nagasaki by way of Cheju Island to Shanghai; from there one branch would extend on land to Nanjing, another to Takao (present-day Kaohsiung) in southern Taiwan via Taihoku (present-day Taipei). In the future, this cable network would further extend from northern Korea to the Soviet border in the north, and from Taiwan to South China and Southeast Asia in the south. In between, the network would cover all important areas in East Asia already under Japanese control.[41]

Given the importance of communication in military, political, and economic activities, design of the network reflected the new priorities in Japan's continental policy. Of all the routes proposed by the Telecommunications Committee, two were entirely new. The so-called Second Japan–Manchukuo cable, which was to extend from Keijō to the city of Mudanjiang (in northern Manchuria) via northeast Korea, would strengthen imperial defenses along the Soviet border as well as help exploit natural resources in northeast Manchuria and northern Korea. Given the perceived importance of Japan-Manchukuo ties, this cable would serve as a back-up route to the first one already under construction. The 1,102-km submarine telephone cable linking Japan and Shanghai via Cheju Island, first proposed shortly after the outbreak of the war in China, was considered necessary because Shanghai would serve as the base of Japan's activities in Central China, but there was no secure telephone connection in that area apart from a wireless route. The Navy was particularly interested because the cable promised to protect telephone conversations from being intercepted by the enemy.

41. MOC Records II, 362.

Map 3 A blueprint of the East Asian cable communications network, 1938
SOURCE: Adapted from "Nichi-Man-Shi renraku kansen keburu tsūshinmō keikakuzu," in "Kokusai denki tsūshin kabushiki kaisha kakujū keikaku yōkō an," August 1938, MOC Records I, Te28n.

The initial design, unveiled in early 1938, featured eight voice circuits between Nagasaki and Shanghai. In addition, there would be one broadcast relay channel, allowing radio programs to be transmitted by the cable. Moreover, the Shanghai–Nagasaki telephone cable could be further extended to Taiwan and thus was the first stage of southward expansion of the entire network. The combined cost of these two cables alone would reach 45 million yen—over one third of the estimate for the entire East Asian cable network. The entire expansion, including constructions within Japan, was estimated to take five years at a total cost of nearly 200 million yen (see Table 7). A total of 9,648 km of cable of various types would be added.[42]

Imperial Design and Technological Choice

In November 1938, shortly after the formation of the Telecommunications Committee, Prime Minister Konoe Fumimaro proclaimed the national policy for a "New Order in East Asia." In particular, his program envisioned that "Japan, China, and Manchukuo will be united by the common aim of establishing the New Order in East Asia and of realizing the relationship of neighborly amity, common defense against Communism, and economic cooperation." Professing no interest in either territory or indemnity for the cost of Japan's military operations, Konoe emphasized that "Japan demands only the minimum guarantee needed for the execution by China of her function as a participant in the establishment of the New Order." Furthermore, Konoe pledged respect for the sovereignty of China and suggested giving "positive consideration to the questions of abolition of extraterritoriality and of the reversion of the concessions and settlements, matters which are necessary for the full independence of China."[43] That Japan was to be the architect and leader of this New Order was understood by all.

Understandably, the choice of technology for the imperial telecommunications network in East Asia had to reflect the character of Japan's new continental strategy. As already discussed in previous chapters, autonomy in international communications had been a basic tenet of Japan's foreign telecommunications policy. Bound by treaties with the GNTC, a Danish submarine cable company, Japan had viewed the ad-

---

42. Denmukyoku, Keikakuka, "Daini Nichi-Man denwa keburu oyobi Nagasaki-Shanhai kaitei denwa keburu shisetsu gaiyō" (n.d.), MOC Records II, 693; "Kokusai denki tsūshin kansen keburu dai-ikki jigyō keikaku an" (August 25, 1938), MOC Records I-Te28n. See also Denmukyoku, Chōsaka, "Tōa chōkyori yusen denki tsūshinmō jigyō keikaku an (miteikō)" (August 11 and 26, 1938), MOC Records II, 744.

43. Lebra, *Japan's Greater East Asia Co-Prosperity Sphere*, 68–70.

vent of wireless communications in the early twentieth century as a heaven-sent answer to its aim of ending foreign dominance of its international cable routes. Traffic over wireless facilities operated by Japan would reduce payments to the foreign cable company. Moreover, wireless communications cost less than submarine cables to build and would encounter no difficulty regarding landing rights in foreign territories. Wireless was thus deemed most desirable in establishing direct communications with countries that were of great importance to Japan.[44] Consequently, a joint Army-Navy-Communications investigation in 1924 recommended that wireless telegraph be given precedence over cable in the expansion of Japan's international telecommunications facilities.[45] As Japan hastened the pace of opening direct wireless communications with many countries in Asia, Europe, and the Americas, by the early 1930s, wireless facilities were carrying about half of Japan's international telecommunications traffic.[46] When he delivered a radio speech at the beginning of 1936, Okumura Kiwao still referred to the coming "Age of Wireless." Indeed, the wireless would continue to play a vital role in Japan's communications with distant destinations such as Europe and Latin America till the end of World War II.

What, then, accounts for the apparent return to cables in Japan's new expansion plans? Although the shift to cable technology in the late 1930s did not mean abandoning wireless, the change in budgetary emphasis, however, raised a number of questions from skeptics. During the Imperial Diet deliberations in early 1939, for instance, a member of the House of Peers asked whether, given the continuing improvement in wireless and other new technological inventions such as television, such a huge project based on costly cables might not lose its value.[47] To this, advocates of a cable network replied that cable was more suited to

44. Nomura Yoshio, "Taigai musen denshin no kakuchō ni tsuite," *TKZ* 332 (April 1936): 145–49; Kobayashi Takeji, "Nihon musen denshin kabushiki kaisha-hō no kaisei ni tsuite," *TKZ* 345 (May 1937): 44–49.

45. "Rikugun daijin, kaigun daijin, teishin daijin ni teishitsu taigai dai musen keikaku ni kansuru sanshō kankei kanri no chōsa hōkokusho" (received on August 12, 1924), JMFA Archives, Microfilm MT 3.6.11.23.

46. Tamura Kenjirō, "Kokusai denwa kabushiki kaisha no sōritsu kettei made," *TKZ* 292 (December 1932): 82–96. By one rough calculation, some 177 wireless frequencies remained unused worldwide.

47. Teikoku gikai, *Teikoku gikai kizokuin gijiroku*, March 20, 1939, 5–6.

Table 7
Construction Plans for the East Asian Telecommunications Network, 1939

Section	Location	Type of	Length (km)	Pairs of cables	Year of construction
Nagoya–Osaka	Japan	C	186	90	1st
Nagasaki–Cheju	Japan–China	LS	470	21	1
Fukuoka–Pusan	Japan	C	225	28	1
Keijō–Wonsan	Japan	C	180	18	1
Mukden–Shinkyō	Manchukuo	C	308	28	1
Shanhaiguan–Tianjin	China	C	283	14	1
Osaka–Matsuyama	Japan	C	374	90	1
Cheju–Shanghai	Japan–China	LS	630	17	2nd
Wonsan–Ch'ongjin	Japan	C	400	18	2
Shinkyō–Harbin	Manchukuo	C	240	14	2
Mukden–Shanhaiguan	Manchukuo	C	420	14	2
Matsuyama–Nagasaki	Japan	C	481	64	3rd
Ch'ongjin–Tumen	Japan–Manchukuo	C	370	18	3
Harbin–Mudanjiang	Manchukuo	C	354	14	3
Mukden–Dalian	Manchukuo	C	396	28	3
Tumen–Mudanjiang	Manchukuo	C	250	8	4th
Beijing–Zhangjiakou	China	C	190	8	4
Tianjin–Jinan	China	C	360	8	4
Jinan–Xuzhou	China	C	320	8	4
Shanghai–Nanjing	China	C	330	14	4

Taejon–Cheju	Japan	C	440	8	5th
Harbin–Qiqihar	Manchukuo	C	350	8	5
Jinan–Qingdao	China	C	410	14	5
Xuzhou–Nanjing	China	C	340	8	5
Taihoku–Tainan	Japan	C	329	18	5
Tamsui–Tainan	Japan	C	71	8	5
Shanghai–Taihoku		LS		12	4th–5th

NOTES: C: carrier cable; LS: loaded submarine cable.

SOURCE: "Kokusai denki tsūshin kansen keburu daikki jigyō keikaku an (2)," MOC Records I, Te28n.

Japan's imperial network, as a cable communications network would provide greater capacity and security, plus stability.

First of all, cable advocates pointed out, the new telecommunications network had to be capable of handling a large volume of traffic. Already, as the MOC indicated in the documents prepared for the Telecommunications Committee, telecommunications traffic in the new imperium was outstripping the capacity of existing facilities. Between Japan and Manchuria, the eight wired and seven wireless circuits together could handle a total standard capacity of 15,600 telegrams per day. In reality, the daily volume was 20,000, which represented a 40 percent increase over a year earlier. Telegraphic traffic between Japan and China had increased 1.3 times within a year. On average 200 telephone calls per day were made between Japan and China, a 2.5-fold increase.[48] As Japan continued to expand and consolidate its political, military, and economic presence on the Asian continent, a sharp increase in telecommunications traffic could be anticipated because of new industrialization plans and continuing Japanese migration.

Although wireless was most appropriate as a pioneer in opening new routes of communication, as Kajii Takeshi noted early on, cables were more suitable when more capacity was needed.[49] Telecommunications Bureau Director Tamura Kenjirō pointed out that, due to the limited number of suitable radio frequencies, wireless communications could no longer meet the requirements of a more powerful communications network in East Asia. As of June 1932, of the 1,878 shortwave frequencies used worldwide, Japan had 212, divided between the MOC (124) and the military (88). Writing in the *Journal of the Communication Association*, Tamura predicted that 47 telegraph circuits and 28 telephone circuits would be required by 1944 to link Japan, Manchukuo, and North China.[50] In view of its great importance, the Japan–Manchukuo route alone, he told a Diet committee, could conceivably require hundreds of channels, as compared to the few that existed then. The new NLC multi-channel network promised greater capacity for the anticipated

---

48. "Nichi-Man-Shi renraku senro no genjō oyobi fukusō jōkyō" (late 1938), MOC Records II, 693.

49. Kajii, "Tōyō denki tsūshinmō yori mitaru Manshū no chii," 170.

50. See Tamura Kenjirō, "Tōa denki tsūshin seisaku to Kokusai denki tsūshin kabushiki kaisha no kakuju ni tsuite," *TKZ* 369 (May 1939): 37.

volume of traffic between Japan and its sphere of influence in adjoining areas. Moreover, the limited frequencies available to Japan had to be reserved mainly for communications with more distant foreign countries in Europe or the Americas. Given the limited number of wireless frequencies available, cable appeared to be the clear winner.

In the meantime, the government had been placing greater emphasis on communications secrecy. Although telegraph could rely on encryption codes, which were widely used at the time, telephony posed a greater technical challenge.[51] In 1936, a young engineer at the central laboratory of the SMR claimed to have invented a device for secret telephone conversations. Because most long-distance telephone links were by wireless, Japan soon applied such home-grown scrambler technology to its most sensitive circuits between the home islands and Manchukuo. This was not adequate, however. Tamura Kenjirō noted that since Japan, Manchukuo, and China were bound by a special relationship, military, political, and diplomatic communication among them required special protection against interception. Given the advances in code-breaking techniques, Tamura told the Diet committee, wireless communications ran the risk of being easily deciphered by the enemy. In this respect communicating by cable had a decided advantage over wireless.[52]

Moreover, stability and safety also favored the cable solution. Although wireless technology continued to improve, the perennial problem of atmospheric interference remained. Depending on the time of the day or the season, the problem of static could be so severe as to put wireless communications out of commission completely. In contrast, a network of metal-shielded cable, buried underground or laid underwater, would be entirely stable and safe. MOC officials were aware of the fact that much of the land-based network would be in territories outside Japan's formal empire, but sabotage could be eliminated by means of underground cables. Even with broadcasting, which was typically associated with wireless, cable technology had its advantages:

51. See "Himitsu denwa no hatsumei," *Manshū nichinichi*, June 3, 1936.

52. Teikoku gikai, *Teikoku gikai kizokuin gijiroku*, 383. Much has been written about the failure of Japan's communications security, despite the fact that most communications traffic intercepted by the Allies consisted of diplomatic correspondence with embassies located outside this network. Japanese engineers began developing a scrambler to be used on wireless telephones to ensure secrecy, but in general it was not satisfactory.

the greater capacity of imperial trunk cables could facilitate regular program exchanges between Japan and other parts of the empire. Local wired networks would be preferable to wireless when it came to protection against enemy air raids as well, since radio signals could serve as a navigation beacon for aircrafts.[53]

Thus the choice of the more expensive cable for the imperial network was not accidental but can be said to reflect the nature of Japan's empire-building project of the 1930s as a whole. The development of non-loaded cables in the early 1930s gave cable a new technological edge over wireless. Although construction and maintenance costs were much higher, the long-distance cable network for East Asia would be more permanent than wireless facilities and therefore more appropriate for Japan's lasting imperial enterprise.[54] As technological choice and Japan's new empire-building agenda became inseparable, the cables were presented as "Japanese-style technology" that could better serve a closely integrated and long-lasting empire.

COMMUNICATION AND IMPERIAL INTEGRATION

The imperial telecommunications network for East Asia represented one of the most ambitious of Japan's techno-imperialist projects. As such, it was not only shaped by Japan's new empire-building agenda, but also reflected the recognition of communications technologies as a force of integration as well as an appreciation of the spatial strategies of other Imperial Powers.

The Power of Communication

At the beginning of 1936, when the construction of the Japan–Manchukuo cable had just begun, Okumura Kiwao delivered a radio speech entitled "The National Strength of Japan." The venue chosen

---

53. On wire-based broadcasting, see Shinohara Noboru, "Daitōa sensō to denki tsūshin gijutsu," *TKZ* 402 (February 1942): 5–6.

54. In this sense, the preference for cable over wireless is not dissimilar to Innis's distinction of space-biased vs. time-biased technology.

for the speech was no accident. First started in Japan in 1925, radio broadcasting had undergone phenomenal expansion and consolidation after the Manchurian Incident. By the mid-1930s, it had become a powerful tool of information dissemination and mass mobilization for the state.[55] In the broadcast, Okumura discussed Japan's national strength from the perspective of transportation and communications, which he described as the movement of people and goods and the transmission of messages and meaning, respectively. He began with a brief review of world history, speaking of the road system developed in the ancient Middle East and Europe and noting that "transportation and communications had been closely linked to the rise of national fortune and development of culture." Scientific progress since the end of the nineteenth century, he went on, had not only given birth to various new means of transportation but also produced extremely large organizations, contributing significantly to the ultimate objective of "the overcoming of distance." Okumura then surveyed the "remarkable development" of transportation and communications in Japan since the Meiji Restoration and concluded that "in this unprecedented era of Japan's modern transformation, we must develop our national strength and culture by further improving transportation and communications in the future."[56]

The topic of Okumura's speech was not entirely new. The Japanese had already embraced the concept of "communications capability" (*tsūshinryōku*) as an indicator of "national fortune" for quite some time.[57] As we have seen, an extensive internal information network was important to the formation of a national community in Meiji Japan. In highlighting transportation and communications as critical measurements of Japan's national strength over national radio, Okumura not only drew from his own experience in Manchuria but also spoke proudly of the great strides Japan was making in strengthening its communications capabilities in the 1930s. By then, telecommunications had also come to occupy an important place in the daily life of Japan's rapidly expanding imperium (see Table 8).

---

55. Kasza, *The State and Mass Media in Japan, 1918–1945*, 72–101. Young, *Japan's Total Empire*, 63–68.

56. Okumura Kiwao, "Kōtsū tsūshin yori mitaru Nippon," *DKS* 6 (February 1936): 19.

57. See, e.g., Teishin kyōkai, Henshūbu, comp., *Teishin jigyō ni kansuru kōen shiryō* (Tokyo: Teishin kyōkai, 1911), 43–46.

Table 8
Communication Indexes in Japan and Its Colonies, 1935

Index	Japan proper	Taiwan	Korea	Southern Sakhalin	South Sea islands	Kwantung territory
Area (km^2)	382,265	35,961	220,769	36,090	139	3,748
Population (000)	69,255	5,316	21,891	323	105	1,637
Number of post offices	11,249	184	873	81	9	265
Area per post office (km^2)	34	195	253	446	15	14
Population per post office	6,156	28,889	25,076	3,981	11,653	6,177
Telegraph offices	8,232	213	878	91	9	211
Area per telegraph office (km^2)	47	169	252	397	16	18
Population per telegraph office	8,413	24,956	24,933	3,544	11,653	7,758
Telegrams sent (millions)	73,860	1,902	7,992	905	254	4,113
Telegrams sent per person	1.067	0.358	0.365	2.807	2.419	2.513
Telephone subscriptions	870,564	16,800	39,763	5,535	392	21,321
Subscriptions per 10,000 persons	125.7	31.6	18.1	171.6	37.4	166.9
Number of public telephones	3,635	34	86	34	2	148
Population per public telephone	19,052	156,353	254,547	9,500	52,500	11,061
Phone usage (million units)	2,906	107	270	21	1	404
Phone usage per person (units)	41.96	20.2	12.35	96.79	29.68	246.7

SOURCE: Chōsen sōtokufu, comp., *Chōsen teishin tōkei yōran* (1936).

Table 9
Speed of Communications from Tokyo, 1937

	Tokyo–Osaka	Tokyo–Keijō	Tokyo–New York
Postal service	18 hours	4 days	15 days
Airplane	2 hours 50 min	8 hours 40 min	127 hours 30 min
Train	8 hours	36 hours	16 days
Ship	21 hours	5 days	28 days
Telegraph—average time	42 min	50 min	39 min
Telegraph—shortest time on record	18 min	26 min	15 min

SOURCE: *TKZ* 320 (May 25, 1937).

NOTES: Pan American Airlines opened the first trans-Pacific route in 1938. Train service is combined with ship when necessary.

Although postal communication continued to be indispensable to the Japanese as well as the native populations, it was electronic communications that opened the most promising horizons for communication (see Table 9). Telecommunications offered several advantages over the various forms of postal service. Since it did not rely on the actual movement of objects, it was less dependent on any particular means of transportation. Okumura may not have been entirely original when he observed:

> The invention of electric communications was a fundamental leap forward in the means of communication of mankind in the past thousands of years. Domestically, [telecommunications] make exchange of meaning extremely speedy and play a role in helping improve social life, speeding up business transactions, and reforming politics and administration both at the center and in the local areas. Internationally, [telecommunications] also promote friendly relations between countries by creating closer political and economic relations. That we feel the world has become small today is largely due to telecommunications.[58]

The power of telecommunications to annihilate distance became even more enhanced as new technologies were introduced and improved. Though still costly, phototelegraphy technology was increasingly available in Japan and played a major role in media coverage of the 1936 Berlin Olympics. The first decade of regular phototelegraphy service in Japan saw it limited to the circuit between Tokyo and Osaka. During the 1930s, events on the Asian continent, especially the war in China, called for rapid expansion of phototelegraphic service as well as improved quality. Temporary service was provided for news agencies during special events, mostly by way of wireless. For example, during a flood in North China, the Japanese blockade of the British Settlement in Tianjin in August 1939, and the border clashes between Japanese and Soviet forces, the Japanese press was able to send photographs back to Japan using phototelegraphy.[59] Before the Pacific War, phototelegraphy service between Japan and the United States had been established as well.

Since sending photo-quality images via telegraphic circuits proved to be prohibitively expensive, it was largely used by news agencies with deep pockets. A similar but less costly method was facsimile (*mosha denshin*), which did not require high resolution or the cumbersome devel-

---

58. Okumura, "Kōtsū tsūshin yori mitaru Nippon," 17.

59. See, e.g., issues of *Tōkyō Nichinichi shinbun* and *Tōkyō Asahi shinbun* for July 1939.

opment process of phototelegraphy. Facsimile was particularly useful in transmitting written script or graphs that did not require shaded resolution. The history of facsimile dated back to the early days of the telegraph, when inventors in America and Europe sought to transmit letters directly.[60] By the 1930s, the Japanese military, interested in using facsimile to transmit meteorological graphs and maps, set up a joint Facsimile Committee. Because facsimile promised to bypass the stage of transcribing the text into telegraphic code, both MOC and Japanese news agencies like Dōmei demonstrated great interest in the technology.[61] In the fall of 1937, Matsumae ordered new research to be conducted on facsimile telegraphy in anticipation of its wide use in the future. Not only would such a service help eliminate human errors in transcribing telegrams, he reasoned, facsimile could also revolutionize telegraph communication in East Asia by transmitting ideographic languages such as Chinese characters.[62] Shinohara Noboru, who succeeded Matsumae as the chief of MOC's Investigation Section, predicted that facsimile would be "most effective in Japan, Manchukuo, and China, where ideographs are in use." Nakayama Jirō, MOC's chief of foreign communications (and the son of Nakayama Ryūji), concurred that facsimile "has considerable value between Manchukuo and China." Remarkably, there was still no formal designated name for the new technology as late as 1940, but all parties seemed to agree on its great potential, particularly for use in Asia.[63]

The completion of the Japan–Korea–Manchukuo long-distance cable in late 1939 was understandably an occasion to showcase new technologies. As a major technological landmark for Japan, it served as a technological confirmation of the newly drawn blueprint of the East Asian telecommunications network. Thanks to the new cable connection, telephone capacity between Japan and Manchukuo increased by 24

---

60. On Japanese interests in early facsimile technology in the late nineteenth century, see Kawanobe Tomiji, *Teregurafu komonjo kō*, 413–68.

61. Okuno Haruo, "Daitōa kyōeiken to atarashii mosha densō," *Tōa denki tsūshin zasshi* (hereafter *TDTZ*) 3.1 (January 1943): 27–36.

62. Toriumi Noboru, "Masha denshin sōchi no kanseisuru made," *TKZ* 429 (June 1944): 17.

63. Facsimile was also referred to as "simplified phototelegraphy" (*kan'i shashin denpō*) or "postcard telegram" (*hagaki denpō*); see Tōa denki tsūshin jimukyoku, *Tōa denki tsūshin kyōgikai dai-2-kai kaigi gijiroku* (1940), 193–98.

channels, and that between Japan and Korea increased by 10.[64] Much of the new technology that was envisioned in the new East Asian Telecommunications Network had already been tested on the Japan–Manchukuo cable. For instance, Japanese newspaper reports took special note of the fact that public teleprinter and phototelegraphy services would begin the following day, using equipment developed entirely in Japan.[65] Teleprinters, which greatly shortened telegram transmission time and reduced human errors in transcription, were in widespread use on Japan's long-distance trunk lines in East Asia. The completion of the Japan–Manchukuo trunk cable in late 1939 greatly improved the prospects of phototelegraphic service in the imperium. Photographs were not the only items sent. The *Manshū Nichinichi* newspaper also welcomed it as a "great leap" in telegram service, since it was error-proof because it omitted the step of transcription into codes. It was particularly recommended for sending letters, tables, graphics, designs, or photographs.[66] Scramblers, used to enhance security in telephone conversations, were installed on telephone cables between Tokyo and Shinkyō.[67]

The technological implication for East Asia of the Japan–Manchukuo long-distance cable became obvious when the Tokyo–Mukden line was extended to Beijing and Tianjin in May 1940, making the 3,000-km route the longest in the world. MOC officials confidently predicted not only a direct telephone link between Osaka and Tianjin in North China in the near future using the Japan–Manchukuo cable, but also "an expanding telephone network that will crisscross the Asian continent like a spider's web." "As the distance between Japan and Manchuria has been completely overcome in such a manner," they pointed out in language that had become all too familiar, "it will make a great contribution to the construction of the New Order in East Asia!"[68] Indeed, as Japan strengthened its communications with its continental dependencies, Japan's empire-building in Asia seemed to enjoy unprecedented

---

64. Teishinshō, *Nichi-Man renraku denwa shisetsu sekkei keikaku kōyō*; *DDJS* 6: 400–401.

65. "Nichi-Man renraku denwa kōji shunkōsu," *TKZ* 375 (November 1939): 120–26; *Manshū Nichinichi*, October 1, 1939.

66. *Manshū nichinichi*, October 1, 1939.

67. Nakamura, "Shūsen," *DJ* 33 (August 1983): 40.

68. Denmukyoku, Gaikoku denshinka, "Nichi-Man denwa 'sabisu' kaizen," *Teishin no chishiki* 3.5 (July 1939): 4.

technological blessings. The completion of the Japan–Manchukuo long-distance cable in late 1939 was a powerful demonstration that a key component of the imperial nerve system was in place.

Imperial Inspirations

As Japan harnessed technologies for integration in East Asia, the experience of Western empires served as inspiration, justification, and eventually targets for its new quest for telecommunications hegemony in East Asia. Japanese admiration for European overseas expansion dates at least to the Meiji period, and European ideas about empire had always found a ready audience in Japan, a relative latecomer to the imperialist game.

Although Germany had been deprived of its overseas empire after World War I, the ideas of German geopolitical thinkers had become popular in Japanese intellectual circles by the late 1930s. In particular, German works on *Raumordnung* (literally, "spatial order") provided a potent theoretical justification for territorial expansion and for Japan's planned redistribution of world resources for autarkic economic purposes. Although communications were often conspicuously absent in the original works, land, sea, and air modes of transportation were considered vital to "reconfiguring space."[69] Many Japanese thinkers did not accept German geopolitical theory uncritically, to be sure. They were careful not to embrace it wholeheartedly, even though it was useful in justifying Japan's mission of constructing a new Asian community. Rōyama Masamichi, one of the leading political theorists and a member of Prince Konoe's brain trust, was insistent about creating Japan's own vision of the East Asian region.[70] Still, *Raumordnung* seemed to provide a scientific rational for reordering the newly expanded imperium.

---

69. Examples include Ichii Osamu, *Tōa kokudo keikaku* (1942); and Kōseikai, *Daitōa kokudo keikaku no kenkyū* (1943). The term *kukan kisei* was used in Manshūkoku, Sōmuchō, Keikakusho, "Sōgō ritchi keikaku sakutei yōkō" (June 1940); and "Sōgō ritchi keikaku ni tsuite" (March 7, 1940), in Sugai Shirō, comp., *Shiryō kokudo keikaku* (Tokyo: Taimeidō, 1975), 1–18.

70. The best study on this subject is Hatano Sumio, "Tōa shijitsujo to chiseigaku," 14–47. See also Miwa, "Japanese Policies and Concepts for a Regional Order in Asia, 1938–1940," 133–56.

The "advantages of followership" enabled Japanese writers on telecommunications policy to borrow from Western sources. Watanabe Otojirō, one of the most brilliant Japanese writers on telecommunications policy during the war, readily evoked examples of how the Western powers had managed their overseas telecommunications. As one Japanese employee at the NCTT noted, strong communications ties between Japanese on the periphery of empire and those at home would prevent the former from losing their Japanese spirit and being assimilated by the vast indigenous populations of Asia. Interestingly, the author of the report attributed the idea of identity loss to a Frenchman.[71]

The British empire, a truly global hegemonic power sustained since the mid-nineteenth century by the twin pillars of a powerful navy and an extensive cable network, garnered the largest share of admiration and envy among the Japanese. Subsidized cable and imperial communications facilities provided exchanges of news at low cost between the various parts of the British empire, with the intended result of creating common bonds of knowledge and interest. At the height of its glory, Britain controlled more than 70 percent of the world's communications cables. This enabled the British to control the flow of news information to their own advantage, whether by subtle manipulation or outright censorship (as during the Boer War). It was no coincidence that the largest news agency in the world was the British-owned Reuters—a fact no other aspiring power could ignore. Like Watanabe, many Japanese writers were both familiar with the history of British hegemony in telecommunications and acutely aware of the ascendance of American influence in the Pacific as well. A Japanese research group on communications, jointly set up by the communications industry and the Japan Broadcasting Corporation, published detailed studies on the rise of British dominance in international telecommunications.[72]

Just as important, Japanese like Matsumae could study Western colonial empires firsthand, as he did during an extensive tour of Southeast Asia at the beginning of 1937. Matsumae's report reflected his interest in the colonial policy of various Western powers in Asia as much as his

---

71. NCTT, Eigyōbu, Gyōmuka, "Ka-Nichi tsūwa no kaitsū ni itaru made" (July 1939), NCTT Records 2028/1493. The name of the Frenchman was F. V. Ferner.

72. See, e.g., such studies as Oka Tadao, *Taiheiyō iki ni okeru denki tsūshin no kokusaiteki bekken*; and idem, *Eikoku o chūshin ni mitaru denki tsūshin hattatsu shi.*

interest in telecommunications operations per se. With thinly veiled envy, Matsumae noted that "in their colonies, Holland, Britain, and France adopted a very conscious communications policy and made great efforts to guarantee the nerve system connecting the colonies and their home countries." He concluded that the Dutch East Indies was becoming the center of telecommunications in the East, even producing equipment locally in case supplies from Europe were cut off.[73] Matsumae was most impressed with the British, however. Despite widespread use of wireless in the world, he noted, Britain's Great Eastern Telegraph Company (GETC) had poured enormous resources into building branch lines that connected the region's submarine cables, so as to maintain Britain's communications link with its colonies by secure means. What showed as one line on the map connecting Singapore to Colombo via Penang was actually five separate cables, as Matsumae discovered to his surprise. In an article published in early 1938, Matsumae described the "spider's web" of the British worldwide telecommunications network "that completely controls world opinion so as to ensure influence in commerce and diplomacy and serve as a powerful weapon in trade and other international business warfare by extending its greedy tentacles into other countries." After outlining his vision for Japan's East Asian cable telecommunications network, Matsumae called for creation of a national policy corporation like Britain's GETC, which ensured the smooth operation of its long-distance submarine cable network.[74]

In a textbook on telecommunications he co-authored in late 1938, Matsumae laid out perhaps the most definitive description of the British model:

> What role do communications play in administrative policy toward colonies—countries consisting of different nationalities? We can see this most clearly in the traditional colonial policy of the British empire. This vast empire, where the sun never sets, occupies one-fifth of the world. Despite various difficulties and wars, it has not lost a single territory but has remained intact. This is due to the fact that it controls a communications network that crisscrosses the

73. Matsumae Shigeyoshi, "Nan'yō shoppō no shisatsu o oete," *Denki tsūshin gakkai zasshi* (hereafter *DTGZ*) (1937): 494–506; idem, "Nan'yō shoppō ni okeru denki tsūshin jigyō," *TKZ* 348 (August 1937): 2–26.

74. Matsumae Shigeyoshi, "Nichi-Man renraku musoka keburu ni tsuite," *Denki gakkai zasshi*, January 1938; reprinted in Matsumae Shigeyoshi ronbunshū kankōkai, *Hatsumei e no chōsen*, 825–49.

earth like a spider's web, pays extraordinary attention to developments everywhere at all times, and strives to influence public opinion through the communications network. This is due to the fact that the state never relents on giving its attention to the most advanced communications technology, at times making a huge sacrifice to build a submarine telegraphic network spanning the seven seas. It was also the first of the Powers to notice the applicability of shortwave wireless and implement it. Maintaining multiple autonomous connections between the colonies and the home country, it has prevented overseas colonies from turning their back [on the home country].[75]

In October 1938, shortly after the Telecommunications Committee was formed, Matsumae Shigeyoshi delivered a speech on "technology and communications policy" at a gathering of telecommunications engineers. He spoke at length about Britain's success in influencing world opinion by controlling a vast information network throughout the world. Matsumae reminded his audience that it was Britain's cables and news agencies that had effectively turned world opinion against Germany during the Great War. Britain's early adoption of new technologies and application of them in national policies, regardless of the moral consequence, spoke volumes about its success in communications policy, Matsumae maintained. Referring to British press criticisms of Japan's actions in China, Matsumae noted that the key to overcoming Britain's anti-Japanese influence was simple: just get the information out ahead of the British. Speed was all that counted! By evoking the British example, Matsumae drove home his favorite theme of welding telecommunications technology with national policy.[76]

Increasingly, as Japan's own horizon expanded in Asia, the Western imperial presence came to be seen as an obstacle to Japan's empire-building mission in Asia. Many proposed concrete measures to counter the Anglo-American presence in East Asian telecommunications. The most senior leading spokesman on communications matters was Nakayama Ryūji, the Japanese equivalent to engineer-turned-publicist Charles Bright in turn-of-the-century Britain. An advisor to the Chinese government in telecommunications matters during the 1910s and 1920s, Nakayama had worked assiduously to establish Japan's influence in telecommunications on the Asian continent. At home, he was a tireless

75. Matsumae Shigeyoshi and Nishizaki Tarō, *Denki tsūshin gairon*, 47–48.

76. Matsumae Shigeyoshi, "Tsūshin no gijutsu to seisaku," *DTGZ* (November 1938): 12–14.

publicist of the importance of telecommunications in Japan's new imperial project in East Asia and a most ardent champion of an aggressive policy. Nakayama argued that the strengthening and expansion of telecommunications must take precedence over everything else. In the past, international communications in East Asia had been oriented toward Europe and America; Nakayama proposed that the center be moved to Japan immediately. For instance, the submarine cables, which were mainly oriented toward Europe and America and concentrated in submarine cable network nexuses such as those in Hong Kong, Singapore, Penang, Batavia, and Manila, should be reoriented toward Japan. Likewise, the majority of wireless communications equipment previously used for communication with Europe and America should be put to use in direct communications with Japan. Finally, he called for immediate restoration, consolidation, and expansion of telecommunications administrations within various areas in Asia in order to maintain security, revitalize the economy, and develop culture.[77] His prescription would soon find its way into official policy

ENVISIONING GREATER EAST ASIA

In the fall of 1939, as the Japan–Manchukuo long-distance telephone cable neared completion, MOC adopted a number of measures to raise public awareness of Japan's new communications policy. To explain Japan's new official position, MOC prepared a pamphlet entitled "Telecommunications Policy in East Asia." Issued by the Cabinet Information Bureau in August 1939, as part of its "Propaganda Material on Current Affairs," this was as close as Japan ever came to a public manifesto on its program for telecommunications hegemony in East Asia.

The pamphlet began with an explanation of the need for an East Asian telecommunications policy:

> At present, it is a task of enormous urgency to establish an East Asian telecommunications network that binds Japan, Manchukuo, and China closely together. It is the great mission of telecommunications to transmit the intentions

---

77. Nakayama Ryūji, "Shin-Tōa kensetsu to denki tsūshin jigyō," 2 pts., *DT* 2.6 (October 1939), and 3.8 (1940); later included in his popular *Sensō to denki tsūshin* (War and telecommunications), which went through three editions after initial publication in early 1942.

of East Asian peoples burning with the desire to revitalize Asia, to overcome the barriers of time and space in continental management, to facilitate intimate mutual contact among various policies of Asian revitalization, and to become the driving force in implementing the national policy.[78]

The pamphlet then surveyed the condition of telecommunications in East Asia.

Like the rest of East Asia, the pamphlet admitted, Japan was late in benefiting from modern communications, its domestic facilities not yet meeting demand. Although wireless had met with some success in Japan's foreign communications, Japan was still far from having caught up with the West. In contrast, advanced countries, realizing the importance of telecommunications, had entered East Asia by operating telecommunications networks, supplying equipment, and securing control through capital, and had then cultivated their firmly entrenched influence there. The pamphlet detailed the history of expansion in telecommunications by each major Western player. Although China had gradually recovered its rights in wireless communications, the pamphlet noted, it still relied on the West for equipment, and its domestic wireless was still primitive. As a result of Western dominance, "East Asian countries do not possess their own ears when they listen, or mouths when they speak."[79]

What, then, should Japan's telecommunications policy for East Asia be? How was Japan to establish a New Order of East Asian telecommunications? The MOC saw two dimensions to correcting the situation: first, Japan's aim should be to establish a "self-reliant East Asian telecommunications bloc." Such an objective, as the Telecommunications Committee resolved, involved three tasks: Japan would promote telecommunications in the "new China," build communication trunk lines between Japan, Manchukuo, and China, and strengthen Japan's telecommunications industry.[80]

Second, Japan must deal with Western influence in East Asian telecommunications. Due to their persistent efforts, Western countries enjoyed a solid hold over telecommunications above and under ground,

---

78. Teishinshō, *Tōa ni okeru denki tsūshin seisaku*, 139. Reprinted in Ishikawa Junkichi, comp., *Kokka sōdōin shi: shiryō* (hereafter *KSSS*), 4.

79. Teishinshō, *Tōa ni okeru denki tsūshin seisaku*, 139–42.

80. Ibid., 143–46.

under water, and even in the air. The pamphlet recognized the fundamentally international character of telecommunications: it obliterated physical distance while enabling instantaneous transmission in time. As cross-boundary human exchanges increased along with cultural progress and as communications capability expanded due to the advancement of technology, it noted, the international character of telecommunications became more and more complicated and obvious. Nevertheless, the pamphlet emphasized, Japan's policy toward these Western telecommunications interests in East Asia must start from the "inevitable necessity of strengthening and expanding the great East Asian Telecommunications Network spanning Japan, Manchukuo, and China." Existing Western interests, even if based on international law, must now be interpreted or handled differently on the basis of the new conditions that prevailed.[81] With such language, the MOC foresaw an Olympian struggle between East Asia led by Japan and entrenched Western telecommunications interests.

The timing could not have been more perfect. On September 1, 1939, as MOC officials were preparing for the much-anticipated completion of the Japan–Korea–Manchukuo long-distance cable, Nazi Germany launched a *blitzkrieg* against Poland; two days later, France and Britain declared war on Germany. The sudden outbreak of war in Europe underscored MOC Minister Nagai's reference to "today's world, full of uncertainties" in his September 30 speech at the opening ceremonies for the long-distance cable. The war in Europe had profound policy implications for Japan's geostrategy in general, and for its designs on Southeast Asia in particular.[82] Developments in Europe provided Japan with an opportunity to strengthen its position in Asia. During the summer of 1940, Germany occupied Holland and much of continental Europe. On June 29, 1940, a week after France's capitulation, Foreign Minister Arita Hachirō broadcast a statement elaborating on Japan's vision of forming "a sphere of co-prosperity and co-existence" that would include the Southern Region. Arita spoke of Japan's principle of each country being "entitled to its rightful place"; of a "natural and constructive system" based on geographical, ethnic, cultural, and

---

81. Ibid., 146.

82. For a general discussion, see Nagaoka Shinjirō, "Economic Demands on the Dutch East Indies."

economic bonds; and of correcting the "unreasonable and unjust" order of the past. Although he did not announce a change in Japan's policy of "noninvolvement" in the European War, he concluded that "countries of East Asia and regions of the South Seas" should be united into a single sphere. The so-called Arita Broadcast was the first public announcement that much of Southeast Asia was to be considered Japan's sphere of influence.[83]

Visionaries of the new imperium gained additional momentum in July 1940, when Prince Konoe Fumimaro became prime minister for the second time. Barely a month later, on July 26, the Second Konoe Cabinet approved the "Outline of Fundamental National Policy," which spelled out the new direction of Japan's external policy in view of developments in Europe. Prepared under Hoshino Naoki, chief of the Cabinet Planning Board and previously the top Japanese official in the Manchukuo government, the outline called for a national defense economy based on the self-reliant economic reconstruction of Japan, Manchukuo, and China. The outline reaffirmed the establishment of a self-sufficient economy with "Japan, Manchukuo, and China as one link," but also encompassing Greater East Asia. It stipulated that the Southern Region was now to be incorporated into Japan's sphere of prosperity, although this was to be achieved by peaceful means. On July 27, the Imperial Liaison Conference adopted a series of new diplomatic and military policy guidelines. At the beginning of August, in a press release, the new Foreign Minister Matsuoka Yōsuke declared that the goal of Japan's policy was to create a Greater East Asia Co-Prosperity Sphere.[84]

The three years after the outbreak of the China War were the heyday of ambitious imperial planning, when the Konoe Cabinet and the "new bureaucrats" led the charge to bring about the New Order both at home and in East Asia. Okumura Kiwao, previously noted for involvement in the establishment of MTT as an MOC official, emerged as a rising "new bureaucrat" (*kakushin kanryō*) in the newly formed Cabinet Research Bureau, which evolved into the powerful Cabinet

83. *Nihon gaikō nenpyō narabini shuyō bunsho* (hereafter *NGNB*), 2: 433–44. See Hatano Sumio, "Arita hōsō (June 1940) no kokunai bunmyaku to kokusai bunmyaku"; and Beasley, *Japanese Imperialism*, 226.

84. Beasley, *Japanese Imperialism*, 226–29.

Planning Board. These newly established "comprehensive policy agencies" (*sōgō seisaku kikan*) took the lead in imperial planning. Established in July 1940, the Second Konoe Cabinet introduced a plethora of plans for restructuring the domestic political and economic system as well as for giving greater substance to the New Order in East Asia.

Imperial planning had to address the spatial reality of the new imperium. In September 1940, Japan signed the Tripartite Pact with Nazi Germany and Italy, which pledged military support if one of the signatories was attacked by a country not yet involved in the war in Europe. In the same month, the Cabinet had adopted an "Outline of National Land Planning," which called for "comprehensive planning of transportation and communications for the home islands, the colonies, and the entire East Asia region."[85] This was Japan's answer to Germany's *Raumordnung*. In early October, the Second Konoe Cabinet adopted an "Outline of Economic Reconstruction of Japan, Manchukuo, and China," the most comprehensive economic program for the New Order in East Asia. The reconstruction consisted of three processes: (1) reorganization of Japan's domestic economy, (2) strengthening of the sphere of self-sufficiency in Northeast Asia, and (3) expansion of the East Asian Co-Prosperity Sphere. To achieve the "organic unity" of the sphere of self-sufficiency on the basis of national defense and geopolitical status, the outline pointed out, required the political, cultural, and economic integration of Japan, Manchukuo, North China, Inner Mongolia, and "certain islands protruding off the South China coast." According to this plan, by 1950 Japan would achieve economic self-sufficiency, with Japan, Manchukuo, and China unified as one. Central and South China, Southeast Asia, and other areas in the south would form the rest of the East Asian Co-Prosperity Sphere, which would support completion of the national defense economy. To accomplish this ambitious goal, the outline called for a "giant leap forward" in strengthening the transportation and communications connections between Japan, Manchukuo, and

85. "Kokudō keikaku settei yōkō" (Cabinet resolution of September 24, 1940), in *KSSS* 4: 1080–82. For an early American work on land planning, see Hoover, *The Location of Economic Activity*, 298–300.

China, in order to promote integration of the economies of the three countries and meet the demands of national defense.[86]

In the same month, the Telecommunications Committee issued another recommendation in response to a third and final policy inquiry from MOC. The recommendation reaffirmed the goals of "consolidating important international communications facilities with the South and other areas," and ensuring the rapid and energetic construction of the New Order in East Asia, along with further accelerating the expansion of "the Japan–Manchukuo–China connection trunk cable."[87]

By then, an information network in East Asia controlled by Japan had become incorporated into these elaborate and ambitious plans for the new imperium. In fact, it is noteworthy that the construction of an East Asian telecommunications network was already under way when most of these plans were formulated. The new imperium envisioned in these imperial blueprints was therefore both a technological and political construct. Or, to borrow a phrase from leading technology bureaucrats such as Matsumae Shigeyoshi and Miyamoto Takenosuke, the "politicization of technology" finally seemed to have borne fruit.

Overseas expansion was not new for Japan, but envisioning imperial integration of a rapidly expanding empire now became materially and intellectually inseparable from the new information technology that overcame time and space with unprecedented speed and efficiency. Japan's wartime empire was not constructed simply according to an ideology of imperial domination; rather, it was blessed with powerful, modern technologies—railway, aviation, and telecommunications—that made it possible to extend Japan's military, political, and economic control over a vast area. In this sense, the new Japanese empire after the 1930s had become a technologically imagined community. In short, techno-imperialism had entered its golden age.

In his classic survey of Japanese imperialism, historian W. G. Beasley observed that Japan conceived of the East Asian Co-Prosperity Sphere as a type of informal empire that served the wartime needs of the Japa-

86. "Nichi-Man-Shi keizai kensetsu yōkō" (Cabinet resolution of October 3, 1940), in *KSSS* 4: 1083–85.

87. Reprinted in *DDJS* 3: 739–41; Yūseishō, *Zoku teishin jigyō shi*.

nese economy. To underscore the haphazard nature of Japanese expansion, he emphasized that "there was never a blueprint towards which statesmen and their officials worked."[88] Certainly there was no single masterplan for Japan's wartime empire that was taking shape in the late 1930s. Nor can we equate various blueprints for Japan's imperial telecommunications network with that of the empire in general. Nonetheless, these plans for the imperial telecommunications network in East Asia speak volumes about the nature of Japan's new expansion in Asia. The envisioned East Asian telecommunications network reflected Japan's ideology of a new regional order, eventually presented as a co-prosperity *sphere* (*ken*) of Asian solidarity with Japan as the leader.

As we have seen, Japan's vision of a New Order in East Asia in the late 1930s was accompanied by a new vocabulary that linked national power and empire-building with mastery of time and space. As Japan's policymakers and propagandists increasingly stressed the need for integrating the vast territories under its control, discussion of Japan's new vision of its imperium in Asia had become inseparable from technologies that "overcame distance"—physical as well as psychological. These technologies would not only strengthen the "bonds" between the regions and peoples in the co-prosperity sphere but also fortify Japan's position as its leader as well. The belief in the transformative power of communications is indicative of nothing less than a conceptual revolution. By the end of the decade, the Japanese were no longer speaking of mere "links" between the imperial center and individual colonies or even between Japan and Manchukuo. Japan was now pursuing a comprehensive "communications network" for the entire East Asia region. Here the change was not just a matter of terminology or scope; it was not even strictly technological. The change from "link" to "network" reflected a leap in the conceptualization of Japan's imperium. Thus, it can be said to be a paradigm shift, signaled by the emergence of what one historian recently called the "modern techno-strategic paradigm."[89]

Discussion of Japan's discourse on space and empire raises the question whether such modernist, optimistic rhetoric was for propaganda purposes only. There is no question that the exploitation in domestic

---

88. Beasley, *Japanese Imperialism*, 233.
89. Griset, "Technical Systems and Strategy," esp. 69–71.

propaganda of Japan's advances in communications technology and their great potential for building Japan's new empire helped build confidence and create enthusiasm in the public and served as a powerful tool for mass mobilization. But that is not all. These concepts were first used for the most part in confidential internal memos within the MOC and were only later deployed to convince other bureaucratic groups as well as the Imperial Diet when legislation and funding were involved. References to future expansion occasionally had to be struck from the Diet record to maintain confidentiality. Moreover, even propaganda could have an impact on policymakers themselves, a phenomenon political scientists call "blow-back." Indeed, an unprecedented confidence in technology seemed to have taken hold among Japan's policy elites by 1940.

As Japanese bureaucrats and engineers came to discover, however, building a network was much more difficult than envisioning it. It is the techniques of building and managing the imperial telecommunications network—both in local areas and as a unified system—to which we turn in Parts III and IV.

A Japanese woman, perhaps a member of the Resident Association, on guard outside the Chinese Northeastern Wireless Station in Mukden after Japanese takeover, late 1931.

Beginning of wireless telephone link between Japan and Manchukuo, August 1933. Prime Minister Zheng Xiaoxu of Manchukuo is the second from left; second from the right is General Hishikari Takeshi, commander of the Kwantung Army.

Wireless facilities at Shinkyō built by the
Manchurian Telegraph and Telephone Co., 1930s

Matsumae Shigeyoshi in Germany, 1934(?).

Japanese technicians on the construction of the Japan–Manchukuo cable.

Local laborers on the construction of the Japan–Manchukuo cable.

Japanese and Danish representatives at the negotiation over Great Northern Telegraph Co.'s status in Japan, 1940. First from left is H. S. Poulsen; Okumura Kiwao is third from left. Tamura Kenjirō, Japan's Director of Telecommunications, is the second from the right.

Inside an Emergency Telephone Exchange in Japan during the final days of the war.

PART III

Control, 1936–1945

Japan's strength of leadership in East Asia must lie, above all, in its effort to control its vast space, and then in directing various people in the Sphere toward a Greater East Asian consciousness.

—Sugitani Hidenosuke, 1942

Control has a double nature. One is the notion of control as exogenous, imposed, abstracted and rationalized. The second is the notion of control as endogenous, as communicative and shared. Their twin histories, the one that of tools, weapons, techniques and structures, the second that of language, of commonality, of self-regulation and nurture, run in parallel.

—G. J. Mulgan, 1991

CHAPTER 6

Negotiating Control at Home

On an inspection tour in 1936 near Kaesong, Korea, Governor-General Minami Jirō had an unpleasant encounter: at a relay station of the new Japan–Korea–Manchukuo cable then under construction, Minami became unhappy with the words "Ministry of Communications" (*Teishinshō*) inscribed on the stone marker. Also disturbed, Yamada Tadatsugu, chief of the GGK Communications Bureau, proposed changing "Ministry" to "Bureau," thus making the construction appear to be a GGK project. An MOC engineer at the scene objected, on the grounds that the facilities belonged not to the Bureau of Communications in Korea, but to the MOC in Tokyo. Upon hearing this, Governor Minami became enraged: "Korea is under the jurisdiction of the governor-general, directly appointed by the Emperor. What authority does the minister of communications have to build facilities here?" In the end, as a compromise, the word "ministry" was dropped from all the markers, leaving only "Communications."[1] Ambiguity thus became the temporary solution.

This episode reveals something more profound than semantics. The MOC-supervised long-distance communications project rekindled an old conflict over jurisdiction between Japan's colonial administration in Korea and the government in Tokyo. The problem was not limited to Korea alone, however. Although Japan had acquired the technological capacity to link all its telecommunications facilities into a single network, as the MOC engineers realized, it lacked the administrative authority necessary to build that network or make it function properly.

1. Murakami, *Ichi gijitsusha no shōgai*, 373–74; Teishin gaishi kankōkai, *Teishin shiwa* II, 250; Matsumae Shigeyoshi ronbunshū kankōkai, *Hatsume e no chōsen*, 33.

The episode in Korea thus raises the bigger question whether a second-tier ministry such as MOC or even the Tokyo government was capable of implementing the ambitious plans of an imperial telecommunications network in East Asia.

Two related developments in the late 1930s seemed to favor the ministry. Through an emphasis on planning and state interference, the government was moving Japan toward a "control economy," aimed at coping with the new world economy and at fortifying Japan's capability for war.[2] A number of Japanese bureaucrats who had worked on economic planning in Manchukuo went on to create "comprehensive policy agencies" such as the Cabinet Planning Board in Japan. The 1939 promulgation of the National General Mobilization Law, in particular, marked a major step toward a full war economy.[3] This trend, together with new plans for telecommunications expansions in the empire, emboldened the MOC bureaucracy. By then, a new breed of "communications men" had come of age, imbued with a sense of mission that they associated with telecommunications and other public utilities in the overall national rejuvenation. Together with their fellow-travelers in other government ministries, they came to reject *laissez-faire* capitalism as well as party politics. Nationalization of the electric power industry, masterminded by former MOC bureaucrat Okumura Kiwao and executed by the MOC, is probably the best known example.

It was against such a political and economic backdrop that officials in the previously lowly MOC adopted a number of high-profile initiatives aimed at restructuring telecommunications at home. But in attempting to assert greater control, they encountered resistance and often generated new conflict as well. As they soon discovered, the technology of empire turned out to be as much about social and political restructuring as about electrical engineering.

---

2. For works on Japan's wartime economy, see Nakamura Takafusa, "The Japanese War Economy as a 'Planned Economy'"; and Okazaki Tetsuji, "The Wartime Institutional Reforms and Transformations of the Economic System," among others, in Eric Pauer, ed., *Japan's War Economy*. An older but still useful work is Cohen, *Japan's Economy in War and Reconstruction*, esp. 10–109.

3. Furukawa, *Shōwa senchūki no sōgō kokusaku kikan*.

CENTRALIZED CONTROL AND ITS ENEMIES

Advocates of Control

Despite the MOC's broad range of activities, as one senior MOC official lamented in the early 1930s, the Japanese public did not understand its actual operations, not just because communications services often took place behind the scenes, but because "MOC employees simply worked studiously out of a sense of responsibility and hardly came forward to the public." This, he regretted, had a harmful effect on the operation of communications services.[4] Still, when Okumura Kiwao graduated from the prestigious Law Faculty of Tokyo Imperial University in 1925, he chose to work for the MOC because of its close connections to "public life and its cultural function." Okumura had reasoned that public utility enterprises (*kōkigyō*) would become more and more important as capitalism experienced a crisis and the "control economy" became necessary to Japan. "As a typical bureaucracy of the public utility administration," he found the MOC attractive.[5] As Okumura had predicted, the situation began to change when the Japanese economy began to move toward greater planning and control. Increasingly, the ministry would come out from under the shadow and raise its profile, and MOC officials often found themselves at the center of the action. Indeed, Okumura Kiwao would emerge as one of the best-known figures of this group of "reform bureaucrats."

The concept of control (*tōsei*) that became popular in Japan in the 1930s, as used in terms such as "control economy," has been attributed mostly to those reform bureaucrats who had been influenced by the Soviet Union or Nazi Germany. Largely overlooked is the fact that control was, above all, a matter of technology: for complex technical systems such as telecommunications networks to function properly, control was essential.

In June 1936, after construction began on the Japan–Manchukuo cable, a section was created within the MOC's Engineering Bureau that began making arrangements for the maintenance of this trunk cable.

4. Makino Ryōzō, *Tokubetsu kaikei to natta tsūshin jigyō*, 4.

5. Okumura Kiwao, *Teishin ronsō*, 1.

This Japan-Manchukuo Telephone Construction Section issued a report that urged a unified maintenance and control system under the MOC for the entire line. The report offered a number of justifications.

First, the report noted that telecommunications, unlike other forms of communications, operated in an instant regardless of physical distance. This called for rapid response in time of disruptions, which could be accomplished only under unified control. Therefore, the report argued, "With the gradual opening of long-distance circuits in Japan, the colonies, and Manchukuo, the maintenance and control of these circuits have become most important in terms of technology." Second, the report claimed that the new non-loaded cable, in contrast to the open-wire line of the Meiji period, was "a unique Japanese technology a step ahead of advanced countries in the West." And precisely because it was so technologically sophisticated, it required maintenance by superior (i.e., MOC) technicians to prevent problems from occurring. Third, the special technological character of long-distance communications had to be taken into consideration. The Engineering Bureau's own test of the 1,200-km Tokyo–Fukuoka section of cable had shown that the frequency of accidents on long-distance cables increased by the square of the cable's length. The unprecedented length of the Japan–Manchukuo cable thus called for a higher level of control. Moreover, the carrier technology used on the non-loaded cables throughout the network meant that a single error would affect all circuits. In other words, the need to prevent mistakes was much greater than with other forms of telecommunications.[6]

The prospect of multiple centers of control over the planned long-distance network was abhorrent from the MOC's standpoint. During the early development, such an arrangement might be appropriate for metropolitan communications or networks between adjunct regions, MOC bureaucrats argued, but it was not suited to a long-distance communications network linking East Asian countries over thousands of kilometers. The Engineering Bureau called for a plan that was "not bound by local interests but based on the common interests in East Asia." It would not be subject to local financial constraints either. Above all, such a network was to be based on a "unification of tech-

6. "Nai-Sen-Man renraku keburu kaisen no hoshu tōsei o hitsuyō to suru riyū" (1936), 1–6. MOC Records I–A1.

nology" (*gijutsu no tōitsu*)—namely, the independently developed NLC technology that had no parallel in the world. "Unification of construction and maintenance is an absolute prerequisite," the Engineering Bureau emphasized, "since the consolidation and operation [of the East Asian network] is impossible under the current multiple agencies."[7]

Unified control and operation would not only reduce accidents and interruptions but provide other economic benefits as well. For example, the MOC calculated that using joint rather than separate cables between Dalian and Harbin would save more than 30 percent in construction and maintenance costs. Moreover, as a result of the "unification of technology," manufacturers would be able to stabilize planning, resulting in technological progress and reduced material costs. Finally, as a result of consolidated manufacturing, Japan would be liberated from the constraints of foreign technology and could actively export its domestic technology.[8]

To strengthen their case for unified control, MOC officials often pointed to foreign models. In early 1937, when Matsumae Shigeyoshi returned from a three-month tour of telecommunications facilities in Southeast Asia, he praised the British model of telecommunications administration in its Asian colonies. Matsumae reported that although telecommunications were under private operation in Britain's smaller direct possessions of Singapore, Malacca, and Penang, they were under government administration in the British protectorates on the Malay Peninsula, and in India all communications matters were under the direct control of the governor of India. As Matsumae observed, "This is a very noteworthy matter, since in communication you cannot have separate administrations. Although such a form [of unified administration] may be taken for granted in an old imperialist country like Britain, few realize immediately that the format used in our country, where Korea and Taiwan have their separate operations, is strange."[9] The MOC report also repeatedly evoked other examples of unified control of their telecommunications, including AT&T in the United States as well as the French and the Soviet systems.

---

7. Ibid.
8. Ibid.
9. Matsumae, "Nan'yō shoppō no shisatsu o oete," *DTGZ*, June 1937, 499.

Because the current level of maintenance was far from sufficient for the 2,700-km cable from Tokyo to Mukden in Manchukuo, the MOC report concluded that "for the higher purpose of future East Asian telecommunications, technical control and maintenance of the Japan–Korea–Manchukuo cable circuits should be entrusted directly to the MOC." In this way, the long-distance circuits inside Japan under MOC jurisdiction could be seamlessly joined to "satisfactorily fulfill the function of our important national policy of building a communications artery." The report also recommended that technicians engaged in maintenance for the entire line follow a single chain of command under MOC supervision. In addition, they proposed creating a powerful new institution and making sweeping revisions to the existing treaties and ordinances governing telecommunications between Japan and its colonies as well as Manchukuo and China. For such a sophisticated system to function fully, MOC's technology bureaucrats called for "coordination and control" (*renraku tōsei*) of its planning and operations.[10] Apparently, construction of the world's longest NLC link gave much-needed confidence to those striving for greater assertiveness.

In 1939, the Telecommunications Committee reaffirmed MOC's reasoning for such a requirement: (1) the cable facility was to be an apparatus of national policy aimed at driving out Western communications interests in East Asia and thus must be planned with a holistic view of the entire region; (2) the long-distance and particularly the highly sophisticated technology used required not only unification of technical measurements but also the control of technicians to realize the capacity of the network fully; and (3) because the cable required considerable investment, due to the dictates of national policy (i.e., military and political needs), certain unprofitable routes must also be built and then incorporated with routes with good returns so as to form a single, financially sound unit across Japan, Manchukuo, and China.[11]

---

10. "Tōa yūsen denki tsūshinmō no kakuchō seibi ni kansuru ken," MOC Records I, Te28n.

11. "Tōa ni okeru tsūshin keburu kansen no seibi ni kansuru ken" (1939); "Tōa ni okeru tsūshin no tōitsuteki seibi no hitsuyōsei ni tsuite" (1939).

Colonial Autonomy Versus Imperial Integration

The new Japan–Manchukuo telephone cable was to run through Japan, the colony of Korea, and the newly created Manchukuo. Although intricately connected by the cable, the three areas had separate administrative apparatuses as far as communications were concerned—the MOC in Japan proper, the Bureau of Communications in Korea, and the Ministry of Communications in the Manchukuo government. As MOC bureaucrats aggressively pursued their designs of unified control, the impact of Japan's technological innovation and new imperial agenda reached bureaucracies outside the home islands as well.

The Government-General of Korea had a special status in Japan's colonial empire. Unlike other colonies, Korea's governor-general was in theory "under the direct supervision of the emperor." It is said that because a GGK section chief had a direct relationship with the emperor, he often felt superior to his counterparts in Tokyo. Although a number of changes after 1919 sought to reduce Korea's colonial autonomy—such as the 1929 establishment of the Ministry of Colonial Affairs, charged with overseeing all five colonies, including Korea—resistance to centralization remained strong.[12] Since the MOC did not have the jurisdiction over Korea that it had over Karafuto or Nan'yō, telecommunications matters involving the home islands and the colonies had been resolved by bilateral agreements between government agencies. In the early 1930s, for instance, the MOC and GGK had agreed that the MOC would pay a fee for the GGK maintenance work on proposed telegraph lines connecting Japan and Manchukuo that passed through Korea.[13]

The new cable network was a different story, however, in large part due to its strategic importance and its highly sophisticated technological nature. As a result, construction of the Japan–Manchukuo NLC

---

12. On the legal framework of Japan's colonial administration and the special status of the GGK, see Edward I-te Chen, "The Attempt to Integrate the Empire: Legal Perspectives," esp. 262–66. On status consciousness, see Teishin Dōsōkai, comp., *Senpai ni kiku*, 666.

13. "Chōsen keiyū nai-Mankan denshin senrō kensetsu hoshuhi nado no futankata ni kansuru ken" (August 22, 1933), MOC Records II-149.

network was not simply an epoch-making technological feat but created unprecedented administrative problems as well. As soon as the construction of the cable got under way in Korea in 1936, the issue of MOC intrusion into GGK territory surfaced. In April 1936, Matsumae Shigeyoshi and Watanabe Otojirō went to Korea as representatives of the MOC and signed an administrative agreement with the GGK over the construction of the new cable. Already some had suggested the need for an imperial ordinance in order to put the MOC-directed construction on a sound legal footing. This was dismissed by the outspoken Watanabe, who commented sarcastically that such an arrangement was tantamount to "proposing marriage to a girl after raping her."[14]

While the Korean portion of the construction was proceeding, the MOC and the GGK continued to be preoccupied with making arrangements for its maintenance. Whereas construction was a one-time event, maintenance required the stationing of MOC technicians permanently along the cable route in Korea and thus had grave implications. In early October 1936, Matsumae Shigeyoshi once again accompanied other MOC engineers, including the chief of the newly created Japan-Manchukuo Telephone Construction Section, on a mission to Keijō to discuss maintenance of the new cable with GGK officials. Matsumae and other MOC engineers repeatedly cited examples of unified control of long-distance networks in advanced countries in the West, and reiterated the necessity for a unified system under MOC control because of the highly complicated technical nature of the NLC. The MOC's proposal proved unacceptable to the Bureau of Communications of the GGK, which insisted that unified maintenance be carried out in the home islands first before being extended to Korea and that Korea be allowed to try doing its own maintenance. The GGK also wanted the MOC to promise to bring about unified control in Manchukuo. Because the GGK was planning to upgrade its Pusan–Keijō telephone

---

14. Murakami, *Ichi gijutsusha no shōgai*, 374. He attributed the remark to Watanabe Otojirō, who had just returned to the MOC after serving in the GGK for a number of years. Yamada was a 1917 graduate of Tokyo Imperial University and joined the MOC the same year. Having previously worked in Korea, Yamada took up the post of GGK Communications Bureau Chief in 1936, after serving for two years as chief of Sendai Communications Bureau in Japan; see Yamada, "Imaida san no omoide," in Imaida Kiyonori denki hensankai, *Imaida Kiyonori*, 776–79.

lines to cables, it welcomed the MOC project. At the same time, the colonial administration in Korea made no secret of its wariness of the intrusion of the home bureaucracy's authority. MOC bureaucrats were well aware of this. As one explained to his colleagues in Tokyo later, "The extension of MOC authority to Korea is a problem of invasion of their administrative jurisdiction, which has always been feared there."[15]

Over the next few days, in a total of eight meetings, Matsumae and other MOC officials stepped up the pressure and engaged in heated arguments with GGK officials. Relentless arm-twisting and also the promise to lease surplus circuits to the GGK finally led Korea's Bureau of Communications to accept the MOC plan. Both sides tentatively agreed to define the cable as "only passing through Korea"; thus defined, it would not affect GGK jurisdiction in any way. The MOC's Engineering Bureau was allowed to set up maintenance offices in Keijō and Pusan, staffed with MOC employees. But the GGK insisted that it would transfer none of its authority over telegraph and telephone administration in Korea to the MOC even for maintenance work; instead, it would ask the MOC to provide "labor-like maintenance activities."[16] Rather than being reimbursed by the MOC for maintenance work on the existing telegraph lines, the GGK would now pay the MOC to maintain the soon-to-be completed long-distance cable. At least on paper, the GGK's jurisdiction remained intact. And both parties agreed to share the costs, beginning in fiscal year 1939.[17] Eventually, an imperial ordinance was issued in Tokyo to legitimize limited MOC operations in Korea.

The tension between the MOC and the colonial administration did not dissipate entirely, however. The conflict between the two was more than a petty rivalry between bureaucracies. There was also a new dimension to the struggle for control of the Japan–Korea–Manchukuo long-distance cable. Japan's new imperial telecommunications network would exacerbate political tensions between the metropole and the periphery throughout the empire.

---

15. Statement by Chief of Inspection Section of MOC Tejima at a Tokyo meeting on November 18, 1936. MOC Records I, Te28n; *DTJGKS*, 17.

16. "Nai-Sen-Man renraku denwa no hoshu ni kansuru uchiawase kaigi gijiroku" (October 5–10, 1936), MOC Records I, A1.

17. "Nai-Sen-Man renraku keburu shisetsu ichi ni kansuru ken" (November 1937), MOC Records II, 710.

In reality, upgrading telecommunications facilities also became an increasingly important priority for the colonial government due to industrialization at home and trade activity abroad. At the 1936 Conference on Korean Industrial and Economic Policy, business leaders from Korea and Japan lamented high telegram rates and scarce telephone service.[18] Expansion of telecommunications in Korea was a major agenda item at the conference convened by the GGK in September 1938 to cope with the new conditions in East Asia following the outbreak of the Sino-Japanese War. The colonial government admitted that inadequate telecommunications facilities in Korea were causing numerous delays and poor service. The conference recommended a wide range of expansions of telecommunications facilities: extension of telegraph and telephone lines, conversion of bare wires to more secure cables, adoption of high-speed telegraphic equipment, and improvement of special communications services for aviation, weather forecasting, and shipping. In the meantime, control was to be strengthened over all forms of communication, by means of censorship and radio intelligence-gathering. Given radio's great role in educating the people, unifying opinion, and stabilizing the public order, the GGK pointed out, broadcast facilities must be greatly expanded, and public radio receivers and high-power radio stations to broadcast to the Soviet Union and interfere with Soviet broadcasts to Asia must be established.[19] In response to the growing business concerns voiced at the conference, in 1939 the GGK established radiotelephonic links between P'yongyang, Pusan, and Keijō in Korea, and Beijing, Tianjin, and Shanghai in China.

The GGK was not the only obstacle toward realizing MOC's plan of unified control, however. It was no accident that the GGK insisted that the MOC bring about unified control in Manchukuo. The Japanese puppet state also objected to the MOC's plan for direct control over the entire cable network. Understandably, none of the major Japanese interest groups in Manchukuo, including the MTT and the Kwantung government, was enthusiastic about the MOC's proposal, outlined in its draft response to the "First Policy Inquiry to the Telecommunications Committee." The Manchukuo government requested that Japan help strengthen the international communications network in Manchukuo

18. Chōsen sōtokufu, *Chōsen sangyō keizai chōsakai kaigiroku*, 555–58.

19. Chōsen sōtokufu, *Chōsen sōtokufu jikyoku taisaku chōsakai shimon tōshinsho*, 55–61.

but not request any funding from it. It insisted that the operation of the long-distance cable—limited to Andong–Mukden–Shanhaiguan—should be "placed under the absolute control of MTT." Like the GGK, the Manchukuo government also demanded an "understanding" that Japan would accomplish unification of trunk line maintenance in the colonies before extending such unification to Manchukuo.[20]

The Kwantung Army, which had previously battled the MOC for effective control of telecommunications in Manchukuo by creating the semi-private MMT, agreed to the principle of coordinating construction and maintenance work. It found the idea of unified control over the East Asian long-distance cable unacceptable because of the "current military requirements in Manchuria" and "the special character of communications administration in Manchukuo." Instead, the Kwantung Army proposed that the MTT be responsible for the entire portion within Manchukuo, from Andong to Shanhaiguan. Although cable manufacturers and construction companies in Japan would be contracted for future work on the network in Manchukuo, the Kwantung Army stipulated, personnel from Japan must work as temporary MTT employees. To fend off a possible MOC complaint that the MTT lacked adequate financial resources, the Kwantung Army suggested that the Manchukuo government subsidize the MTT in this endeavor.[21]

Unlike the colonial bureaucracy in Korea, the powerful Kwantung Army was unlikely to yield to MOC pressure. As a result, demarcation of bureaucratic authorities approached absurd levels. Even the relatively minor issue of exactly where the MTT portion of the cable (belonging to the MTT) began and the MOC portion (government property) ended would become a matter of serious contention. Manchukuo officials insisted that the dividing line be in the middle of the Yalu River separating Manchukuo and Korea. Since this was technically impossible, an MOC engineer recommended connecting the two lines at the Andong Telephone Office on the Manchukuo side of the border. This would have allowed the MOC to do maintenance work

20. Various views expressed by Japanese interests in Manchuria are included in Kantōgun, Sanbobu, Dai-4-ka, "Tōa chōkyori tsūshin denran shisetsu no tōitsu seibi ni kansuru ken" (November 28, 1938), MOC Records II-95.

21. "Tōa chōkyori tsūshin denran shisetsu no tōitsu seibi ni kansuru ken" (November 20, 1938), transmitted from Kwantung Army Chief of Staff (Isogai Kensuke) to Chief of Kwantung Bureau (Ōzu Toshio) (December 1, 1938). MOC Records II-95.

on a small stretch of Manchukuo territory. Neither side gave in, so the matter was taken all the way to the minister of communications.[22] Such problems of coordination must have been frustrating to officials in Tokyo. Thus it is not difficult to understand why, in a speech to electrical engineers in Tokyo in 1939, Matsumae openly complained about the lack of unification of facilities in Manchukuo.[23] Japan's difficulties in establishing an empire-wide telecommunications network in the late 1930s made it clear that imperial integration would not be attained as easily as technical experts in Tokyo had hoped it would. These problems must also have convinced Matsumae and others in the MOC of the need to fortify the argument for unified control and to seek new solutions.

THE ITC: CONTROL THROUGH RATIONALIZATION

Merging Wireless with Cable

Bureaucratic resistance in the colony and the puppet state to centralized control was not the only obstacle facing MOC officials, however. The MOC bureaucracy also had to cope with the problem of financing an undertaking of this unprecedented scale.

The trend of relying on private capital to fund Japan's international wireless communications began in the 1920s and continued until the beginning of the 1930s. In 1930, the MOC requested a budget to establish wireless telephone connections between Japan and Taiwan, but the Ministry of Finance rejected it because of government budget constraints. Believing international wireless telephone service "a question of national face," MOC officials granted permission for the establishment of International Telephone Company (ITC) in 1933. This was a private company created specifically for overseas wireless telephony with Japan's colonies and other Asian destinations, and hence a tri-

22. Murakami, *Ichi gijutsusha no shogai*, 378–79. The engineer was reprimanded by the MOC for "bypassing the rules," which was one of the reasons he left MOC and joined the newly established NCTT.

23. Matsumae Shigeyoshi, "Tsūshin no gijutsu to seisaku," *DTGZ*, November 1939, 551.

umph of many years of private initiative to operate telecommunications. Thanks to the enthusiastic public response, ITC shares were oversubscribed by 57 times the original number of 50,000 shares.[24]

Technological as well as strategic developments soon raised new questions about the institutional arrangement. To cope with the increasing threat to their extensive cable network posed by the wireless, the British convened the Imperial Wireless and Cable Conference in 1928. The conference recommended a unified system of cable and wireless communications to best serve British national and imperial interests. The following year, major British cable and wireless holdings were merged into a new operating company named Imperial and International Communications, Ltd, which was renamed Cable and Wireless in 1934.[25]

Such developments would soon affect Japan. As the MOC began planning for the Japan–Manchukuo long-distance cable as well as future telephone cable links between Japan and Shanghai and Taiwan, the two forms of telecommunications—cable and wireless—would similarly come into direct competition. By then, shortwave technology had largely replaced the longwave technology, which required greater power output and capital investment beyond the government's capacity. Moreover, the domestic telegraph service tended to lose money whereas the telephone posted profits, but the opposite was true of international wireless service. As a result, the ITC was not in a position to rapidly expand international service.

The MOC found a solution through consolidation. In July 1935 the MOC suggested that the Japan Wireless Telegraph Company (JWT) and ITC merge. Director Hirazawa Kaname of the MOC Telecommunications Bureau told company executives that since the MOC was going to operate long-distance telephone service by cables to Manchukuo and Taiwan, such ITC services would be in direct competition and had to be discontinued. In November 1935, senior executives from both companies were summoned to the MOC minister's office for more detailed instructions. The minister thanked them for their contribution to Japan's international communications autonomy and for reducing

---

24. Kokusai denwa kabushiki kaisha, *Kokusai denwa kabushiki kaisha jigyō shi*, 67.
25. Barty-King, *Girdle Round the Earth*, 203–27; see also Pike and Winseck, chap. 10.

Japan's payments to foreign companies. He then suggested that not only should wireless telegraph and wireless telephone be consolidated, but cable communications should be considered as well. He cautioned that Japan could not be satisfied with the status quo in its overseas submarine cable connections. He then added that, in view of the "special relationship" between Japan and neighboring China, a merger of the two companies would better prepare Japan to expand telecommunications into China "on a significant scale" in the future.[26]

Neither company was thrilled about the proposed merger. In particular, the ITC complained that the MOC order was tantamount to breaching the government's own promise made just a few years before and thus was damaging to the interests of its private shareholders. The president of JWT did not oppose absorption of the ITC because this would enable his company to expand into international wireless telephone service, a request denied a few years earlier. He was unprepared, however, to consider inclusion of cable telegraphy in the new venture. Not surprisingly, both companies were opposed to the prospect of increased government control through MOC appointment of executives in the new company. And both opposed the suggested reduction of the government's annual financial subsidies.

In the end, both companies gave in under MOC pressure and agreed to the merger, although considerable differences remained. It took lengthy negotiations, followed by the intervention and mediation of prominent business leaders for the two companies to iron out stock price appraisals and agree on the organization of the new company. The ITC, the only completely privately owned telecommunications company in Japan, was absorbed by the special semi-governmental JWT after a mere four years in existence. The new company, called the International Telecommunications Co. (also ITC), was formally launched in March 1938. The biggest winner of all, as these business leaders had rightly feared, was the MOC. Although all other top executives came from the two companies, Ōhashi Hachirō, a rather colorless former MOC administrative vice minister, became its president. The government also increased its share of the company's revenue from 10 to 25 percent, citing its rising operational costs as well as the higher profit margin when the wireless companies switched from the

26. Kokusai denwa kabushiki kaisha, *Kokusai denwa kabushiki kaisha jigyō shi*, 129–30.

costly longwave to shortwave transmitters.[27] The newly merged ITC continued to provide facilities for NHK overseas broadcasting, which began in 1935 under a government subsidy. The broadcasting would be expanded in subsequent years, with a new transmission station completed in 1940. The merger not only solved the potential conflict between Japan's own international wireless telephone and telegraph services, it also gave the government greater control of overseas telecommunications.

ITC for the East Asian Network

The ITC merger turned out to be only the first step, however. No sooner had the merger between the JWT and the ITC been consummated than the plan for an East Asian telecommunications network called for a complete reorganization of the new company. Several factors seemed to be involved, primary among them the fact that finances continued to be a problem for the MOC. Technological justification also played a big part, as did political convenience.

By the beginning of the 1930s, it had become clear to MOC officials that the solution to its chronic financial problems lay in some degree of control over the ministry's revenues. Hence the earlier scheme of a Special Account for Communications Service (SACS) gained new popularity. As a senior MOC official explained to a group of Tokyo businessmen shortly after its passage in March 1933, it became the "only medicine, even if it is not a panacea."[28] Under this scheme, the MOC agreed to pay up to 82 million yen to the national treasury each year—the equivalent of the MOC's total annual revenues in 1932. In return, it was allowed to keep whatever remained for telecommunications expansion both at home and abroad. The SACS was widely viewed as a victory for the MOC, having resolved with one stroke a 30-year-old problem. It not only strengthened the MOC bureaucracy vis-à-vis the Ministry of Finance but also gave the MOC more autonomy vis-à-vis

27. For a cost analysis of using longwave versus shortwave in international telecommunications, see Kokusai denki tsūshin kabushiki kaisha shashi hensan iinkai, *Kokusai denki tsūshin kabushiki kaisha shi* (hereafter *KDTKKS*), 442–53.

28. Makino Ryōzō, *Tokubetsu kaikei to natta tsūshin jigyō*, 20.

the Diet.[29] As it turned out, however, the SACS alone could not solve all the financial problems facing telecommunications expansion, especially with the ambitious new East Asian telecommunications network. The financial burden remained significant and was made worse by a compulsory additional annual contribution to military operations after 1938, under the Provisional Military Special Account.

Added to the financial problems at home were, of course, the political complications of operating an empire-wide network outside Japan proper, as we have seen in the cases of the colony of Korea and Manchukuo. Political realities in China also posed a challenge, since the appearance of an independent and sovereign China under Japan's leadership was considered an important principle of Japan's Asia policy. All these situations called for innovative organizational solutions. Unified control over the East Asian telecommunications network—the dream cherished by MOC engineers and bureaucrats—would have to be modified or realized in some other way. Although favoring a greater role for the government, these advocates of state control were nonetheless fully aware of its limits, especially in financial terms. As a first step, they sought to rationalize telecommunications operations within Japan. In a clever move, the MOC accomplished this via a semi-private enterprise.

As early as November 1937, the MOC Engineering Bureau had drafted a proposal suggesting the establishment of a private institution—named the East Asian Long-Distance Cable Facility Company—for future construction and maintenance work for the new cable network. Enjoying corporate status in all countries concerned, the company would receive appropriate protection from the governments of those countries, although its capital would come mainly from Japan.[30] That such an idea came from MOC engineers was not surprising, given the difficulties they had with the GGK the previous year. Such an approach gained currency when the Telecommunications Committee began formulating a comprehensive policy for East Asia. By the beginning of 1939, the MOC had drafted an "Outline for Consolidating the East

29. The official biography of Ōhashi Hachirō, MOC vice minister at the time, attributed the resolution in part to the end of the party politics that began with the Saitō Cabinet of 1932; see Ōhashi Hachirō denki hensan iinkai, *Ōhashi Hachirō*, 161–62.

30. "Tōa kokusai chōkyori yusen denki tsūshinmo tōsei yōkō an" (November 24, 1937), as reprinted in *DTJGKS*, 206–7.

Asian Cable Communications Network," with details on how to implement the new East Asian telecommunications policy outlined by the Telecommunications Committee.[31] As part of its First Policy Recommendation to the Telecommunications Committee, issued in early 1939, the outline suggested the principle that "construction and maintenance of communications cable inside our country that forms part of the East Asian telecommunications network be entrusted to an appropriate private institution [throughout] the home islands and the colonies, for use by the government":

> Since this communications network requires immediate consolidation by means of large amounts of funding and abundant construction capability, it is necessary to entrust it to an institution that can raise the needed capital easily and that possesses sufficient construction capability. In view of the above, it is considered ideal to adopt the principle of entrusting the unified construction and maintenance of the East Asian telecommunications network to an appropriate private institution for the use of telecommunications operators in the areas concerned.[32]

In supplementary documents, MOC again detailed the rationale for "unified construction and maintenance of the East Asian telecommunications network." To fulfill this unprecedented mission, the committee considered a single private company to have the following advantage: since some of the routes were to be built for military purposes and thus would not generate enough commercial traffic, a private enterprise could balance them with profit-generating routes.[33]

Instead of creating a new company from scratch, the MOC had now found a new use for the recently consolidated ITC. By mid-1939 the MOC completed the planning of construction and maintenance for the unprecedented endeavor of an East Asian long-distance telecommunications network. It had to project needed capacity, material, and funding. Originally conceived as the sole operator of international wireless communications in Japan, the new ITC was to be given the mandate of

---

31. MOC, "Tōa tsūshin kēburu mō seibi yōkō" (May 1939), MOC Records II, 693; "Tōa denki tsūshinmō seibi yōkō setsumei shiryō" (May 1939), MOC Records II, 86.

32. "Denki tsūshin iinkai tōshin" (January 16, 1939); "Tōa denki tsūshinmō seibi yōkō" (January 1939), MOC Records II, 693.

33. MOC, "Tōa denki tsūshinmō no tōitsuteki seibi no hitsuyōsei ni tsuite" (n.d.), MOC Records II, 693; "Tōa denki tsūshinmō seibi yōkō setsumei shiryō" (May 1939), MOC Records II, 86.

expanding into overseas cable communications as well. Specifically, it was to be responsible for the construction and maintenance of the planned East Asian long-distance trunk cable. Expansion of the cable network would be the top priority of the company for the next five years, at a record total cost of 122 million yen. The government had obviously realized that such an ambitious scheme required a large amount of capital and thus was impossible with government funds alone.[34]

Although the new plan put great emphasis on cable expansion in areas that came under Japanese control, expanding wireless facilities for international communications over a five-year period was also considered an urgent need. In view of international developments in Europe and Asia, the new plans called for strengthening overseas broadcasting, expanding communication links with cities in Southeast Asia. Moreover, duplicating overseas wireless communications centers in both Tokyo and Osaka was considered necessary in time of war whereas previously Tokyo had been responsible for communication with the United States, the South Seas, and the Far East, and Osaka had dealt with Europe. Other assignments included sufficient worldwide communications for the upcoming Tokyo Olympics in 1940. The ITC was to take up these tasks as well.

The government also strengthened its "protection" of the ITC by granting it many privileges. These included raising the ceiling on bonds the company could issue to three times the limit set by Commercial Codes, backed by government guarantees; increasing its share of the company's capital but giving preference to privately owned shares in terms of dividend payments; and various tax exemptions for a period of ten years, beginning in 1940. As a tradeoff, however, the ITC's autonomy vis-à-vis the government was further eroded. A government supervisor (*kanri-kan*) was to be installed in the company; with the power to veto company resolutions and to fire executives, he would oversee the ITC's borrowing and its business plans. The MOC's influence would now extend to include company finance as well as personnel policies.[35]

---

34. *KDTKKS*, 28–29.

35. Tamura Kenjirō, "Tōa denki tsūshin seisaku to Kokusai denki tsūshin kabushiki kaisha no kakujū ni tsuite," *TKZ* 369 (May 1939): 38–42; Hanaoka Kaoru, "Tōa kokusai chōkyori denki tsūshinmō no seibi to Kokusai denki tsūshin kabushiki kaisha no kakujū," *DT* 2.4 (1939).

In the deliberations at the 74th Imperial Diet in early 1939, the MOC justified the ITC's reorganization on the grounds that the MOC would run into difficulties operating telecommunications outside the home islands. As then-Director of the Telecommunications Bureau Tamura Kenjirō emphasized, the East Asian network made it necessary to enter foreign countries in order to build and maintain the new cable network. Given current international conditions, he noted, it would be extremely difficult for the government to do this because it would be tantamount to building Japanese government facilities on foreign territories. The ITC would be the ideal vehicle of expansion into foreign countries. Financial considerations also loomed large. Tamura argued that it would be easier for a private company to raise the capital needed for the kind of rapid expansion anticipated and to make construction economical.[36]

Even some Diet members considered the MOC's approach high-handed and raised objections in committee deliberations. One member of the House of Representatives voiced a common complaint, accusing the government of not doing nearly enough at home despite the poor condition of telephone service in Japan. He also objected to government interference with business management: "Not just in Japan but all over the world, when government officials enter companies, there are bad results." Others questioned the financial soundness of such an ambitious expansion plan. It is interesting that some House of Peers members noted the extremely large allowance for debt issuance and questioned what future expansion plans would be. Apparently a very sensitive subject, Tamura requested that no records be kept of his comments since he would be revealing the government's "draft plans."[37]

Interestingly, Tamura responded that because of the peculiarities of the new technology and the inadequate technical expertise of the private sector, the government had to provide technological guidance. Technological necessity thus provided a convenient justification for increased control, as even the dissenting Diet member had to concede. He pointed out that he had no objection to the involvement of government technicians but was opposed to government bureaucrats "with

---

36. Teikoku gikai, *Teikoku gikai soggijiroku dai-74-kai: Shūgiin* (March 6–11, 1939), 380–409; Teikoku gikai, *Teikoku gikai soggijiroku dai-74-kai: Kizokuin* (March 18–20, 1939).

37. Teikoku gikai, *Teikoku gikai soggijiroku dai-74-kai: Kizokuin* (March 6–11, 1939), 7.

bachelors of law degrees" and other administrative officials descending on the company from government (*amakudari*).[38] Remarkably, due to this member's persistence, the House of Representatives actually added an amendment to the law prohibiting government officials from being appointed company executives within five years of their retirement. Whatever the minor objections raised, by the late 1930s, opposition to overseas expansion in favor of improving domestic service had lost much of its earlier effectiveness. The government had little difficulty getting its reorganization plan approved by both houses in the Imperial Diet.

Apart from seeing the new ITC reorganization law through the Imperial Diet, MOC officials had to negotiate with other Japanese authorities affected by the plan. In an effort to ward off the GGK's concerns, careful language was crafted putting the reorganized ITC under the jurisdiction of the "government"—a more ambiguous term than "the minister in charge," as in the government draft. This implicitly acknowledged the role of GGK and also cleared the way for the colony of Taiwan to be included in the future. To make it more acceptable to operators such as the MTT and NCTT, the MOC prepared a memo outlining the financial benefits of using these facilities.[39] However, when MOC officials visited North China, they still encountered some local resistance to turning facilities over to the Tokyo-based ITC. Such reluctance was mostly financial: the NCTT emphasized the need to ensure its own revenue and asked the MOC to consider providing a subsidy to compensate for its loss of revenue from giving up the cables. The NCTT also wanted no interference with its autonomy of operation, even when it had to commission the ITC to undertake maintenance for the sake of unification.[40] The Japanese military authorities in China expressed stronger reservations about the MOC's plan, demanding more time to study the matter and consult with the Kwantung Army. In particular, the officers complained of "feeling pressed with decisions that

38. Teikoku gikai, *Teikoku gikai soggijiroku: Shūgiin* (March 8, 1939), 3.

39. MOC, "Man-Shi no un'eisha ga Kokusai denki tsūshin kabushiki kaisha no setsubi o shiyō surukoto no rieki" (May 27, 1939), MOC Records II, 86.

40. MOC, "Tōa denki tsūshinmō no seibi ni kansuru ken ni kansuru kyōgi yōkō" (June 17, 1939), MOC Records II, 643.

Table 10
ITC Capital Composition, 1938–43

Share type	Category	Number of shares	Price per share	Paid-in capital
Old shares	Government	46,000	50	2,300,000
	Private	454,000	30	13,620,000
New shares	Government	700,400	50	35,020,000
(1940)	Private	53,600	22.5	1,206,000
SUBTOTAL		346,000	22.5	7,785,000
		1,100,000		44,011,000
2nd new shares	Government	58,000	50	2,900,000
(1943)	Private	58,000	12.50	725,000
SUBTOTAL		116,000		3,625,000
TOTAL	Government	804,400		40,220,000
GRAND TOTAL	Private	911,600		23,336,000
		1,716,000		63,556,000

SOURCE: *KDTKKS*, 457.

have already been made in Japan proper," adding that, given the shortage of material, priority must be given to national defense.[41]

The MOC must have been relieved when the ITC reorganization was completed in July 1940. Some 600 government employees from the MOC and GGK joined the reorganized company. By then, the new company became the sole operator of long-distance communications by wireless and, increasingly, by cable as well. The MOC transferred to the ITC the entire Japan–Manchukuo long-distance cable, including the newly completed non-loaded cables between Tokyo and Nagoya and between Fukuoka and Andong. As a result, the company's total capital increased to 80 million yen, and the value of government's contribution increased from over 14 to 62 percent. The number of government shares became approximately equal to the number of private shares (see Table 10). After 1941, the ITC tried to raise additional capital by issuing bonds totaling 130 million yen over the next five years. Moreover the ITC also served as a conduit of investment into other telecommunications national policy companies in Japan and

41. "Tōa denki tsūshinmō seibi keikaku an ni taisuru genchi gun no iken" (June 18, 1939), MOC Records II, 780.

abroad, with 12 percent of its total capital invested in a dozen or so subsidiaries.[42]

The institutional arrangement created by the MOC continued to have its critics, to be sure. Writing in 1941, Oka Tadao, a researcher with the private Communications Research Association, cited the fear that such a middle-of-the-road solution between government and private operations, although supposed to combine their respective strengths, would end up nullifying them both.[43]

Considering the enormous difficulties it had once faced in bringing about unified maintenance, however, the MOC deserved some credit for having found a marvelous solution. In particular, MOC bureaucrats aptly turned technology to their advantage in their relationship with the private business. By emphasizing its own technological superiority, the MOC gained more control of the ITC—but control that deliberately fell short of assuming complete financial responsibility. The creation of the ITC also rode the new wave of "national policy companies" that emerged after the Manchurian Incident. Before 1931, there were 21 such companies; between 1935 and 1939, more than a hundred were set up in Japan, Manchukuo, and China, with Manchukuo alone boasting 50. Unlike the MTT and other companies created in China, however, the ITC was devoted to maintenance as well as operation of long-distance telecommunications facilities on the cutting edge of expansion. Its area of operation, therefore, was to be as wide as possible. Through direct operation or capital investment in other parts of the empire as well as in the rest of Asia, the ITC would thus become the powerful tool of the MOC, functioning as the core component of a unified communications system in the East Asian sphere.[44]

COMPLETING NATIONALIZATION

From Cooperation to Conflict

Both consolidation of Japan's own overseas telecommunications service in the ITC and Japan's new imperial telecommunications policy

---

42. *KDTKKS*, 456–57, 465.

43. Oka Tadao, *Taiheiyōiki ni okeru denki tsūshin no kokusaiteki bekken*, 490.

44. *KDTKKS*, 457.

had another impact as well. The MOC's success in consolidating its control over overseas telecommunications service and manufacturing industries not only gave the Japanese greater confidence but helped put on the agenda the issue of ending foreign control over Japan's external communications. Although there were always financial reasons for seeking total independence from foreign business interests, Japan's move toward a semi-war economy increasingly made the presence of a foreign company like GNTC on Japanese soil a security problem.

In 1935, the MOC sponsored a feature film entitled *Wireless Breakthrough*, telling the story of a lowly postal clerk named Yamakawa whose lifelong dream of personal success (*shussei*) is to become the head of wireless communications on a cable ship. One day Yamakawa rescues a young woman named Fumiko from harassment at the hands of rascals in the cosmopolitan port city of Yokohama. Yamakawa then has Fumiko spy on a suspicious foreign company named Cobra Trading. To demonstrate the power of new technical advances, Fumiko reports a secret of Cobra's that she has uncovered to Yamakawa via the mobile wireless telephone on a train, while Yamakawa rushes to the large commercial city of Osaka by airplane. Using newly developed phototelegraphy, Fumiko also manages to send a sketch of Cobra Trading's secret hideout. The film ends with the young couple successfully foiling an attempt by evil Westerners to sabotage Japan. To reach a wider audience, the MOC also had the entire film broadcast on radio throughout the country.[45] The MOC must have been quite satisfied with a film that brought together Japan's technological accomplishments, positive images of "communications men" through their depiction by a handsome actor, and a heightened sense of urgency about guarding Japan against foreign infiltration.

This last point was particularly poignant for MOC. Contrary to the popular impression, the era of "unequal treaties" in Japan did not end completely following its victories over China and Russia at the turn of the twentieth century. One last vestige of that treaty system—GNTC's telegraph station in Nagasaki—had been an object of envy and displeasure for the Japanese government since it was built. Great Northern established strict office rules and punished errors with heavy fines but oth-

---

45. *Toppa musen*, script: Suzuki; director: Murata Minoru; cinematography: Aojima, with sponsorship by the MOC, *TKZ* 326 (October 1935): 146–65.

erwise gave its employees considerable freedom. It was able to provide high-quality service that emphasized speed and accuracy. Danish employees received much higher pay than their Japanese counterparts, but Japanese employees nevertheless enjoyed the highest salaries found in Japan, exceeding even those of ranking Japanese government officials. For instance, a Japanese messenger carrying telegrams for the Nagasaki office of the GNTC was paid roughly the same salary as a Japanese official in charge of an entire telegraph office. In fact, the GNTC's Japanese employees were so well paid that some became regular patrons of expensive restaurants and geisha houses.[46] As time went by, the GNTC presence in Nagasaki became a painful reminder of Japan's humiliating dependency in the bygone Meiji era. One ranking MOC official later recalled his indignation while visiting Nagasaki in 1914 at being reminded of Japan's "third-class status" in telecommunications even after the country as a whole had become first-class. As the country sought greater autonomy in its industry and economy after World War I, many foreign-owned companies either withdrew or were bought by Japanese companies but GNTC remained nearly untouchable.

The MOC sponsored the film as part of an attempt to raise public awareness of the role of wireless in international communications as well as domestic security concerns. After the Manchurian Incident, the Japanese government introduced a number of measures to tighten communications censorship at the request of the military. It eventually created a new category of officials, called telecommunications officers (*denshinkan*), within the MOC's Foreign Telecommunications Section. Initially charged with censoring international news telegrams dispatched by foreign correspondents in Japan, their task soon broadened to include monitoring international telephone conversations, deciphering foreign telegrams, and even broadcasting foreign news from Dōmei, Japan's own consolidated "national policy" news agency.[47] As the northern Kyu-

46. Kokusai denshin denwa kabushiki kaisha shiryō sentā, *Aoi umi o hashiru*, 42–43.

47. According to one Japanese official involved in communications censorship, the telegram from U.S. President Roosevelt to Emperor Hirohito was among those delayed under this system. To prevent leakage of national/military secrets, the MOC later installed a special apparatus on international telephones so that the censor could interrupt the conversation immediately. Shortly after the outbreak of the Pacific War, a telecommunications officer cut off a telephone conversation between an army commander in

shu area became a magnet of heavy industries closely associated with military production, concerns over the existence of a foreign communications establishment in their midst were raised with increasing frequency. Under these circumstances, the GNTC Nagasaki office was an obvious target for the MOC. It was perhaps no coincidence that Tamura Kenjirō, a driving force behind the elimination of the GNTC's rights in Japan, had been a director of communications in nearby Kumamoto prefecture and had had ample opportunity to observe the GNTC operation at close range.[48]

Economics continued to provide a powerful rationale for excising the foreign presence in Japan's international communications. Under its agreement with the GNTC, Japan had to pay the company one third of the income generated by all submarine cables between Japan and China, including traffic on the Dalian–Chefoo and Qingdao–Sasebo lines, which were not owned by GNTC. In 1935 alone, Japan paid the GNTC more than 10 million yen in such fees, although the amount fell to less than 6 million yen in 1938 (see Fig. 1). The reduction was partly due to the fact that, at the encouragement of the MOC, the Japanese business community made greater use of wireless in its international communications. Major commercial centers such as Kobe and Osaka set up international wireless promotion associations.[49] Even after the government decided to use land and submarine cables as its trunk imperial telecommunications network in Asia, wireless continued to be an indispensable means for communicating directly with foreign countries, especially those located far away. Although Japan's international communications via wireless had made much progress and occupied an increasingly large share of its international communications traffic, there were still gaps: for example, the British Commonwealth countries, such as Hong Kong, Australia, South Africa, and Egypt, had no wireless links with Japan because of Britain's colonial communications policy.

---

the field and, unknown to him, Prime Minister Tōjō; Shirao Tataki, "Onikko ni natta denkinkan," in Teishin gaishi kankōkai, *Teishin shiwa* II: 572–85.

48. Tamura Kenjirō's comment in a roundtable discussion on May 24, 1940, concerning the takeover of the GNTC in Japan; "Zandankai: 'Daihoku denshin kaisha' no sesshū o kataru (1)" *TKZ* 382 (June 1940): 17.

49. Nihon musen shi hensan iinkai, *Nihon musen shi* 6: 559–60.

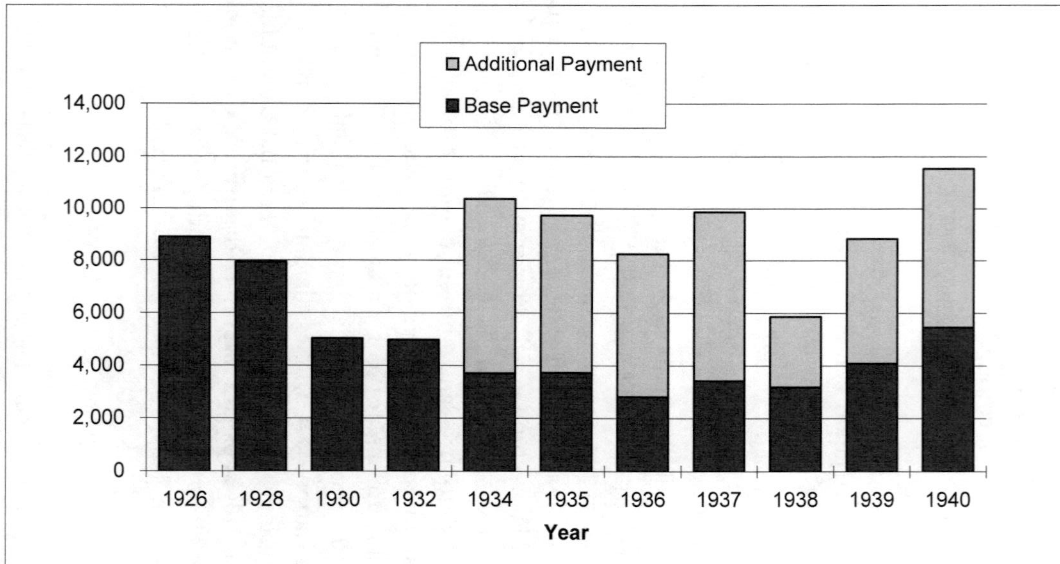

Fig. 1 Japan's payments to foreign cable companies, 1926–40. Additional payments compensated for devaluations of the yen (SOURCE: Nihon musen shi hensan iinkai, *Nihon musen shi* 5: 560).

In fact, it was the MOC's vigorous promotion of wireless in the 1930s that intensified the conflict. Relying solely on submarine cables in an age of increasing competition from wireless, the GNTC was determined to stay in the business. It not only protested repeatedly to the Japanese government but responded aggressively in an effort to recover lost business. In doing so, it overstepped the legal boundaries. On December 10, 1936, the MOC sent a letter to GNTC Far Eastern Manager H. S. Poulsen in Shanghai, alleging that the company was violating its concession by directly dealing with customers by handling messages regarding repetitions, inquiries, etc. It added that judging from the fact that code messages were frequently exchanged between the GNTC officials and the Nagasaki station, the above information seems likely to be true.

The language of the letter was vague as to the consequence, however:

> In view of the long-established relations with your Company, we hope that there might not occur any such irregularities, [which are] nothing short of infringement of the Concession. But if by any chance it may prove true, the matter will be of a grave nature, and so it is requested that you take proper steps at once not to allow them any longer to interfere with the Administration's proper business.[50]

Because the MOC left the allegation unclear, Poulsen simply acknowledged receipt of the letter and assured the ministry that "strictest instructions have been issued."[51] The lack of an explicit denial must have confirmed the suspicions of Japanese officials. The Japanese government began stationing censors in the GNTC Nagasaki station, which served as a check on its activities.[52] Evidence suggests that the MOC also began searching for a permanent solution to this issue in earnest in 1937, when it first conducted research into earlier exchanges between

---

50. Director General of Telecommunications (Hirasawa) to General Manager in the Far East (December 10, 1936), File 610, GNTC Archives, Riggsarchiv, Denmark (hereafter GNTC Records–Copenhagen).

51. General Manager in the Far East to Director General of Telecommunications (Hirasawa) (December 17, 1936), GNTC Records–Copenhagen 610.

52. Kobayashi Takeji, then MOC Chief of International Communications, later commented that GNTC appeared to admit the allegations; "Zandankai: 'Daihoku denshin kaisha' no sesshū o kataru (3)" *TKZ* 383 (August 1940): 50.

the Ministry of Foreign Affairs and the GNTC.[53] It then examined various scenarios for a legal basis on which to eliminate what some Japanese had termed a "longtime cancer in international communications." The flurry of Japanese telecommunications expansion in China probably delayed attention to it for another two years.

Toward a Showdown

As Japan began to implement the new telecommunications policy in East Asia, entrenched foreign interests in telecommunications constituted a major stumbling block. The propaganda pamphlet entitled "Telecommunications Policy in East Asia" noted that the GNTC still took 25 percent of Japan's foreign telecommunications income. In a section devoted to "measures vis-à-vis the Western presence in East Asian telecommunications," the pamphlet argued that given the new political developments in East Asia, existing Western rights now had acquired new meanings and must be interpreted differently in international law as well. Japan should "seize opportunities afforded by political, economic, and diplomatic events" in dealing with such foreign interests.[54]

The outbreak of war in Europe shortly after this pamphlet offered such an opportunity. Emboldened, MOC officials had reached the conclusion that the only way to "eliminate the trouble at its roots" was to terminate the concession granted by the minister of communications.[55] To do so, the MOC decided to accuse the GNTC of illegal business practices in Japan. In justifying its high-handed measure to other government agencies, it listed the company's "detrimental effects" from the perspective both of national defense and of Japan's overseas communications policy. It also cited the "possibility of espionage," given the GNTC's control over overseas communications, even though an internal memo admitted that it would be difficult to substantiate this despite constant surveillance of the company. After the outbreak of the Sino-Japanese War, even with tight surveillance and censorship, there was serious concern about the location of a foreign communica-

53. Horiuchi (Vice Minister, MOFA) to Tomiyasu (MOC Vice Minister) (February 6, 1937), MOC Records II, 369.

54. Teishinshō, *Tōa ni okeru denki tsūshin seisaku*, 140–41, 145–46.

55. "Daihoku denshin kaisha ni kansuru ken" (n.d.), MOC Records II-332.

tions facility near military installations and heavy industries in northern Kyushu. The fact that GNTC agents had visited many "important firms throughout Japan" was suspect, as were reports of frequent meetings between GNTC representatives and British and American consular officials.[56] The MOC also justified the need for immediate action against the GNTC by citing indications that the company was considering preemptive measures, such as a proposal to shore up cable traffic and wireless service between Japan and Denmark. Since the GNTC's rights in China would not expire until 1944, the company seemed to have every reason to delay any change in its status in Japan to the same expiration date as that of Anglo-American landing rights in China. Therefore, Japan must prevent the formation of a united front of all three foreign countries by tackling Danish interests ahead of Anglo-American rights.[57]

Determined to end the GNTC's presence in Japan, the MOC made careful preparations. In a classified document issued in March 1940, it prepared a list of 32 separate questions related to the ending of GNTC rights, such as: Are there any concerns overseas about countermeasures to our moves? Will Denmark bring the case to arbitration? Will Denmark bring a civil suit? The MOC was even prepared for temporary interruption of telegraphic service to areas such as Hong Kong, Northern Borneo, and British West Africa, where GNTC lines were used.[58]

Before confronting the company, the MOC had to obtain full cooperation from other government agencies, which was by no means assured. Although the effort to end GNTC privileges in Japan enjoyed increasingly wide support in the Japanese government, there was disagreement over how to achieve such a goal. The Army, Navy, and Home ministries had no objection to the MOC approach, but the Ministry of Foreign Affairs was much more reluctant. Neither Foreign Minister Arita Hachirō nor Director of the Eurasian Bureau Nishi Haruhiko was supportive of an aggressive policy. Officials in the Treaty Bureau of the Ministry of Foreign Affairs even demanded that negotiations over the

---

56. Teishinshō, "Daihoku denshin kaisha mondai shitsugi ōtai shiryō" (March 1940), MOC Records II, 332.

57. Ibid.

58. "Daihoku denshin kaisha ni kansuru ken," undated, MOC Records II, 332.

GNTC be carried out through their ministry, not the MOC. Through persistent efforts, the MOC gradually persuaded key officials including Vice Minister Tani Masayuki and Yamada Yoshitarō, a section chief in the Eurasian Bureau. It was not until March 12, 1940, that a consensus was reached within the Japanese government. As a compromise, the MOC agreed not to revoke the GNTC's landing rights immediately but to adopt a gradualist approach.[59]

Completely unaware of the MOC's new strategy, the GNTC was still on the offensive. In a memo dated March 11, 1940, the Danish legation in Tokyo made another protest to the Japanese government over the latter's "illegal propaganda against the cable." The protest called attention to a complaint by a Danish firm in Kobe that employees of the "[Japanese] Telegraph Administration do not even refrain from using intimidation in their efforts to persuade business firms to send their telegrams by wireless instead of by wire."[60] This came too late, however. Within days, the MOC struck back. On March 18, MOC officials handed a carefully worded letter to GNTC Far Eastern Manager H. S. Poulsen, who happened to be in Japan to discuss local taxation problems. It accused the company of continuing illegal business practices in Japan, such as direct contact with customers, and demanding a reply within three weeks. Caught off guard, Poulsen asked for an extension to five weeks, citing the Easter holiday in Denmark. A compromise of four weeks was reached.[61] In a cable to GNTC headquarters in London after returning to Shanghai, Poulsen resignedly invoked the "old maxim of [the Great] Eastern [Telegraph Company] that [it was] rarely good policy to fight a powerful concessionary government on its own ground."[62] Great Northern's board of directors in London shared his pessimism and likewise deemed it futile to put up a fight against a government determined to regain its control over foreign companies. It

59. "Zandankai: 'Daihoku denshin kaisha' no sesshū o kataru (1)," *TKZ* 381 (June 1940): 18; Nippon denshin denwa kōsha kaiteisen shisetsu jimusho, *Kaiteisen hyakunen no ayumi*, 346–52.

60. The Danish report alleged that a Japanese official from the Kobe Central Telegraph Office pressured a Danish firm to send telegrams via Japan's wireless service or otherwise "be prepared to receive bad service on their telegrams"; MOC Records II, 327.

61. Teishinshō denmukyoku, *Tai Daihoku denshin kaisha tokkyōjō fuku tokkyōjō jōkan kaitei keika* (June 1940), MOC Records II, 328.

62. Poulsen to London (April 3, 1940), GNTC Records–Copenhagen, 610.

was not at all surprising that the GNTC beat a retreat on all issues. The timing could not have been better for the MOC: its pressure on the company coincided with a deterioration of the Danish position in the European war. When Poulsen returned to Tokyo a month later for further discussion, Director Tamura Kenjirō tactfully informed him, before negotiations began, of the German invasion of Denmark, which had taken place during Poulsen's journey.[63]

With both sides having shown their cards, the final negotiation went surprisingly smoothly. Under the new agreement signed in April 1940, the GNTC would receive a "license" rather than a "concession" from the Japanese government for its business activities in Japan. Its Nagasaki facilities would be taken over by the MOC on June 1, the date the new agreement would take effect. Beginning in 1941, telegrams between Japan and Manchukuo or China would no longer be handled by the GNTC. The company's landing rights in Nagasaki, the basis of its operations in Japan since 1871, would be relinquished on April 30, 1943. The MOC considered this as a compromise on Japan's part, due to the fact that Denmark, a neutral country occupied by Japan's ally, Germany, needed to save face. Whether a graceful exit or not, the agreement ended Japan's dependency on foreign telecommunications interests. It also ended a major symbol of the "informal imperialism" that had been forced on Japan. Following the settlement, an exultant Tamura Kenjirō proudly proclaimed that "a cornerstone has been placed for the construction of the New Order in East Asian Communications," which "bears witness to a powerful Japan making great strides."[64]

Realizing the technological vision of the new Japanese empire involved much more than the work of planners and engineers. It called for creating new organizational structures as well as transforming existing

63. "Zandankai: 'Daihoku denshin kaisha' no sesshū o kataru (1)," *TKZ* 381 (June 1940): 16–29.

64. Tamura Kenjirō, "Daihoku denshin kaisha menkyōjō no kaitei ni tsuite," *TKZ* 381 (June 1940): 2–9. Communications over the GNTC's cables were stopped during the Pacific War and not reopened until 1948. During the postwar period, the company made a comeback and handled about 5 percent of Japan's total international telegraphic traffic; see Hanaoka Kaoru, "Taihoku denshin kaisha no eigyōken," in Teishin gaishi kankōkai, *Teishin shiwa* II: 296–298.

administrative arrangements. As a result, Japan's empire-building project in the late 1930s was accompanied by, and in turn hastened, structural changes at home and in the colonial empire.

As we have seen, the new empire-building mission not only created the need for greater control, justified as a technological imperative, but also injected much vigor into the MOC bureaucracy, which emerged as an active agent in this transformation. In telecommunications, the bureaucrats in the MOC attempted to establish a greater degree of control in operations, maintenance, and manufacturing: (1) they argued strongly in favor of unified central control over the construction and maintenance of the long-distance communications network then being built in Northeast Asia; (2) they called for a greater concentration of telecommunications manufacturing, on the grounds of efficiency and rationality; and (3) they pursued an aggressive course of action to recover control over the remaining part of Japan's external communications, which had been under foreign control since the early Meiji period.

The MOC's record of achievements at home was quite impressive. Years of insufficient funding and the Ministry of Finance's firm grip on the budget compelled MOC bureaucrats to find alternatives to direct state control, although the new requirement of contributing to the emergency military budget after the beginning of the war in China undermined the benefits of such arrangements. Politically, the resistance of the colonial bureaucracy as well as the military forced the MOC to modify its original conception of a highly centralized telecommunications network under its own management. To adjust to these new realities, both financial and political, and carry out its mission, the MOC created and then consolidated the ITC as its own.[65]

In unifying overseas cable and wireless operations, Japan's solution through the ITC was similar to the Cable and Wireless Company in Britain, to which even MOC officials referred in Diet discussions.[66] There were some important differences, however. Japan adopted the ITC at a time when its telecommunications were expanding into new

---

65. "With the realization of increased capital, it is the pride of the Company to become the monopoly in telecommunication in the near future, both in name and in reality" (Noda keizai kenkyūjo, *Senjika no kokusaku kaisha*, 113). For a postwar assessment of the ITC's character by the U.S. Occupation authorities, see SCAP CCS, "Disposal of ITC" (November 27, 1946), 4, RG 331, Box 3188, National Archives, Washington, D.C.

66. Barty-King, *Girdle Round the Earth*, 203–10.

territories, whereas Cable and Wireless was fighting a largely defensive, rearguard battle to protect cable companies. The creation of such a company thus represented a compromise between the direct state control favored by the central bureaucracy and the private operation favored by the business sector or the separate operation favored by the colonial bureaucracy. As such, the ITC would be a useful tool for further expansion in telecommunications in Asia. It is important to note that the transformations at home, even if owing to deeper internal roots, were also affected by developments in the new empire and beyond.

In ending the GNTC's 70-odd years of control over Japan's external telegraph communications, the MOC could claim an almost complete victory. Even there, however, it had to compromise by giving the GNTC another three years before it had to relinquish all its holdings in Japan. In this case, MOC had to bend to pressure from the Ministry of Foreign Affairs. The Japanese state was thus far from being a monolith and did not achieve its goals without encountering resistance. If the second half of the 1930s saw major changes in telecommunications in Japan, they did not always occur in ways the MOC bureaucracy desired. Resistance from the colonial bureaucracy and the field army often forced concessions. The MOC's attempt to consolidate the telecommunications manufacturing industry produced little, with the exception of various control organizations. In this sense, control at home, even when accomplished, was a negotiated victory at best.

CHAPTER 7

Consolidating Control in China

In the fall of 1939, the exhibition "Communications for Asian Development" opened in the posh Mitsukoshi Department Store in downtown Tokyo. Organized by the Ministry of Communications' own Communications Museum, the exhibition was a joint effort by several dozen governmental agencies and private or semi-private companies, all involved in communications and transportation operations on the Asian continent. The six-part exhibit began with a section entitled "The Revitalization of China"; Japan's colonies of Korea and Taiwan and its puppet state of Manchukuo were presented in two separate sections. The numerous models, demonstration boards, and manufactures in more than a hundred thematic pieces covered the postal service, telecommunications, aviation, shipping companies, and other communications services in the vast area under Japan's control. The exhibit was impressive not only for its comprehensive scope but for its display of advanced new technologies such as the television, tape telex telegraph, and telephones specifically designed for rural use, all featured in on-the-spot demonstrations. The dozen or so films shown at the exhibit included such titles as *The Expanding Electric Waves*, *Bonds with the Continent*, *Non-loaded Cable Carrier System*, and *Telegrams at Dawn*. The exhibition, which toured six other cities in Japan, was an apparent success. In Osaka alone, more than 130,000 people visited the exhibit during a two-week period, which had to be extended for two extra days.[1] Given the large number of visitors to this exhibition, it can be

1. Teishinshō, Teishin hakubutsukan, *Kōa teishin tenrankai shi.* In addition to Tokyo, the exhibition traveled to Sapporo, Kobe, Fukuoka, Hiroshima, and Kokura.

assumed that the MOC enjoyed some success in raising public awareness of the role of telecommunications in Japan's new continental enterprise and boosting public confidence in Japan's leadership capabilities in Asia. The accompanying brochure proudly declared that "at present, epoch-making changes toward an ideological, political, and economic New Order are taking place in a part of Asia twice the size of the Japanese Empire, with a population of 170 million." Indeed, it asserted, "A giant step has been taken through Japan-Manchukuo-China cooperation to denounce both capitalism dependent on the West as well as Bolshevism, in order to ensure a sphere for Asian nations under Japan's leadership."[2]

What the exhibition did not reveal was that although telecommunications *technologies* facilitated Japan's control on the continent, who actually controlled telecommunications *operations* in occupied China was highly contested. In both occupied China and Manchukuo, as this chapter shows, Japan's efforts to consolidate telecommunications operations were far from complete or satisfactory.

CHINESE RESISTANCE

Unlike Japan's earlier takeover of telecommunications in Korea or even in Manchuria, Japan's attempt to control telecommunications in China took place in the midst of a protracted and often bloody war. When Japan set up the "national policy companies" to operate telecommunications, first in Manchuria in 1933 and later in China proper in 1938, it did not deem it necessary to consult the local population. None of these telecommunications enterprises could work, however, if Japan were unable to overcome Chinese resistance and secure Chinese cooperation.

To achieve the former, the Japanese military often resorted to brutal tactics. In a telegram to the MOC dated June 1, 1938, the Japanese communications official in Chefoo reported that "our troops adopted forceful measures," and many Chinese were "brought in and shot every day." This chilling report of massacres of Chinese POWs, sent in a specially coded telegram, must have been just the tip of the iceberg.[3]

---

2. Ibid.

3. Tellingly, the phrase "hobakushi kitaride mainichi jūsatsushi" was coded; see MOC Records I, A269.

To minimize an interruption of telecommunications service in areas that came under occupation, the Japanese initially had only to continue Chinese practices as much as possible. Still, when Japanese were sent to take over Chinese telegraph and telephone offices, they encountered various forms of resistance, from apathy to resentment to outright opposition. As one Japanese engineer in North China admitted, with rare candor:

> We Japanese probably are justifiably proud of taking up the task of rebuilding East Asia. We probably have displayed condescension and suspicion toward the Chinese. On the other hand, ordinary Chinese, having seen their country and compatriots suffer in this Incident, can easily become pessimistic. Even a small irritation will probably produce a strong reaction [from them]. Can we say that they do not harbor fear and hatred toward the Japanese?[4]

No match to the Japanese in military force, many Chinese in Central and North China resorted to using neutral foreign countries to resist Japan's takeover attempts. This was the case with the Shanghai International Radio, built by the American firm RCA in the late 1920s and operated by the Chinese government. Its continued communications service with overseas destinations posed a direct threat to Japanese efforts to restrict outgoing information from Shanghai, and it was considered a source of anti-Japanese propaganda. However, since its radio transmitter and business offices were located in the International Settlement and French Settlement, respectively, their occupation by Japanese authorities proved extremely difficult. Anticipating this, the Japanese carried out careful advance preparations in late 1937. Japanese wireless technicians from the JTTCC, using multiband receivers in a nearby apartment, familiarized themselves with the International Radio's reception patterns so that they could take over its operations. Other Japanese, disguised as customers, learned about the International Radio's rates and rules of operation. To assist the effort, other officials and engineers from the MOC arrived in Shanghai on Christmas Day. By December 28, a 42-member Japanese operation team assembled in the city, ready to act. Additional arrangements were made with the

4. Shirozaki Fumio, "Kahoku densei sōkyoku ni hakensarete," *TKZ* 362 (October 1938): 70.

MOC to re-direct Shanghai's international traffic in case of disruption of wireless service.[5]

The opportunity finally came on January 3, 1938. The Japanese decided to act when Chinese employees began removing equipment by truck from the International Radio office in Sassoon House. After consulting with the Japanese military authorities, MOC officials demanded that they be allowed to take over the building. Meeting them was an American manager of RCA who claimed to be in charge of the International Radio on the basis of an agreement with the Chinese. He refused to turn over the keys and lists of employees and property to the Japanese. Several Japanese were left in the building to maintain a watch. Before dawn on January 5, the Japanese were notified that the Chinese were cutting wires and preparing to flee. By the time other Japanese arrived, they found most Chinese employees gone and most rooms locked. Faced with no other choice, the Japanese had to take over the operations by themselves. To give the public the impression that the International Radio was still functioning, the Japanese retained the former Portuguese secretary to the American manager and, at her insistence, hired her sister as well to answer telephone inquiries. At the same time, the Japanese consulate issued statements justifying the Japanese action. Thanks to a business arrangement with MacKay Cable Company, a rival of RCA, the Japanese managed to reopen communication with Manila only a week later.[6] The Japanese takeover thus proceeded without causing a major disruption of international telegraph business in Shanghai.

As this episode surely reminded the Japanese, the long-standing foreign settlements in Chinese cities could complicate their plans. Before Japan occupied these foreign enclaves when it declared war on the United States and Britain at the end of 1941, they proved to be a major headache because the Japanese had to tread carefully to avoid confrontation with the Western countries involved. Foreign settlements easily became a safe haven for anti-Japanese Chinese activists. For example, after the Chinese Government Radio operating in the French Settlement in

---

5. "Shanhai kokusai dentai sesshū tenmatsu gaiyō," in *Sanbochō kakka ni taisuru Higashihan setsumei shiryō* (1938), in *Jihengo ni okeru kyū Kōtsūbu densei kikan sesshū keii narabini misesshū bubun no sesshū hōsaku ni kansuru chōsa kenkyū*, 19–26.

6. "Shanhai kokusai dentai sesshū tenmatsu gaiyō," in *Sanbochō kakka ni taisuru Higashihan setsumei shiryō* (1938).

Tianjin was closed, Japanese intelligence sources reported that Wang Ruoxi—the Chinese director of the Tianjin Telegraph Office and a ranking member of the Nationalists' local branch in Tianjin—secretly moved all the equipment into the British Settlement and continued operation. To make matters worse for the Japanese, this radio was in contact with the Nationalist stations in Hankou, Chongqing, Chengdu, and Xi'an and had thus become a "bastion of anti-Japanese activities." The secretive Wang had previously eluded Japanese searches. Following some leads in a new search, the Japanese *kempeitai* (military police) finally broke into a secret hideout at 6 A.M., arrested Wang, and found a small transmitter and a few batteries. The Japanese then found the radio equipment at a different location inside the American section of the settlement. The British member of the Joint Municipal Council protested that without the permission of the American consulate, it had been illegal to search the American section in the first place. Much to their chagrin, Japanese authorities had to halt their search and turn Wang over temporarily to the British, with the hope that he would be handed over to the Japanese after the Japanese head of Special Services returned from Shanghai. This never happened, and Wang managed to flee Tianjin.[7]

Another prolonged battle in Tianjin was fought over control of the Chinese Government Telephone Bureau, whose telephone branch exchanges were located in the International Settlement. Taking advantage of the exchanges' location and encouraged by Communist Party members, the top Chinese officials of the Telephone Bureau refused to cooperate with the Japanese or with representatives of the puppet Chinese government. Although the Japanese and Chinese collaborators offered incentives for Chinese employees to leave the besieged bureau, they also resorted to such measures as damaging the facilities and openly seizing Chinese employees outside the International Settlement. In April 1938, they even abducted the chief Chinese engineer of the Telephone Bureau inside the settlement. The engineer later died at the hands of the Japanese, as did a number of other Chinese employees who were arrested.[8]

---

7. Maejima (Director, Tianjin Bureau) to Wada (Director, NTCC General Affairs), MOC Records I, 271.

8. Zhu Zhen'gang, "Kangzhan chuqi Tianjin dianhuaju de 'kangjiao' douzheng," *Youdian wenshi tongxun* 23 (June 1995): 14–17.

To deal with this and other instances of Chinese defiance in Tianjin, the Japanese authorities established a blockade of the International Settlement that developed into a major diplomatic brawl between Britain and Japan. Under relentless Japanese pressure, the International Settlement in Tianjin took over operation of the Telephone Bureau from the Chinese in July 1938. British, French, and Italian employees of the Settlement Council then placed it under the authority of the Joint Municipal Committee of the International Settlement. After the war broke out in Europe in 1939, Japan took advantage of the fact that both Britain and France were much weakened and exerted new pressure. It was not until September 1940 that the Settlement Council finally transferred telephone operations to the pro-Japanese city government, which then turned them over to the Japanese-controlled NCTT.[9]

Both in Manchuria and China proper, acquisition of private telephone operations was essential not only to Japanese control of local areas in China, but also to the telecommunications companies' ability to create an economy of scale in the lucrative telephone business. In Manchuria, within a year or so, the MTT purchased relatively large local telephone companies in a number of cities, increasing its own subscription base by 3,667. The Chinese owners or shareholders were paid very little for their assets. As one Japanese employee who had participated in these buyouts admitted long after the war, the MTT would pay next to nothing for these outdated facilities. Although these buyouts made good business sense for the company, he noted, they ended up alienating many local Chinese businessmen, who apparently agreed to the MTT's terms under Japanese pressure but cursed Manchukuo and hated Japan in private. Quite often, he recalled, Chinese owners simply fled the day after signing an agreement. In his view, even three or five times the price would have amounted to less than the cost of building a new office after the Japanese takeover and would have secured the cooperation of powerful Chinese. However, the MTT vetoed his suggestion.[10]

---

9. *GKDTS*, 9: 228–29; Hokudenkai, *Kahoku denden jigyō shi*, 186–92. On Japanese preparation for the takeover, see "Tianjin eikoku sokai denwa sesshu ni kansuru ken," NCTT Records 2028/1124.

10. Maseda Masue, "Min'ei denwa baishu yowa," in Manshū denden tsuiokuroku kankōkai, *Akai sekiyō*, 27–32.

In Central China, Japan tried a different tactic. In August 1940, the CCTC began negotiating with the Wuxi Telephone Company, a relatively large firm with 1,400 subscribers. The process took more than eighteen months because of disagreements over price. The Japanese blamed the owners of the Wuxi Telephone Company, who, like most Chinese businessmen at the time, were not familiar with the concept of property depreciation. In the end, the CCTC had to increase its payments to several Chinese executives before reaching an agreement in early 1941.[11] Consolidation of local operations was not just a simple business transaction, as these Japanese realized.

Consolidation of private telephone operations in occupied zones also brought the Japanese into direct confrontation with the Chinese in rural areas where Japanese military control was often much weaker than in the cities. The amalgamation of local telephone operations in Jinan in North China showed that Chinese resistance could be quite effective, even without the backing of foreign interests. In February 1938, several Japanese employees from the North China Telecommunications Administration (NCTA) Bureau arrived in Jinan to take over telephone operations from the financially struggling Jinan Telephone Company, a privately owned Chinese enterprise founded in 1915. The value of all the equipment, with a capacity of 8,000 subscribers, was set by the Japanese at 300,000 yen. By May, when final price negotiations began with the Chinese, the improved management had already begun to show results. The Chinese side, headed by the general manager, argued that the company had been undervalued and refused to sign the contract at the price of 300,000 yen. At the suggestion of one of the Japanese negotiators, military police were brought in to pressure the Chinese. After spreading a rumor about "dangerous elements" in the Jinan Telephone Company, Japanese soldiers surrounded the company. The Chinese shareholders were forced to sign the agreement prepared by the NCTA Bureau.

The next morning, the Chinese manager and employees disappeared. Moreover, they had appealed to the Japanese Special Service Agency in the area, declaring the agreement invalid because it had been signed under duress. Wary of escalating the tensions with local Chinese elites, the Special Service Agency intervened. Thus, by the end of June, negotia-

---

11. Tōa denki tsūshin jimukyoku, *Tōa denki tsūshin kyōgikai dai-3-kai kaigi gijiroku*, 33; *DDJS* 6: 471–72.

tions had to start again from scratch. At the behest of the Jinan Special Service Agency, a new evaluation committee was organized, with ranking NCTA Bureau Japanese officials such as Asami Shin and Yamamoto Hideya joining from Beijing. The negotiations reopened in the middle of the summer and dragged on for more than a month, with both sides fervently arguing their cases. Finally, at the intervention of the Special Service Agency, both sides agreed on the price of 450,000 yen. Frustrating the Japanese, the Chinese shareholders had been able to raise the company's value by 50 percent. The purchase was finalized in January 1939.[12] As the Japanese came to realize, a high-handed approach did not always work, especially when the Chinese could exploit disagreements among the Japanese themselves.

Securing Chinese cooperation proved especially difficult in Central China, the stronghold of the Nationalist Chinese government. In the Shanghai area, for instance, the Nationalist government issued orders to arrest and prosecute any Chinese suspected of collaborating with the Japanese occupation. Xu Bing, a Chinese who had worked at the Shanghai International Radio, chose to work with the Japanese after the takeover. Appointed by the Japanese to head its secretariat and soon promoted to chief of the Business Section, Xu subsequently reported receiving death threats issued by the Nationalist government in Chongqing.[13] Xu's fear was understandable. In May 1939, the Chinese chief of the Radio's Acceptance and Delivery Office was assassinated in Shanghai, accused of being a Japanese spy.[14] In the same month, when the CCTC in Shanghai advertised its first public recruitment of Chinese employees, it had to conceal the fact that it was a Japanese-controlled company. Moreover, it held the screening examination inside the International Settlement in Shanghai for fear of disruptions.

The CCTC management certainly had the uneasy relations with the Chinese population in mind when it established an elementary school outside Shanghai, near its huge wireless station. Named after the company, the Futian (Fukuda) Elementary School opened in 1939 with Japanese-language teachers and funds supplied by the company. It soon ran

---

12. Hokudenkai, *Kahoku denden jigyō shi*, 192–94.

13. MOC Records I-205.

14. "Japanske Kontrol Med Det Kinesiske Telegrafvasen," GNTC Records–Shanghai File 418.

into financial difficulty and had to close after just one year. CCTC did not give up, however. In early 1941 it obtained support from Japan's Asia Development Board and the military and sought to reopen the school in order "to spread elementary education, promote Sino-Japanese friendship and mutual harmony, and to contribute to the building of a New Order in East Asia."[15]

WESTERN COMMUNICATIONS FACILITIES

Besides the international settlements, the presence of foreign cable and telephone companies in China proper posed a much bigger problem for the Japanese than had foreign cable companies in Manchuria, where Japan had traditionally enjoyed a special position. In China, extensive foreign interests in the form of treaty rights such as settlements, as well as business and cultural institutions, continued to exist even after Japanese forces occupied much of coastal China. Although the Chinese government had prohibited foreign operation of telecommunications in the interior of China, foreign ships plying the Yangtze River, for instance, were fitted with wireless communications facilities. However, it was international settlements housing radio stations that Japanese authorities tried to eliminate, for business as well as political reasons. As one Japanese executive of the CCTC lamented, settlements had become safe havens for the parasitic existence of foreign communications facilities.[16]

During the early phase of the Japanese occupation, talk of an Open Door in China was still in the air. In the December 1937 "Outline on Policies Toward Telecommunications in China," MOC officials even recommended that foreign capital might be utilized in the formation of new telecommunications companies as long as it did not undermine Japan's control. If this scheme did not work, Japan would build a communications network to compete with that of the foreign companies. In addition, Japan would strive to gain control of internal Chinese communications and sever their contacts with foreign communications

---

15. "Sili Futian yinsheng xiejing xiaoxue zhi an," R48-21-271, Shanghai Municipal Archives.

16. Kumon Yō, "Chūshi keizai no tokuisei to denkitsūshin," *TDTZ* 2 (January 1942): 32.

networks so that the latter were limited to the international settlements.[17] The suggestion of absorbing foreign capital while maintaining Japanese supremacy, of course, turned out to be wishful thinking.

In contrast to their earlier competitive stance in Manchuria, foreign cable companies were actually quite cooperative after the outbreak of the conflict in China. This was hardly surprising, given that cable companies could operate only on the basis of landing rights and fixed facilities easily controlled by an occupation army. On August 12, 1937, the day before Shanghai became engulfed in the war, the Tokyo representative of the American firm Commercial Pacific Cable Company asked the MOC chief of the Foreign Telegraph Section "to kindly pass" a check for 50 yen "to the proper authorities of the Imperial Japanese Army Headquarters, to be added to the general subscriptions being donated for the comfort of the soldiers participating in the trouble abroad."[18] After hostilities extended to the Shanghai area in mid-August, all three foreign cable companies based in the city—Commercial Pacific, Great Eastern of Britain, and the GNTC of Denmark—continued to operate. However, Japan's physical control of Shanghai and vicinity threatened their position. These cable companies, in order to continue business, made concessions to the Japanese authorities by agreeing not to pay terminal fees to the Nationalist Chinese government, as they had done before. Instead, they agreed to pay the funds into a new account set up at the Shanghai branch of Japan's Yokohama Specie Bank.

Whatever illusions some Japanese entertained, they soon came to view foreign interests in China as a hindrance to their own relations with the Chinese. Gradually, undermining and eventually driving out third-country interests became one of the foremost objectives of Japan's policy of establishing telecommunications hegemony in China. As commercial rivalries for telecommunication traffic continued, the Japanese authorities in Shanghai found the situation rather serious. A Japanese survey of the traffic destinations of the three companies on March 31, 1938, revealed that communications with Hong Kong occupied nearly 25 percent of all traffic from Shanghai, second only to traffic

---

17. "Shina densei taisaku yōkō" (December 21, 1937), MOC Records I, 253, 8–9.

18. J. Reifsnider to A. Tachibana (August 12, 1937), in MOC Records I, 167. Despite his disclaimer that "this small gift comes from me personally and has nothing to do with the Company whatsoever," the goodwill gesture, written on Commercial Pacific stationery, was not likely to be lost on the Japanese government.

with the interior of China (39 percent). Communication with Japan, Manchukuo, and North China took up about 13 percent, as did traffic with America and Europe. These results gave the Japanese some hope that if they could quickly restore the Chinese-owned International Radio Station, they could easily absorb some of the lucrative business from the foreign cable companies.[19] The foreign companies did not sit idle, however. The GNTC, which had been operating in Japan and China since 1870, continued to interfere with Japan's objectives as it had in Manchuria. The firm hired Japanese representatives and managed to maintain good business relationships with Japanese companies and banks in Shanghai, to the dismay of Japan's own telegraph office in the city.[20] Even three years after the establishment of the CCTC, these cable companies still had a daily volume of over 3,000 telegrams, compared to more than 10,000 sent and received by the entire CCTC.[21]

The continued presence of foreign telecommunications enterprises in China posed a direct threat to the newly established CCTC and NCTT, particularly as they sought to expand their business base and increase revenues. In Shanghai, with a population of more than 5 million, the Japanese faced another headache: the American-owned Shanghai Telephone Company (STC) posed an obstacle to efforts by the Japanese-controlled CCTC to expand its foothold in China's business center. Founded in 1881 by British businessmen, the STC had been bought by the American firm International Telephone and Telegraph (ITT) in 1930. In the same year, it obtained a 40-year concession from the International Settlement Committee to provide telephone service, on the condition that it convert all lines to automatic exchanges within two years and continue to offer quality service. The STC quickly grew into a large, modern telephone company, with more than 50,000 subscriptions, the majority of them with automatic connections by 1932. Before the war broke out in August 1937, the STC had already established long-distance connections with 40 cities in China as well as with Hong Kong and the United States. Its telephone subscription base

19. "Zai Shanhai Daihoku, Daitō, Shōtai san kaisha hasshin denpō chakuchibetsu tsūshū chō" (April 20, 1938), MOC Records II, 319.

20. Matsunaga Hangorō, "Shanhai ni okeru denki tsūshin no kakuchiku," *DT* 4.16 (October 1941): 16.

21. Datefumi, "Chūshi keizai no tokuisei to denki tsūshin," *TDTZ* 2 (January 1942): 33.

peaked at 63,355 in 1938.[22] Even one business-minded Japanese employee at the CCTC candidly admitted in a Japanese journal that since the STC was run with American-style efficiency, there were "many things worth learning about its business and facilities."[23]

After its establishment in 1938, the CCTC initially refused to deal with the STC. This strategy soon backfired, given the STC's much larger subscription pool. The more practical-minded Japanese Local Residents Association wanted to continue to receive STC phonebooks, despite the MOC's formal policy of cutting off contact with the American-owned company in Shanghai. The Japanese formulated a different strategy. When STC President James E. Fullam visited Tokyo for business related to its parent company, ITT, he was met by Minister of Communications Nagai Ryūtarō as well as CCTC President Fukuda Kon, who went to Tokyo specifically for the meeting. They were joined by a number of important business figures including the general manager of the Sumitomo group, Ogura Masatsune, and Kajii Takeshi, the former MOC engineer who now headed the Sumitomo subsidiary NEC. Japan had seriously considered purchasing the company's entire operation in Shanghai, but as it did not have enough funds, it proposed purchasing the part of the STC north of the Soochow Creek, an area with the largest concentration of Japanese residents. The STC, on the other hand, was more interested in simply linking the two systems in Shanghai.[24] The two sides were too far apart to reach any agreement.

While Japan was expanding its own telecommunications network in East Asia, it continued to apply pressure to these foreign companies operating out of international settlements in China. The uneasy relations between the Japanese-controlled CCTC and foreign companies

---

22. Kōain gijutsubu, *Shanhai denwa kaisha no gaikyō.*

23. Nakatani Kiyoshi, "Shanhai shinai denwa ni tsuite," [MTT] *Gyōmu shiryō* 11.2 (February 1944): 201–17; Sakaki Kazuo, "Shanhai denwa kaisha ni gaiyō," *DT* 5.19 (4/1942): 47–50. For the technical aspects, see Nihon denshin denwa kōsha, *Jidō denwa kōkan 25-nen shi* III: 457–64.

24. See remarks by Fukuda Kon in "Kachū den sōgyō zadankai," *KDT* (1943): 99–100. The Japanese settlement in Shanghai was located north of the creek. In his letter to the ITT, STC President James E. Fullam conveyed the impression that "they [the Japanese] are looking for foreign financial cooperation in the development of their plans for China." Fullam's letter, dated November 4, 1938, is appended to Frank C. Page to Raymond O. Mackey (November 25, 1938), RG59, 894.75/5, National Archives and Record Administration, Washington, D.C.

ended with the outbreak of the Pacific War. The CCTC finally assumed operation of Western telecommunications companies on behalf of the military. After the Wang Jingwei government joined the war effort in 1943, their facilities were formally appropriated by the Chinese government and were considered part of the Chinese government's contribution to the CCTC in July 1944. Similarly, the Japanese military took over telephone operations in Canton, run by Sino-American joint venture China Electric Company, which was now considered as enemy property.[25] Only then did Japan succeed in eliminating the last stronghold of foreign-held communications in areas already under Japanese control.

VARIETIES OF COLLABORATION

The Japanese occupation of China had the support of many Chinese from the start. Wang Keming and his Provisional Government in North China worked closely with the Japanese, and the Restored Government under Liang Hongzhi in Central China did the same. In late 1939, they were joined by the high-profile Chinese political leader Wang Jingwei, who came to head the new government in Nanjing with Japan's blessing.[26] As Japan consolidated its occupation of China, many residents in occupied areas came to accept the fact that the Japanese were there to stay. Continuing to make a living seemed an important reason for many Chinese to work under the Japanese.

Securing Chinese cooperation was essential for the operation of telecommunications companies. Not long after the NCTT was founded, two Chinese employees who had worked at the Beijing Telephone Office submitted a letter to NCTT President Inoue Otsuhiko emphasizing the importance of "genuine harmonious cooperation between the Chinese and Japanese."[27] One of them was Luo Jin, who became perhaps

25. *DDJS* 6: 474–76. After the Japanese invasion in October 1938, Japan was able to install advisors in the company.

26. For a recent collection of essays on wartime collaboration, see Larry Shyu and David Barrett, eds., *Chinese Collaboration with Japan, 1932–1945: The Limits of Accommodation* (Stanford: Stanford University Press, 2001). See also John Boyle, *Japan and China at War* (Stanford: Stanford University Press, 1972); and Gerald Bunker, *The Peace Conspiracy* (Cambridge, Mass.: Harvard University Press, 1972).

27. Shirozaki Fumio, "Kahoku densei sōkyoku ni hakensarete," *TKZ* 362 (October 1938): 70.

the most prominent example of Chinese working closely with Japanese in the NCTT. Luo had been handpicked by Nakayama Ryūji—then serving as adviser to the Chinese government—to study in Japan in the late 1910s. A graduate of Kuramae Technical High School in Tokyo and fluent in Japanese, Luo soon became a favorite of the Japanese and was placed in charge of the Beijing Telephone Bureau.[28] Shi Tong, another Chinese who studied electrical engineering in Japan about the same time, became the head of the Engineering Division in the NCTT Beijing branch office.[29]

Chinese who enjoyed such visibility like Luo were the exception, however. Lower-level Chinese managers were often caught in a bind. Whereas the Chinese Nationalist government had encouraged Chinese telecommunications operators and technicians to move to inland areas with the government itself, administrative clerks were simply released from duty with severance pay.[30] For them, continuing to work was very much a practical matter. The Japanese clearly knew this, and they, too, could be practical when it came to cultivating Chinese ties. In early 1940, shortly after Japanese forces had driven away Chinese guerrillas there, the NCTT dispatched a team to the eastern part of Shandong province to take over local telecommunications operations. The question for the Japanese team became what to do with Chen Zerong, the 42-year-old Chinese head of the Shidao Telegraph and Telephone Office. Chen, an eight-year veteran at the office, which had been under the jurisdiction of the Nationalist government's Ministry of Communications, was described in an internal Japanese report as a "gentle person, with profound understanding of the current situation." Shortly before Japanese forces entered Shidao, Chen had been pressured by anti-Japanese guerrillas to turn over communications equipment to prevent it from falling into Japanese hands. Chen apparently balked and tried to placate the guerrillas by handing over only one telephone set. After his office came under Japanese military control and ceased to operate, Chen reportedly paid his employees with his own money and consequently found himself in financial distress. Thanks to what the Japanese report described as a "stellar reputation among the local

28. Matsuoka, *Nakayama Ryūji*, 157; Hokudenkai, *Kahoku denden jigyō shi*, 203.

29. A short biography of Luo was printed in *Hokuden* 3 (November 1939): 11.

30. Mei and Song, eds., *Bainian dianxin zhu huihuang*, 89.

residents," as well as recommendations from Japanese officials stationed in the area, Chen was allowed to keep his job, which paid 150 yuan per month. Grateful for the Japanese gesture, Chen petitioned the NCTT group to assign restoration work to his Chinese employees and others in the area.[31] Keeping his men and their families fed seemed the biggest concern for a local Chinese manager like Chen.

Apart from middle- and lower-level managers, there were tens of thousands of ordinary Chinese who worked in the Japanese-controlled telecommunications companies. Unlike the MTT, telecommunications companies in occupied China employed more Chinese than Japanese. When the NCTT was founded, it already employed over 3,000 Chinese—most of whom were former Chinese government employees—along with less than a thousand Japanese. For one reason or another, many Chinese sought employment in these companies. In the city of Jinan, nearly 400 young Chinese women took the examination to become telephone operators for the NCTT.[32] Due to the shortage of qualified Japanese technicians, which worsened during the war, these companies had to resort to hiring and training more Chinese employees. As one of the CTCC's Japanese executives admitted at a business meeting in 1940, the departure of skilled Japanese technicians in China would be expected due to Japan's domestic situation. Therefore, in addition to "thorough use of Japanese technicians already in China," the CTCC had to enable many Chinese technicians to work independently as soon as possible.[33] The NCTT set up a training school in 1940, headed by a Chinese. By the end of the war, 1,240 Chinese telegraph operators and 670 repair technicians had graduated from the school.[34]

The Japanese-controlled telecommunications companies in China launched vigorous efforts to inculcate their workforce as well as the public with the new Asian ideal of solidarity. Their professed ideals and window-dressing notwithstanding, all three major telecommunications companies in China and Manchuria—the NCTT, CTCC, and MTT—

31. Sesshū chōsa 1-han, "Chin Taku-ei saiyō ni taisuru ikenshō" (June 1940), NCTT Records 2028(2)/42.

32. "Kajin joshi wamuin yōsei ki," *Hokuden* 8 (April 1940): 10.

33. [CCTC,] *Dai-3-kai eigyō shochō kaigi* (June 15–17, 1940), CTCC Records.

34. Hokudenkai, *Kahoku denden jigyō shi*, 118.

like other Japan-China "joint ventures" created during the war, were anything but truly cooperative enterprises between the Japanese and the Chinese. In reality, not only were the majority of ranking positions held by Japanese, but even when a Chinese was given a relatively high position, it was his Japanese lieutenants who were actually in charge. With only a few exceptions, Chinese received far less pay than their Japanese counterparts even though they performed essentially the same jobs. Even one Japanese executive of the NCTT admitted with concern that the living standard of Chinese employees had deteriorated and that well-qualified Chinese engineers were underpaid.[35] This was in contrast with their much-envied status and pay before the war.[36]

For ordinary Chinese, working for the Japanese during the war could turn into psychological trauma. In an article published in a Japanese magazine, one Japanese telephone-exchange operator who had worked in Shanghai before the Pacific War recalled that some of her Chinese coworkers had applied for the job in order to make a living and had to conceal the identity of their employer. Occasionally these Chinese women would burst into tears in the midst of connecting phone calls. As this Japanese woman found out, Chinese operators working for the American-owned Shanghai Telephone Company, safely ensconced in the international settlement, would often ridicule her Chinese coworkers for working for the Japanese.[37]

Managing a binational workforce was no small challenge. Many of the Japanese who took up positions in these "joint ventures" were former MOC employees. For them, it was already a major change to switch from government offices to national policy companies. Working along with Chinese employees, as the author of one MOC memo admitted, "increasingly complicates the personnel component" of these companies. Mixing Japanese and Chinese employees "with fundamentally different thoughts" had to be a temporary if inevitable measure, the memo suggested. After a considerable period of time, the MOC predicted, these companies would have to become self-sufficient in human resources by creating a workforce with "firm beliefs in the

35. Murakami, *Ichi gijutsusha no shōgai*, 402–3, 412–13.

36. For anecdotes about telecommunications workers in prewar China, see Zhang Jian, ed., *Lao dianhua*, 50–53, 71–79.

37. Kondo Ai, "Denwakyoku yonen no kaikō," *TDTZ* 2.6 (June 1942): 39.

liberation of East Asian peoples." Since a basic industry such as telecommunications could not be operated by Japanese alone, it noted, the vast majority of Chinese must be relied on; the Japanese would provide guidance, and the Chinese would fill the actual posts.[38] One of the prize-winning essays in a contest sponsored by the NCTT in 1940 similarly pointed out that it was inevitable that more and more Chinese would be recruited into the company. This would not be easy, its Japanese author concluded, citing the generally low level of education in North China and the employment demands of various development companies there. Still, the essayist argued, the Japanese must be used strategically—to supervise Chinese employees, to protect military secrets, and to carry out mobilization plans during emergencies.[39] Obviously, Japanese in Tokyo and in China alike had set a clear limit on the oft-touted "Japan-China cooperation."

Given such mind-sets on both sides, friction between Chinese employees and Japanese managers was quite common. Many Chinese employed in these companies simply used delay and other tactics to vent their discontent or to cope with deteriorating economic conditions. For example, Japanese managers at the CCTC discovered that some Chinese telephone operators were receiving extra fees from certain business users for making connections for them. In November 1940, for instance, a strike broke out among Chinese messengers at the Shanghai International Radio, now part of CCTC; it ended with the firing of two Chinese employees and a salary reduction for others. The Japanese chief was also reprimanded.[40] Whatever their causes, instances like these undermined Japanese control and often led to confrontations between Japanese supervisors and Chinese employees. In July 1945, Chinese telephone operators again staged a strike to protest poor treatment and low pay at a time of runaway inflation in Shanghai. This time, disruption of telephone-exchange service forced the CCTC man-

---

38. MOC, "Tōa denki tsūshin seisaku" (n.d.), copy made available by Professor Hikita Yasuyuki.

39. Mikami, author of a third-class prize essay in Kahoku denden kurabu, *Tairiku ni okeru tsūshin seisaku o ronzu*, 182–88.

40. [CCTC] *Gongsibao* 252 (January 11, 1941). For a Chinese account of Japanese abuses and Chinese resistance, see Mei and Song, eds., *Bainian dianxin zhu huihuang*, 87–96.

agement to meet their demands that Chinese employees receive extra rice.[41] Under such circumstances, economic grievances and tacit political resistance reinforced each other.

TECHNOLOGIES OF HARMONY

Given their vast potential to annihilate distance, communications technologies seemed to hold much promise for closer ties between Japan and China. The "Hokuden March," the NCTT company song written by a Japanese, depicted a harmonious relationship between the Japanese and the Chinese thanks to modern communications:

> The country of cherry blossoms,
> The land of orchid fragrance,
> Bound together, by the culture
> of communication,
> In the new tide embracing Asia,
> Is the shining New Order.[42]

When wireless long distance telephone service started between Nanjing and Tokyo in October 1939, the now ritualized exchange of personal greetings by politicians was described in the newspaper under the deadline "A Feeling of Japan and China Brought Closer."[43] If international telephone exchange was the preferred technology of the day, facsimile (*mosha denshin*) was the desired technology of the future between Japan and China since it could eliminate a major barrier between the two languages, which shared many Chinese characters.[44] Finally, here was the technology that would turn the cliché "same script, same race" (*dōbun dōshū*) into reality.

After its establishment in 1938, the NCTT showed an interest in such technological developments and drew up plans for adapting Chinese-language telegrams to facsimile. Wada Yoshio, the NCTT's director of general affairs, predicted in the company newsletter, *Hokuden*, that such

---

41. Mei and Song, eds., *Bainian dianxin zhu huihuang*, 91–97. The authors attribute the incident to instigations by underground Communist Party members among the Chinese workers at the CCTC.

42. For the complete lyrics, see Hokudenkai, *Kahoku denden jigyō shi*, 128.

43. *Yomiuri*, October 9, 1939.

44. *Manshū nichinichi shinbun*, October 1, 1939.

a service would greatly ease communication between Japan, China, and Manchukuo. Although abolishing Chinese-language telegrams altogether and replacing them with Japanese *kana* telegrams was the most reliable method from the perspective of an East Asian telecommunications network, Wada noted, there were many difficulties in such a drastic change, and for the time being, facsimile technology would fill the need.[45] During a visit to Japan, Wada met with Matsumae Shigeyoshi and Kajii Takeshi, both also involved in facsimile development, and suggested developing a new system of "postcard telegrams." Customers would purchase ready-made facsimile-paper postcards, write their message, and then deposit it in a mailbox. These cards would be collected at 20- to 30-minute intervals and transmitted from telegraph offices via facsimile. If the cost could be brought down, Wada predicted, people would even be able to send messages from their homes as easily as they would make telephone calls. This would greatly popularize the use of telegrams.[46]

The use of facsimile generated enough interest to be put on the agenda of the annual meeting of Japanese telecommunications operators from East Asia in 1940. The Shanghai-based CCTC predicted a "large demand" for this service in China, since it would not only absorb ordinary telegrams but also create new demand. NCTT officials proposed conducting experiments of facsimile transmission between Japan, Manchukuo, and China. Although price was to be a decisive factor, feasibility was also determined by demand for the equipment as well as paper. The NCTT's Watanabe also seemed optimistic: the company had already written the facsimile into its five-year plan. He suggested yet another benefit of using such a technology in China. "If we can deploy such equipment embodying the essence of Japan's technology," he predicted, "we can first of all demonstrate Japan's technology to the Chinese people, who make up 90 percent of the population in China, and at the same time open up a new field of Chinese-language telegrams."[47] The Japanese newspaper *Yomiuri* enthusiastically welcomed the start of a service that would "secure the bonds of common script"

---

45. Wada Yoshio, "Hokuden shunjū (9)," *Hokuden* 9 (May 1940): 19.

46. Wada Yoshio, "Hokuden shunjū (10)," *Hokuden* 10 (June 1940): 14.

47. Tōa denki tsūshin jimukyoku, *Tōa denki tsūshin kyōgikai dai-2-kai kaigi gijiroku* (1940), 193–98.

between Japan, Manchukuo, and China. It predicted that the "postcard telegrams" would bring about a lively "exchange of blood based on common script."[48] Due to budgetary and technical concerns, however, MOC's engineers cautioned that any further implementation would have to wait for the results of tests to be carried out between Japan and Manchukuo in 1941. It was not until the end of 1943 that Radio Tokyo reported that "a new postcard telegraph service" had begun between Tokyo and Shinkyō, which would transmit a facsimile reproduction of handwriting or a sketch the size of a postcard.[49]

Ultimately, such ambitious plans were slowed by the lack of mass-produced, reliable facsimile equipment. Although a number of Japanese telecommunications manufacturers set up factories in China, none was advanced enough to engage in such research, let alone production. According to Japan's idea of the proper division of labor within its sphere of influence, technological development should take place in the center—namely, Japan—but even at home, developments were slowed down. Although the NEC had produced a prototype of facsimile equipment as early as 1928, it was not yet economical for mass production. The Japanese considered using American facsimile technology, but the outbreak of the Pacific War of course eliminated that source. Japanese technicians in the MOC Research Institute strove to turn their prototype into a usable model. It was not until 1944 that another Japanese electronics manufacturer, Adachi Electric Company, produced the actual equipment. The first commercial line with facsimile equipment opened in November 1944.[50] Dōmei's research lab, as well as electronic manufacturers like NEC, finally produced workable models in early 1945.[51] Given its high cost, facsimile had a long way to go before replacing conventional telegraph service.[52]

---

48. *Yomiuri*, October 5, 1940.

49. *Yomiuri*, November 7, 1943; Radio Tokyo (Japanese), December 1, 1943, in Office of Strategic Services, Research and Analysis Branch, comp., *Transportation and Communication in Japan*, 201.

50. *GKDTS*, 10: 517.

51. Tsūshinsha shi kankōkai, *Tsūshinsha shi*, 949–51.

52. Although facsimile saved transmission time by eliminating transcription to codes and thus eliminating human errors, it used broader bands and thus had to match teleprinting in volume. Okuno Haruo of the MOC's Electrical Institute did the calculation: some 548 *kanji* could be sent by facsimile per minute through one entire circuit. If each

The Japanese also explored others ways to break down the communication barrier. They sought to minimize the difference between Chinese and Japanese telegrams in order to reduce the workload and need for equipment. The CCTC in Shanghai actively explored the possibility of unifying Chinese-language telegrams with those in Japanese. It recognized that while the "superior Japanese-language telegram system" would expand to entire Greater East Asia, it was not yet "desirable to simply abolish the Chinese-language telegrams under the current circumstances." One recommendation was to change the format of Chinese-language telegrams, which were usually written horizontally, into the vertical Japanese format. The CCTC reasoned that if Chinese-language telegrams (30 percent of outbound traffic) and Japanese-language telegrams (60 percent) could be combined, it would greatly simplify service and reduce material costs. Moreover, the vertical Chinese telegram not only followed China's own tradition, but was in keeping with the growing use of the Japanese language in the Greater East Asia region and thus facilitated "fusion of Chinese and Japanese culture." The vertical Chinese telegram, CCTC suggested, could be the first step toward a Chinese telegram system based on *kana* codes. Although the CCTC devoted considerable time to studying the matter and proposed its trial adoption, other operators were less enthusiastic. The MTT raised two major objections: since a vertical Chinese telegram would look very similar to a Japanese telegram, their different rate structures would smack of discrimination in the eyes of the public. Moreover, since at the beginning of 1942 only 1.6 percent of its telegrams handled were in Chinese, and because European-language telegrams, also written horizontally, could not be eliminated in the near future, MTT came to the conclusion that the effect of streamlining would be very limited.[53] The proposal for vertical Chinese telegrams was thus shelved.

---

kanji equals two *kana*, the 1,100 *kana* per minute exceed the regular Morse Code method, assuming 80 *kana* are sent on each of the 12 channels that can be operated on a single circuit. However, facsimile could not match the teleprinter, which could send as many as 3,220 *kana* on all 12 channels; Okuno Haruo, "Daitōa kyōeiken to atarashii mosha denshin," *TDTZ* 3 (January 1943): 27–36.

53. "Kabun denpō no jūsho ni kansuru kyōdo kenkyū no ken" (September 1942), and MTT denmu buchō to NCTT eigyō buchō (January 14, 1942), in MOC Records I-FC-A321; Tōa denki tsūshin jimukyoku, *Tōa denki tsūshin kyōgikai dai-4-kai kaigi gijiroku*, 76–

More ambitious was the effort to revise the existing Chinese telegraphic codes. It had long been recognized that the Chinese-language telegraphic codes, invented by a Frenchman and popularized by the Danish GNTC during the 1870s, were cumbersome. In the 1920s, a Chinese railway official devised a system of using 52 phonetic symbols, which was used successfully for railway telegraphic communication in Manchuria. Due to the great variation of Chinese dialects, the phonetic telegram (*guoyin dianbao*) was not adopted for public telegram service.[54] The idea of revising Chinese telegraphic codes gained renewed momentum during the Japanese occupation largely for the purpose of increasing revenue for the national policy companies. For revenue-minded Japanese executives, the Chinese population represented a vast underdeveloped telecommunications market waiting to be exploited.

In Manchukuo, the Chinese-language telegram service was particularly underutilized, the largely Chinese population notwithstanding. A one-day survey of all telegrams handled by the MTT in November 1935 revealed that Japanese-language telegrams accounted for more than 90 percent, whereas Chinese-language telegrams were a mere 6 percent (the remaining were in European languages). In other words, although only one in every 75 people in Manchuria was Japanese, collectively they managed to send fifteen times as many telegrams as all the Chinese added together. Although the extension of Japanese-language telegraphic service to all areas with a sizable Japanese population made a huge difference in the number of telegrams sent, the lack of Chinese-language telegrams was not due entirely to poor access. Thirty-five percent of all telegraph offices that handled Chinese telegrams had fewer than five such telegrams each day.[55]

Although the low literacy rate among the Chinese population as well as their lower living standards were often cited as leading causes of depressed telecommunication use, much had to do with the tariff structure based on the cumbersome Chinese-language telegram codes. One

---

77. According to CCTC's calculation, a vertical Chinese telegram charged at the same rate as a Japanese telegram would lead to a 26.6 percent loss of revenue per telegram.

54. On the creation of telegraph codes in China, see Baark, *Lightning Wires*, 84–85. Tomida Kenichi, "Sōritsu sanshūnen ni atari wagasha no shōrai ni kibosu," *KDT* 25 (September 1941): 27.

55. Maeda Naozō, "Manshū ni okeru denshin denwa jigyō no keiei," *TKZ* 330 (February 1936): 153.

Chinese employee of MTT who learned how to send telegrams in Japanese at an MTT training institute in Harbin recalled that his telegraph office expanded from 8 employees to about 40 under the Japanese administration. Japanese-language telegrams had an advantage over Chinese-language telegrams, he admitted, because they were cheaper. Japanese-language telegrams cost 6 *fen* per word (consisting of five *kana*), and the recipient's entire name and address counted as only two words. In contrast, Chinese-language telegrams were charged the same 6 *fen* for each character, with the name and address of the recipient treated the same way as the text. As a result, one could send a Japanese-language telegram for 60 or 70 *fen*, whereas a similar Chinese-language telegram cost three times as much. He recalled that Chinese businesses such as department stores, shops, and factories, as well as individuals, preferred to send telegrams in Japanese if they could. In the end, Japanese-language telegrams made up over 80 percent of all traffic in Manchukuo, while the share of Chinese-language telegrams was under 20 percent and English-language telegrams were gradually abandoned.[56]

A disparity between Chinese and Japanese telegrams also existed in China proper. One local survey in North China revealed that while the Chinese sent far fewer telegrams than the Japanese, they sent many more express-mail letters (85 percent of total) and nearly half of all regular letters.[57] A three-day survey conducted by NCTT, for instance, showed that 41.6 percent and 43 percent of telegrams sent and received in North China, respectively, were in Chinese. On certain routes, Chinese-language telegrams even surpassed those in Japanese.[58] But Japanese dominated the external traffic in occupied China as in Manchuria. In October 1940, the NCTT surveyed 14,474 telegrams sent and received between North China and Japan (including Korea). All but 2 percent were in Japanese.[59] A similar survey of telegraphic traffic be-

56. Tao Ye-rong (recorded by Liu Tianguang), "Rijun qingzhan hou de Jiamusi dianbaoju," *Weiman wenhua* (Changchun: Jilin renmin chubanshe, 1993), 350–57.

57. Tanaka Sei, "Denpō senden jin no kyoka o nozomu," *Hokuden* 11 (July 1940): 8.

58. NCTT, Eigyōbu, Eigyōka, Eigyō chōsa-kakari, "Denpō kōryū ni kansuru chōsa" (1. Naikoku denpō) (May 1940), NCTT Records 2028/1352. Chinese-language telegrams exchanged with Central China accounted for 55.8 percent (outgoing) and 60 percent (incoming) of the total.

59. NCTT, Eigyōbu, Eigyōka, Eigyō chōsa-kakari, "Denpō kōryū ni kansuru chōsa" (Nikka) (May 1941), NCTT Records 2028/1352.

tween North China and Manchukuo about the same time revealed that 80 percent were in Japanese, which the NCTT attributed to the "overall dependence on Japan in Chinese-Manchurian affairs," in addition to what it described as "the lack of appreciation of telegrams by the Chinese population" and the inconvenience of Chinese-language telegraphic codes.[60]

Although Japanese running telecommunications in Manchukuo first raised the possibility of revising the Chinese telegram codes to encourage increased usage, little progress was made. This was partly because the Japanese population in Manchukuo was large enough to provide a stable revenue base. Companies in China proper were different. For one thing, their telegraph services were running at a deficit; hence they were more interested in "developing Chinese use of telegrams" to increase their revenue.[61] In a prize-winning essay in a contest sponsored by the NCTT in 1940, one Japanese employee even proposed that Chinese-language telegram service be made the cornerstone of the company's business strategy. The bottom line was economics: since the company's external telegram rates were kept artificially low to facilitate communication with Japan proper, he argued, its real profit must come from Chinese-language telegrams within the NCTT area.[62]

The NCTT, which considered itself the "leader of telecommunications in China," also had bigger ambitions. In addition to the large Chinese population in China, there were an estimated 7 million overseas Chinese in Southeast Asia who were potential customers. For all these reasons, Japanese executives in China considered Chinese-language telegrams an important component of all communications traffic. The reform of Chinese-language telegraphic codes was necessary, the NCTT declared in early 1942, for the "reconstruction of China's economy and culture as well as the development of telecommunications in the East Asia Co-Prosperity Sphere." In addition to the fact that Chinese-language telegrams made up more than half of its traffic, the NCTT was aware of the fact that millions of overseas Chinese living in

---

60. Ibid., (Nichiman) (May 1941), NCTT Records 2028/1352.

61. Wada Yoshio, "Hokuden shunjū (13)," *Hokuden* 13 (September 1940): 14.

62. Sonoda Takeo, in Kahoku denden kurabu, *Tairiku ni okeru tsūshin seisaku o ronzu*, 71–72. Issued by the NCTT Club as an extra of the September 1940 issue of *Hokuden*.

Southeast Asia had just come under Japanese control.[63] Both the political future and the economic prospects seemed promising.

In January 1942, encouraged by the outbreak of the Pacific War, the NCTT set up a special committee devoted to revising the Chinese telegram codes. Headed by Luo Jin, the Japan-trained Chinese engineer who was then section chief in the company's Business Department, the committee consisted of five Chinese employees. Instead of improving the phonetic telegram, Luo and his associates sought a different solution. Of the 8,800 characters in the Chinese telegraphic code, they discovered, only about 3,000 were commonly used. Under this new system, each of these 3,000 Chinese characters would be transcribed into a simpler code of either two or three symbols, instead of four Arabic numerals as in the old Chinese telegraphic code. The remaining 5,000 characters would be represented by more symbols. This method would drastically reduce the number of Morse strokes needed for each of those 3,000 commonly used characters from an average of 46.4 strokes to about 29 (see Table 11). The new system thus promised higher efficiency and the possibility of rate reduction of more than 50 percent. Moreover, Japanese *kana* would be used as basic symbols in the new code, thus further integrating the Chinese and Japanese telegraph systems. For his efforts, Luo was rewarded the rare distinction of reporting (in Japanese) this development of a new Chinese-language telegram system to an all-Japanese gathering of telecommunications officials and executives in Tokyo in 1942.[64]

After nearly two years of investigation, the committee produced a Codebook for Japanese-style Chinese Telegrams (*Heshi Huawen*, or *Washiki Kabun*). Using three *kana* for each Chinese character, the new system promised to reduce the number of necessary Morse strokes by an average of eight and simplify the use of telegraphic equipment. After adopting the Japanese-style Chinese system, the committee predicted, 98.5 percent of all telegrams handled by the NCTT could be processed

63. "'Kabun denpō seido kaisei junbi iinkai' setchi ni kansuru ken" (drafted January 12, 1942), NTCC Records 2028/1642. Machida Itsuyoshi, "Daitōa denpō no kakushin (2)," *KDT* 40 (February 1943): 12–13.

64. Tōa denki tsūshin jimukyoku, *Tōa denki tsūshin kaigi dai-4-kai kaigi gijiroku* (September 1942), 30–34.

Table 11
Proposed Revision of Chinese-Language Telegrams, 1942

	Current Chinese system	Option A	Option B	Japanese telegram
Each character represented by	4 numerals	2 symbols	3 symbols	
Number of symbols needed for 3,000 common characters		56 (*kana* + numeral)	15 (*kana* only)	
Average number of Morse strokes needed for each character	46.4	29.2	29.6	10.9* for 1 letter

*Five Japanese characters (*ji*) counted as one word (*go*), the basic unit in Japanese telegrams (at 10 *sen*).

SOURCE: Based on Luo Jin's report in Tōa denki tsūshin jimukyoku, *Tōa denki tsūshin kaigi dai-4-kai kaigi gijiroku* (September 1942), 30–34.

with the same Japanese-style hole puncher, a significant simplification. The codebook itself was reduced from 96 pages to 21. In early 1944, the NCTT carried out a three-month experiment with its version of a Japanese-style Chinese telegram between Jinan and Jining, two large cities in Shandong province. The result was far less encouraging than the NCTT had expected, however. A Japanese report drawn up in early 1945 concluded that the new system was "theoretically epoch-making but still needed further study in actual application." One major problem was lack of familiarity with Japanese *kana* among the Chinese public so that Chinese telegrams had to be translated into *kana* by employees of the company. Chinese businessmen who were used to sending messages in commercial codes, the report admitted, would be reluctant to have their telegrams transcribed at the telegraph office for fear that their contents would be leaked.[65] Several Chinese employees at the Jinan Central Telegraph Office also submitted a letter to NCTT headquarters listing the disadvantages and inconveniences of the new sys-

65. "Washiki denpō shinpen shikō ni kansuru ken" (drafted April 4, 1944); Kōno Kunio, "Fukumeisho" (January 20, 1945), 2, NCTT Records 2028/1642.

tem. In addition to the extra time needed to transcribe Chinese into *kana* codes, and to the lack of secrecy in such a process, they pointed to the high probability of error due to the very short codes.[66] However, the NCTT report acknowledged the need for replacing the Chinese-language telegram—based on Morse code, an artifact of European culture—with an "East Asian Chinese-language telegram," and noted that if used on automatic telegraphic circuits, it would improve efficiency considerably. Therefore, the project was worth improving so that "even those Chinese who were conservative by nature would be willing to abandon their old telegram system." Given the difficulty involved, however, the report recommended postponing application of the Japanese-style Chinese telegram system in the East Asia region until after Japan's victory in the Pacific War.[67] With about six months left before Japan's surrender, such a day seemed more and more remote.

ENTRENCHED JAPANESE INTERESTS

In 1942, Watanabe Otojirō, the head of the NCTT Business Department since its inception, published a book entitled *Telecommunications National Policy and Telecommunications Enterprise.* Having previously served in both the MOC and in Korea, the 40-year-old Watanabe was perhaps the most astute Japanese expert on telecommunications management. Watanabe considered prewar Chinese telecommunications administration a total failure, because of such problems as disunity of operations, low communication demand, unstable finances, and poor management. Understandably, he applauded the drastic changes that had taken place under Japanese control in telecommunications in China since 1937, which he attributed to the advanced management and technology brought to China by Japan. Watanabe had one serious complaint, however: the continued division of telecommunications operations into several geographical areas on the basis of the principle of "administration through political divisions" undermined the mission of telecommunications.[68]

---

66. Zeng Jiarong, Ma Shiguang, and Guo Yubao, "Dui Heshi Huawen dianbao zhi ganxiang" (n.d.), NCTT Records 2028/1642.

67. Kōno Kunio, "Fukumeisho" (January 20, 1945), 32–33, NCTT Records 2028/1642.

68. Watanabe, *Denki tsūshin kokusaku to denki tsūshin jigyō*, 298.

Chinese resistance and foreign establishments were not the only obstacle to Japan's effort to further consolidate telecommunications operations in China. Japanese-controlled national policy companies often encountered problems from other Japanese institutions as well. Some were largely a result of conflicting economic interests, whereas others were based on political considerations.

The problem with amalgamation of Japanese telecommunications began in Manchukuo, where different administrative zones existed alongside powerful quasi-governmental corporations. In eastern Manchuria, the Jiandao area posed a special problem for MTT expansion. This area bordered Japan's colony of Korea and had a largely Korean population, and the GGK had been operating the telephone exchanges as well as telegraph service, including Korean-language *han'gŭl* telegrams. Although the Jiandao area was supposed to be incorporated into the MTT, only after the intervention of the Kwantung Army and through the mediation of the Korea Army did negotiations begin in 1935 between the MTT and the GGK over transfer of GGK telegraphic facilities. Initially, the GGK insisted on deferring the transfer until after the MTT had unified all other telecommunications facilities in Manchukuo, while the MTT demanded immediate relinquishment, citing the unreliability of existing service and the time needed to improve the quality of operations. The negotiations were by no means always amicable, but under pressure from the military, the GGK gave in once the MTT promised that it would undertake unification measures immediately after the transfer. The GGK also agreed that telegrams between Jiandao and Korea would continue to be treated as domestic telegrams, although at a different rate, and that Korean *han'gŭl* telegrams would be continued, with the town of Tumen added. GGK employees at the Jiandao Telegraph Office were allowed to stay.[69]

The well-established Japanese administrative spheres in southern Manchuria—the Kwantung Leased Territory and the SMR Zone—

---

69. "Kantoshōnai ni okeru Chōsen sōtokufu shokan no denki tsūshin shisetsu o Manshū denshin denwa kabushiki kaisha ni jutōsuru ni itaru made no keii," MOC Records I-196; "Zai-Kanto Chōsen teishinkyoku shokan denshin bunshitsu no Manden kaisha e juto Hoka kyōgikai no ken," MOC Records I, 186; W, "Sen-Man no kyogai no tsūshin shisetsu," in Teishin gaishi kankōkai, *Teishin shiwa* II: 210; *DDJS* 6: 407. Ironically, the use of *han'gŭl* was banned in Korea just a few years later.

posed a much larger problem. Despite the agreements reached over the status of the telecommunications facilities in these two areas, actual implementation was anything but smooth. Most Japanese agreed in principle that all existing communications facilities would perform most efficiently and economically when fully integrated. Since the MTT would almost certainly run in the red initially, some argued, it would be fiscally wise to operate it together with the Kwantung Communications Bureau, which was turning a profit. But because Japan had not agreed to relinquish any of its established interests despite the creation of Manchukuo, some Japanese officials in Manchukuo proposed that it would be better if the MTT were incorporated in Japan so that it could take over Japanese facilities in the Kwantung area. Many officials there were cool toward the merger, however. For one thing, it was estimated that the Kwantung Territory would lose annual revenues of more than 1 million yen if its telecommunications were to merge with the new company. The Ministry of Colonial Affairs, which was directly in charge of administering the Kwantung Territory, requested that the decision be postponed until it discussed matters with officials in the field.[70]

The SMR Zone, consisting of narrow strips of land adjacent to the famed railway that had been under Japanese jurisdiction, was another bone of contention. Although many MTT employees and managers came from the SMR, the latter refused to relinquish its control over telecommunication operations in the SMR Zone. It was not until November 1937, four months after the outbreak of the conflict in China, that Japan finally announced the abolition of extraterritoriality and other special rights in Manchuria. The agreement also formally transferred communications rights held by the SMR to the Manchukuo government. Even then, there was a remaining problem: international telecommunications. Since Manchukuo was not recognized by most nations and was barred from international conventions on telecommunications, communications service between Manchuria and most foreign destinations was classified as Japanese traffic so as to avoid possible rejection. In this way, the MTT was to provide such service on behalf of the MOC.[71]

70. "Manshū ni okeru tsūshin jigyō ni kansuru ken" (August 9, 1932), in *Shōwa zaiseishi shiryō*, microfilm (Microfilm 135–1).

71. Okazaki Seichi, "Minami-Manshū tetsudō fuzokuchi ni okeru tsūshin gyōseiken no chōsei riyū ni tsuite," *TKZ* 352 (December 1936): 2–9. On the effect on postal ser-

Even after territorial consolidation was accomplished, the MTT's control over telecommunications in Manchukuo was incomplete. In August 1938, the Manchukuo government, together with the MTT, launched a major consolidation of all telecommunications facilities in Manchukuo, with the stated goal of "economically rationalizing" their construction and operations.[72] Many Japanese enterprises and organizations in Manchukuo continued to operate their own telecommunications facilities. Given the long presence of Japanese enterprises in the area, it was not surprising that many of them maintained mini-telecommunications networks for their own internal use. The sharp increase in Japanese business activities in Manchukuo further increased the number of these private facilities. In late 1937, there were 47 special telegraph circuits with a total of 8,233 km of lines operated by five large companies such as SMR. In addition, Japanese-controlled security, aviation, and meteorological services had their own systems, with thousands of kilometers of communications lines either installed or planned. The situation with telephones was worse, with 4,034 telephones in 68 private branch exchanges (PBX) for special use by railways, business, and government agencies. The SMR alone owned 122 toll lines over a total distance of 20,924 km.[73]

As military preparation and economic development in Manchukuo entered a new phase in the late 1930s, overlapping communications facilities would become a serious drain on resources. From the MTT's perspective, even the private telephone exchanges installed for internal use by large business enterprises and government agencies were often used for non-business communications, thus siphoning off potential revenue for the company. In many localities, as the MTT discovered, police telephones were occasionally used by the public, further aggravating its problem of losing revenue.[74] As a temporary measure, the MTT strove to reach a special agreement with each user. The Manchukuo News

---

vices, see "Chigai hōken no tehai to Nichi-Man yūbin," *Teishin no chishiki* 2.1 (January 1938): 8.

72. Yūsei sōkyoku, "Zen-Man denki tsūshin shisetsu no tōgo seibi yōkō" (August 1938) and "Zen-Man denki tsūshin shisetsu no tōgo seibi o hitsuyō tosuru riyū" (August 1938), MOC Records II-270.

73. "Sen'yō tsūshin shisetsu, kansho jimu sen'yō tsūshin shisetsu chō" (October 1937), MOC Records I-183.

74. *DDJS* 6: 394–95.

Agency, for instance, operated an extensive wireless network in Manchuria and was in direct communication with Japan. In early 1937, the MTT reached an agreement with the agency: in return for transfer of all wireless facilities to the MTT, the company gave special permission to the news agency to send telegrams not related to its news-gathering operations.[75] In contrast to the cases of the Fushun Coal Mine and Shōwa Steel Works, which operated in limited geographical areas, connecting the SMR's PBX with the MTT telephone network proved most difficult. Partly because of that, the MTT moved from a single telephone rate with unlimited use to a unit rate system so as to prevent PBX systems from hurting the company's revenue.[76] After considerable negotiation, the SMR finally agreed in 1939 that its PBX system would not be used for non-railway-related communications.[77] Only then did MTT more or less complete its consolidation of telecommunications in Manchukuo.

In China proper, similar battles were fought among the Japanese over the consolidation of telecommunications facilities, although perhaps on a smaller scale. The conflicts of interest among different Japanese groups remained a serious problem. The situation in Inner Mongolia was quite serious: of the 18,471 km of telegraph lines in the area, about 8,000 km were devoted to railway communications, mostly along the railway from Beijing to the important city of Baotou. Since the area was vast and thinly populated, investment in new telegraph lines would be economically unsound. In December 1937, the Japanese Army took the unusual measure of ordering the incorporation of the railway communications system into the soon-to-be established Inner Mongolian Post and Telecommunications Administration in the puppet regime. This unprecedented move encountered strong opposition from the SMR, which was operating the railways. The SMR mounted a vigorous campaign, both in China and in Tokyo, to derail the amalgamation. In the end, a compromise was reached. Although railway telecommunications was supposed to be under the jurisdiction of the Inner Mongolian government, the newly established Inner Mongolia Telecommunications Facility Company (IMTFC) was limited to the construction and repair of

75. Ibid., 390–91.

76. Yanagina Masaya, "Manshū denwa zakki," in Manshū denden tsuiokuroku kankōkai, *Akai sekiyō*, 55–62.

77. Manshū denshin denwa kabushiki kaisha, *Denden tokuhon*, 401.

toll lines, leaving other aspects of railway telecommunications largely intact. What was considered the test case of unified control over telecommunications facilities in Inner Mongolia thus had mixed results: the measure made it easier to procure needed material for repairs, but the trunk lines of communications in Inner Mongolia were merged into a single line that was vulnerable to disruption.[78]

Consolidation of telecommunications in other parts of occupied China was also subject to political as well as to business rivalries among the Japanese. Competing interests among different Japanese groups prevented the emergence of a single telecommunications enterprise in occupied China, just as political considerations made it impossible to unify different puppet regimes under one roof. For instance, it was primarily political expediency that made Inner Mongolia distinct from China proper, justified in terms of their ethnic makeup and importance to Japan's defense needs. As each company sought to maximize its interests, the argument also emerged that the amalgamation of telecommunications operations was undesirable from a business perspective. The lack of funding and of sound business prospects in Inner Mongolia, according to such views, could be harmful to the NCTT's own operations. As a result, even advocates of unified operations in North China were not really keen about merging their operations with those in Inner Mongolia, despite the geographical proximity of the two areas.[79]

Japan's entrenched treaty rights in China were another major contentious issue. Recall that in Prime Minster Konoe's 1938 speech proclaiming "a New Order in East Asia," he pledged to respect the sovereignty of China and suggested giving "positive consideration to the questions of abolition of extraterritoriality and of the rendition of the concessions and settlements, matters that are necessary for the full independence of China."[80] In reality, however, as soon as it became apparent that the conflict in China would be protracted, many MOC officials considered Japan's military occupation a great opportunity to settle the age-old question of strengthening Japanese telecommunica-

---

78. *GKDTS*, 10: 668–72; *DDJS*, 6: 451.

79. "Mōkyō denki tsūshin setsubi kabushiki kaisha to Kahoku denshin denwa kabushiki kaisha no gappei ni tsuite," MOC Records I-23.

80. "Konoye on the New Order in East Asia," in Lebra, *Japan's Greater East Asia Co-Prosperity Sphere*, 68–70.

tions rights in China once and for all. In this context, the relationship between existing Japanese telegraph offices in China and the new telecommunications companies became delicate. The Japanese telegraph offices in China, the MOC officials pointed out, were one of the most important concrete forms of Japan's communications rights in China. Even if Japan were to avoid acquiring completely new rights in newly occupied areas, it was nonetheless appropriate to expand on the existing rights. They questioned the wisdom of suggestions that these existing rights be transferred to a pro-Japanese regime in China because in the past such regimes had not lasted. Rather than abandoning rights acquired through strenuous efforts, these officials argued, only after the future of the pro-Japanese regime in China became ascertained could Japanese telegraph offices be gradually merged with the new telecommunications enterprises to be set up in China.[81] As a result, despite its reputation as a champion of consolidating operations in China, the MOC favored keeping existing Japanese government installations in China separate from the new telecommunications companies.

The continued existence of these Japanese government telegraph offices in areas serviced by the new telecommunications companies posed a business problem for the latter. Many Japanese working in the "national policy companies" found such continued separation detrimental to the companies' revenue. At an annual gathering of telecommunications operators in the East Asia region in late 1940, Hirada Kōzō, a senior executive of the CCTC, listed unlawful foreign telecommunications facilities in Shanghai as the foremost obstacle to be overcome. After praising the MOC's elimination of the GNTC's concessions in Japan, Hirada pleaded with the MOC to take steps to end "the competitive relationship between the Japanese government telegraph office in Shanghai and our company."[82] This was understandable, given the heavy volume of traffic between Japan and Central China handled by the office and the rather precarious financial situation of the CCTC.

Little progress was made, however, even after signing of the Treaty of Alliance between Japan and the Wang Jingwei regime in January 1943.

---

81. Denmukyoku, Gaishinka, "Zai-Shi teikoku tsūshinken no kakujū kyoka o hitsuyō tosuru riyū" (February 1938), MOC Records I, 245.

82. Hirada Kōzō's report in Tōa denki tsūshin jimukyoku, *Tōa denki tsūshin kyōgikai dai-2-kai kaigi gijiroku*, 35.

In December 1942, the Japanese government had adopted "Basic Principles Concerning a Settlement with China" and had announced (1) its intention to return all its exclusive concessions in China, (2) its plan to relinquish all extraterritorial rights enjoyed by the Japanese in China, and (3) its assent to Chinese recovery of the international settlements.[83] In that context, the Japanese government finally agreed to incorporate its telegraph offices in China into the Japanese-controlled telecommunications companies. The measure became necessary not only because of politics but also because of the increasing emphasis placed on streamlining operations and improving the efficiency of telecommunications operations. The 80 or so Japanese government employees working in telegraph offices in Shanghai, Qingdao, and Chefoo could work in other capacities at a time when the shortage of skilled personnel in the imperium was acute. Moreover, once cable and wireless communications were well integrated, those Japanese telegraph offices that had been set up only to operate submarine cables would become antiquated. Perhaps more important for Japan, since the telecommunications companies in North and Central China were firmly under Japanese control, the separate Japanese offices that had been crucial to Japanese telecommunications in China became redundant. After the submarine cables ceased operations in Shanghai in August 1943, conditions in Central China made it increasingly difficult to continue to send telegrams by wireless alone. Yet it was not until June 1945 that the Japanese Telegraph Office in Shanghai finally transferred all its operations to the CCTC, with the exception of its Acceptance Service and the wireless connection between Shanghai and Osaka. Its 102 employees, 54 of whom were Japanese, were reduced to a mere 12.[84]

Those MOC officials were not the only ones who favored preserving existing Japanese installations in China, however. The Japanese military considered Japanese government telegraph offices in China to be of strategic importance, although they often differed over the types of control needed. The port city of Qingdao in North China provided one such testing ground. Early in the war, the Japanese Army insisted that the NCTA take over all facilities except the operation of the Qingdao–

83. For an English version, see Foreign Affairs Association of Japan, *The Japan Year Book, 1943–44*, 985–86.

84. *GKDTS*, 11: 285–87.

Sasebo submarine cable. The Navy opposed this vehemently on the grounds that separate Japanese control of the cable was vital. To counter the Army's pressure, the Navy requested support from the MOC and prevailed.[85] Even after the 1943 Treaty of Alliance with the Wang regime was signed and the MOC reversed its position, the Japanese military in China did not want to take any chances by relinquishing direct Japanese government control of these key assets. In an order issued in November 1943, the China Expeditionary Army expressed support for the principle of amalgamation but added a number of strict conditions: (1) the telecommunications companies would not be allowed to use former Japanese-managed facilities for purposes other than communicating with Japan; (2) senior employees as well as those handling military telegrams at these telegraph offices must be Japanese; and (3) any changes in personnel must be approved by the Japanese government.[86] Given the reluctance on the part of the military, it is not surprising that the process of returning assets to China through the Japanese-controlled telecommunications companies met with repeated delays. The Japanese government continued to hold on to some of its telegraph offices until the end of war. In North China, for instance, the Japanese government finally transferred the Chefoo Telegraph Office to the NCTT in May 1945 but held on to the Qingdao office until Japan's surrender.[87] Even in the last days of the war, therefore, Japan would not give up direct control over one of the most vital links in the empire. The much-desired consolidation was never completed.

Judged by Japan's basic goals, Japanese control over telecommunications in China was remarkably successful. Within months of the outbreak of war in China, Japan had established several regional telecommunications companies by adapting the MTT model. Firmly under Japanese control, these national policy companies restored telecommunications operations in their respective areas, rationalized telegraph and telephone operations,

85. Director, Qingdao Telegraph Office to Tachibana (February 1, 1938), MOC Records I, 269.

86. Chief of Staff, China Expeditionary Army, "Zai Shi Nihon denshinkyoku to tsūshin kaisha kan chōsei ni kansuru ken" (November 17, 1943), NCTT Records 2028(2)/40.

87. *DDJS*, 6: 415.

and expanded connections with Japan. Given the fact that fighting never ceased in occupied areas in China, it was remarkable that all these companies overcame the financial odds and stayed afloat throughout the war, with varying degrees of success.[88] In 1940, all three major telecommunications operators set up in China and Manchuria seemed to be in reasonably good shape and were paying 6 percent returns to their investors.[89] To solicit grand visions for its future expansion, the NCTT even hosted an essay contest in that year on the subject of a new communications policy for the Asian continent.[90] Later that year, with two successful years of operation, the company produced a number of outlines on future communications expansions for North China and beyond. Targeted over 5-, 10-, and 30-year periods, respectively, they reflected the ambitions of the "national policy company." Billed as part of the East Asian land planning endeavor, the most ambitious among them envisioned a total investment of 1,625 million yen over a 30-year period. Total telephone subscribers would reach three quarters of a million. North China would feature "powerful wireless bases" as part of the East Asian communication system, as well as cables extending to Manchuria and Central China, and to Central Asia and India.[91]

To a great extent, these Japanese-controlled telecommunications managed to secure Chinese cooperation, even though they never completely eliminated resistance. They strove to expand their customer base among the Chinese population, relying on new technologies and other innovations. Yet, there were also severe limitations to Japan's efforts of consolidation of telecommunications in China. As Watanabe pointed out, consolidation had succeeded only to some extent within fairly separated areas on the Asian continent. Although Watanabe recognized the

---

88. One of the peculiar aspects of the war was that postal service between Japanese-occupied areas and Nationalist-held areas continued throughout; see Forman, *Changing China*, 285.

89. See, e.g., replies by the Asia Development Board's director of Economic Affairs in a 1940 Imperial Diet session; Kōain seimubu, comp., *Dai-75-kai Teikoku gikai Shina kankei shitsugi otoshu* (1940), 646, 675–76.

90. Kahoku denden kurabu, *Tairiku ni okeru tsūshin seisaku o ronzu.*

91. Kōain Kahoku renrakubu, "Kahoku denki tsūshin 5-ganen keikaku hōshin an" (September 5, 1940); "Kahoku denki tsūshin 5-ganen keikakusho" (September 18, 1940); "Kahoku denki tsūshin 10-ganen keikaku yōkō"; "Hokushi kokudo keikaku no ichibumon toshite no sogo tsūshin keikaku yōkō" (November 7, 1940), all in NCTT Records 2024(2)/235.

need to adapt to the special circumstances of each area under Japanese occupation, he reminded his fellow Japanese of the desirability of forming a single seamless telecommunications network throughout China and the importance of cooperating in unifying its operating methods.[92] The political reality as well as the entrenched Japanese interests made that unlikely. The task of integrating these local systems into a single empire-wide telecommunications network had to be attempted elsewhere.

92. Watanabe, *Denki tsūshin kokusaku to denki tsūshin jigyō*, 298.

CHAPTER 8
Gaining Control in Southeast Asia

Speaking at a gathering of communications engineers in May 1937, Matsumae Shigeyoshi described a strange phenomenon he noticed during his recent trip to Southeast Asia: whereas a letter from Britain or the United States might take as little as three or four days to reach Singapore, mail to and from Japan, which was "really just next door if you look at a map," would take up to a month. The difference was the result of different types of transportation, but it had less to do with technology and more with politics: steamship was the only means of postal communication between Japan and Singapore, whereas both Britain and the United States had established regular air routes to their respective colonial outposts in Asia.[1] Matsumae's point was that modern technology alone could not reorder space; rather, it had to be backed by what he called "communications policy" to fully realize its space-adjusting potential.

Matsumae's trip to Southeast Asia was part of the effort to define such a communications policy. A similarly unsatisfactory situation existed with telecommunications: Japan could communicate with only major cities in Southeast Asia through foreign-owned cables and occasionally by direct wireless. However, telegram rates between Japan and the region were much higher than those between Japan and Europe or those between European countries and their colonies in Southeast Asia. Moreover, Japan was unable to exchange telegrams in Japanese

1. Matsumae Shigeyoshi, "Nan'yō shoppō no shisatsu o oete," *DTGZ* (1937): 504–5.

characters (*Wabun denpō*) with the tens of thousands of Japanese residents in the region.[2] Moreover, Japanese exports in telecommunications equipment in the region faced increasing difficulties due to competition and protectionist policies.

If Japan's disadvantage was obvious, the solution was not. The problem was primarily political; the area consisted largely of colonies of Britain, the Netherlands, and France, as well as territory within the U.S. sphere of influence. Japan's attempted expansion into the region in the 1930s would meet with suspicion and resistance from these powers. Bridging the gap between technological potential and political reality in Southeast Asia was thus a major task for Japanese communications officials in Tokyo during the late 1930s. Given the complex political realities, extending Japan's imperial telecommunications network into Southeast Asia would be even more challenging than it had been in China, even after Japan's military victories after 1941 accomplished what had been impossible by diplomatic and economic means.

SOUTHERN INTEREST REKINDLED, 1930–39

Nan'yō and Nanpō

In the prewar Japanese lexicon, there were a number of regional concepts that applied to the vast area lying to the south of Japan and China. The oldest was "Nan'yō" (the South Seas), which was used in the Tokugawa period to describe the entire region. The term gradually came to refer only to the South Pacific islands that Japan administered after World War I, now sometimes called the Inner South Seas (Uchi Nan'yō). Another general term, "Nanpō" (the Southern Region), tended to emphasize the continent and other large landmasses, reaching as far as Burma and Australia and New Zealand. By the second decade of the twentieth century, some Japanese also began to use the term "Southeast Asia," even before it became established in the West.[3] To some extent,

2. For a survey of Japan's telecommunications links with Southeast Asia before and during the war, see *DDJS* 6: 479–535.

3. For a study that traces the development of the regional concept, see Shimizu, "Southeast Asia as a Regional Concept in Modern Japan." See also Hikita, ed., *Nanpō kyōeiken*, 4–5.

the lack of a single well-defined term reflected Japan's shifting perception of its own relationship with the entire region; indeed, Japan's "Southern Advance" (*nanshin*) would emphasize different destinations at different times.

Located over 1,500 miles from Japan proper, the 1,400-plus Nan'yō islands became the most far-flung and dispersed territories under Japanese administration. Maintaining communications with and between the islands was vital but difficult. Japanese shipping companies, supported by government subsidies, carried most mail and other supplies. By 1940, Japanese planes regularly flew to the islands. The regular ferry service could not meet the need of more urgent correspondence: even the Nan'yō Agency in Palau received surface mail from Japan only twice a month in the mid-1930s. As a result, telegraph and later telephone services played a crucial role. The Imperial Navy took over the existing German-built cables and communications facilities on the islands such as local telephone exchanges. These facilities, together with the post offices and wireless stations were later transferred to the Nan'yō Agency.[4] Since the island of Yap had been the center of Germany's cable network in the Pacific, the Japanese redirected the captured Yap–Shanghai submarine cable to Okinawa and thus linked it to Japan's domestic network. This became the only Japanese cable in the area. Maintaining secure and steady communications to the islands assumed such importance that in 1923 the MOC named its third and newest cable ship the *Nan'yō maru*. With a displacement of 3,600 tons and a capacity to carry 900 nautical miles of deep-sea cables, *Nan'yō maru* was the largest cable ship in Japan and reflected the new spatial realities of empire.[5]

In comparison, the Japanese presence in the area known as the Southern Region (present-day Southeast Asia) had a much longer history, predating the arrival of European colonial powers. When Japan embarked on industrialization at the end of the nineteenth century, Southeast Asia became an important market for Japanese manufactured

---

4. *DDJS* 6: 352–65; Takazaka Kiichi, "Nan'yō guntō ni okeru tsūshin gyōmu," *TKZ* 329 (January 1936): 93–104.

5. Nippon denshin denwa kōsha kaiteisen shisetsu jimusho, *Kaiteisen hyakunen no ayumi*, 212–19; K. R. Haigh, *Cableships and Submarine Cables* (London: Adlard Coles, 1968), 382–83.

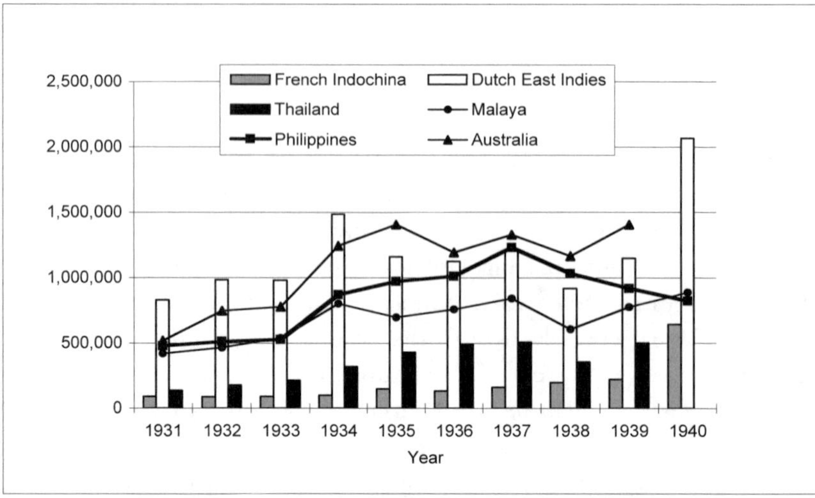

Fig. 2 Japan's telegraphic traffic with the Southern Region, 1931–40 (SOURCE: *Denmu nenkan*, 1942).

goods. At the same time the region's rich natural resources, particularly rubber, tin, and petroleum, became increasingly indispensable to Japan's growing industry. In the wake of the worldwide economic depression, Japanese interest in the Southern Region intensified further during the 1930s (see Fig. 2). The Imperial Navy had its own designs.[6]

Beginning in the mid-1930s, the MOC launched a series of studies of the telecommunications operations in Southeast Asia. In early 1937, Matsumae Shigeyoshi, then head of the Investigation Section in the MOC's Engineering Bureau, embarked on an extended tour of the entire region. From January to April, he surveyed telecommunications conditions in the Philippines, the Dutch East Indies, Singapore, British Malaya, Thailand, and French Indochina. Although to his regret his schedule did not permit him to stay at any one place for more than ten days, Matsumae gathered a considerable amount of information. After returning to Japan, he spoke widely about his observations. Matsumae reported in great detail on the state of telecommunications in

6. See Shiraishi and Shiraishi, *The Japanese in Colonial Southeast Asia.* For an overview of the 1930s, see Peattie, "Nanshin: The 'Southern Advance,' 1931–1941," in Duus et al., *The Japanese Wartime Empire*, 189–242.

the region. In the Commonwealth of the Philippines, which was to become independent in seven years, the government ran the telegraph, whereas the more lucrative telephone business was in the hands of America's telecommunications giant, AT&T. He noticed that RCA, which operated the Philippines' radio communications with the United States and Japan, had been very hospitable to him because the Japan traffic accounted for a large portion of the company's profits—30 percent. In the Philippines, the Americans were relatively open to foreign products as long as the price was right, although government taxes—30 percent on imported industrial goods—put non-American imports at a disadvantage. The Netherlands, Britain, and France, according to Matsumae, pursued a communications policy that would strengthen the crucial links between the colony and its home country.[7]

Matsumae's interests apparently went beyond telecommunications facilities, for he also paid close attention to the colonial policies of various countries, their images of Japan, and the presence of Japanese interests there. In the area of broadcasting, according to Matsumae, Japan faced much discrimination. Newspapers under the control of the British, he noticed, published the broadcasting programs of European countries but not that of Japan. Matsumae also alleged that French warships in French Indochina deliberately used electric interference to make it impossible to hear news broadcast from Japan. In addition, he was outraged by the negative portrayal of Japan in geography textbooks used in Philippine schools. As a result, his report shed light on European and American colonial practices in Southeast Asia, as compared with Japan's own. Matsumae was critical of Japan's own lack of initiative, accusing the MOC of being "seclusionist" about overseas expansion in this area.[8] Understandably, he returned to Japan with a new sense of urgency about Japan's position in Southeast Asia. MOC studies such as Matsumae's 1937 report provided the basis for a new, proactive communications policy toward Southeast Asia.

---

7. Matsumae, "Nan'yō shoppō no shisatsu o oete," *DTGZ* (1937): 494–506; "Nan'yō shoppō ni okeru denkitsūshin jigyō," *TKZ* 348 (August 1937): 2–26.

8. Matsumae, "Nan'yō shoppō no shisatsu o oete," *DTGZ* (1937): 494–506; idem, "Nan'yō shoppō ni okeru denkitsūshin jigyō," *TKZ* 348 (August 1937): 2–26.

Renewed Presence

In July 1937, two months after Matsumae's return from Southeast Asia, war broke out with China. As Japanese sought the ever-elusive final victory in China, Southeast Asia took on additional importance in terms of public diplomacy as well as strategy of encirclement. To begin with, there were millions of overseas Chinese, particularly in British Malaya and in the Philippines, who rallied behind anti-Japanese causes. Western powers in the region became unfriendly to Japan, especially after reports circulated about Japanese atrocities in China. In early 1938, the Cabinet Planning Board set up a Council on Southern Policies, which drew up a series of policy recommendations for Japan's approach to the Southern Region. One called for stepping up Japan's "information campaign" vis-à-vis Southeast Asia to counter the "misunderstandings" spread by foreign countries and the Chiang Kai-shek regime since the outbreak of the China War. But radio broadcasting alone was often unstable due to atmospheric interference. Besides stationing Japanese correspondents in major localities with sizable populations of Japanese residents, the proposal recommended that Japan provide more news stories to the area by regular telegraph service. Japan would supply news to Japanese newspapers, as well as foreign newspapers, free of charge. The proposal emphasized the critical importance of speedy transmission of news to Southeast Asia, "even if it beats news from China and elsewhere by only an instant."[9] Efficient communications links with Southeast Asia had become vital to Japan's diplomacy in the realm of public opinion.

The prolonged war in China also prompted Japan to take the first concrete step toward advancing into Southeast Asia. In February 1939, Japanese troops occupied Hainan, a large tropical island off the southern China coast not far from northern French Indochina. Although the move was part of a larger attempt to cut off foreign aid to the Chiang Kai-shek government and win the war in China, Japan had other calculations as well. The Imperial Japanese Navy had long harbored an interest in Hainan as a launching pad for further extending its operations

9. Kikakuin, Sōmubu, "Tai-Nan jōhō kyokyū no ken" (February 23, 1938), in *KSS* 6: 297. For a Western assessment of increased Japanese propaganda activities in Southeast Asia before the Pacific War, primarily radio, see Robertson, *The Japanese File*, 83–110.

into Southeast Asia. Discovery of high-quality iron deposits on the island also added another incentive. Suddenly Hainan became a hotbed of Japanese military and economic activity. Two months after the island came under Japanese occupation, the Navy asked the MOC to dispatch several engineers to join ITC employees in restoring telecommunications facilities on Hainan and laying a new cable that would link the island to the city of Guangzhou (Canton), then already under Japanese occupation.[10]

By this time, the ITC was working closely with the military to build and operate telecommunication facilities in newly occupied areas. It also proved to be a useful source of information in the field for the MOC, which was eager to stay abreast of developments throughout the region. For instance, the Taihoku branch of the ITC sent reports to the MOC on developments on Hainan as well as in Southeast Asia. According to one such report, the Government-General of Taiwan (GGT) had earmarked 300,000 yen for investigating conditions in southern China and Southeast Asia. As a result, two GGT officials, one of whom was a technician, were dispatched to French Indochina.[11] Just as the GGT had played a major role in Japanese occupation of Xiamen just across from the Taiwan Strait, so Japanese officials in Taiwan considered the island to be the gateway to Hainan. Not only did all goods and personnel have to go through Taiwan to reach Hainan Island, but all communications with Hainan were handled through a single wireless circuit between Haikou and Taihoku. Given Taiwan's location and history as Japan's first overseas colony, the GGT saw its mission as carving out a sphere of influence in the region and serving as a natural leader in Japan's Southern Advance.[12]

Japan's growing interest in Southeast Asia had a visible impact on the South Sea islands. In 1936, the Japanese government adopted a ten-year development plan for Nan'yō that included expansion of wireless facilities at a budget of 1.22 million yen. Expansion of wireless facilities

10. *KDTKKS*, 51–52, 114.

11. "Futsuin ni taisuru jōhō" (April 1941), MOC Records II-461.

12. On Taiwan's role in Japan's Southern Advance, see Gotō Ken'ichi, "Taiwan to Nan'yō," in Ōe Shinobu et al., eds., *Kindai Nihon to shokuminchi* 2: 147–50; Adam Justin Schneider, "Business of Empire: Taiwan Development Company and Southern Advance," Ph.D. diss., Harvard University, 1998.

in the area was not just for meeting the growing demands of public communications and broadcasting, but also to support increased Japanese aviation and shipping activities in the South Pacific. In 1939, Japan began regular air service to Palau via Saipan. As the international situation became tense and the need for communications links assumed greater urgency, the time for the plan to be completed was shortened to three years, and the budget was increased to 1.61 million yen. The ITC constructed and maintained a 10-kW shortwave transmitter on Palau, which was used for both wireless communication with Japan and radio broadcasts beamed at Southeast Asia. Since the Nan'yō Agency was in charge of the telephone service and NHK was responsible for broadcasting, the same facilities were shared by the three organizations.[13] In May 1941, wireless telephone connection was established between Palau and Tokyo. The two older colonial outposts, the South Sea islands and Taiwan, thus began to assume new strategic roles in Japan's Southern Advance.

OPPORTUNITIES AND OBSTACLES, 1939–41

The outbreak of war in Europe in September 1939 presented Japan with new opportunities in Southeast Asia. A week after the war in Europe started, the MOC wasted no time in drafting "Measures Concerning Advances into the South in Telecommunications During the Turmoil in Europe." A clear indication that the Japanese were consciously exploiting the events in Europe for their own interests in Southeast Asia, the document proposed "establishing and strengthening our communication rights" in Southeast Asia by opening new, direct wireless connections with the region. It also called for increasing the export of telecommunications equipment and construction services, and under some circumstances operating under license or through investment. In this way, Japan could establish "leadership authority" (*shidō ken*) through technology export and capital investment. To avoid arousing the suspicions of the Western colonial governments, the MOC called on non-

13. *DDJS*, 6: 361–62.

governmental entities such as the Telecommunications Association, ITC, and JTTCC to assume these tasks.[14]

As the war in Europe continued over the next year, the MOC worked to formulate a "communications policy toward the Southern Region." The rapid turn of events in the European War—Germany's invasion of Scandinavia and especially the fall of the Netherlands and France—greatly hastened the tempo of its deliberations. Given the new political uncertainties in French Indochina and the Dutch East Indies, the MOC suggested adopting measures that would induce their respective governments to privatize both domestic and international telecommunications, thus making it easier for them to "become dependent on Japan's communications influences." Specifically, Japan should strive for "strong influence" over technical and design matters that were the basis of telecommunications facilities.[15] It was around this time that the MOC prepared several confidential plans on the feasibility of setting up a regional communications network in Southeast Asia that would later be incorporated into the Japan-Manchukuo-China bloc.

By late July 1940, on the eve of Japan's proclamation of a Greater East Asia Co-Prosperity Sphere, MOC officials had drafted perhaps the most ambitious and comprehensive plan ever made for telecommunications expansion in Asia. The plan called for the completion of ten major land and submarine trunk cables as well as a large number of wireless facilities over the next five years (see Map 4). Together they would form a so-called East Asian Stability Sphere Telecommunications Network. Its geographical scope would encompass Northeast Asia, Southeast Asia, and the South Seas mandate islands and would reach as far as Australia. Non-loaded cables would be used on land in Manchuria, China proper, and Indochina, but two underwater telephone cables and three underwater telegraph cables would be laid in Southeast Asia. In addition, all major cities in Southeast Asia and Australia would be connected by powerful wireless facilities as well. The total cost, which also covered the construction of three new cable ships, would be a staggering 600 million yen—almost five times more

14. MOC, "Ōshū dōran ni tomonau denki tsūshin no Nanpō shinshutsu ni kansuru sochi (an)" (September 9, 1939), MOC Records II, 461.

15. "Tai Nanpō tsūshin seisaku yōkō" (July 30, 1940), MOC Records II, 244.

Map 4 Planned cable routes in the East Asian Stability Sphere, 1940
(SOURCE: "Tōa anteiken denki tsūshinmō keikakusho," July 20, 1940, MOC Records II, 244).

than the estimated cost of the East Asian Trunk Cable Network drawn up a year and a half earlier.[16] With such an ambitious blueprint, visions of Japan's communication empire reached their apex, at least on paper.

At the end of August 1940, the deliberation over expanding telecommunications into Southeast Asia moved to the Cabinet Planning Board. It concluded that the South China Sea Western Circular Cable proposed by MOC, running from Takao in southern Taiwan to Batavia in the Dutch East Indies via Canton, Hainan Island, and Saigon, was to

16. "Tōa anteiken tsūshinmō keikakusho" (July 20, 1940), MOC Records II-244.

be given the highest priority. Compared with other routes, as the MOC noted, this telephone and telegraph cable was the "most feasible based on the current situation." The MOC took into consideration both political and economic concerns. A proposed cable landing in Singapore was dropped to avoid complications with the British authorities. The MOC predicted that this new trunk cable would draw traffic not only from Japan but also from China, given the large overseas Chinese population in Southeast Asia. MOC officials based their calculations on actual traffic between Japan and Southeast Asia in 1939, plus the annual increases of 40 percent on the Japan–Manchukuo and 10 percent on the Japan–China circuits. According to MOC calculations, the cable would generate 12 million yen in revenues for the ITC in its first year of operation from telegram, telephone, and phototelegraphy services, as well as from leases on telephone circuits.[17]

In November 1940, the Cabinet approved MOC's "Communications Policy Toward the Southern Region," with only minor changes. At the same meeting, the Cabinet also approved MOC's "Outline of Maritime Policy for the South," calling for acquiring various maritime rights in the area as well as opening new shipping routes to the region from Japan.[18] Japan now seemed ready to launch a major offensive in Southeast Asia to realize its "communications policy." The MOC's plan appeared rather modest and well-tested: it recommended using the ITC as the tool for expansion at the outset and setting up joint ventures with local companies in each country in the future.

Japan's advance into Southeast Asia after the outset of the European War generally was cautious and avoided direct confrontation with the British and the American presence and instead took a three-pronged approach at the weak links of European colonial power and at the only independent country in the region: although the Dutch East Indies was to be the main target for asserting Japan's pressure, Thailand and

---

17. See appendix to "Tai-Nanpō tsūshin seisaku yōkō" (July 30, 1940), MOC Records II, 244.

18. "Nanpō tsūshin seisaku yōkō an" (September 13, 1940), Cabinet approval on November 15. "Nanpō kaiun taisaku yōkō" (drafted September 9, 1940, approved on November 5, 1940), MOC Records II, 461. See "Nanpō kakuchitai ni taisuru honpō teiki kōro" (August 13, 1940), MOC Records II, 461. Perhaps because of the lack of documentation, a recent Japanese study concluded that the draft proposal never reached the Cabinet level; see Hikita, ed., *Nanpō kyōeiken*, 25*n*44.

French Indochina were to serve as stepping-stones in Japan's Southern Advance.[19] Contrary to Japan's expectations, however, European colonial powers would not yield to its pressure. In fact, Japan's actions not only met with strong resistance but also hardened American and British resolve to curtail Japan's ambitions in Southeast Asia.

The Dutch East Indies

The South China Sea Western Circular Cable, which would end in Batavia, received the highest priority at MOC and the Cabinet Planning Board for good reasons. The island of Java, where Batavia was located, boasted the best communication facilities in the entire region. When the ITC and JTTCC were striving to win the bid to install automatic telephone exchanges in Batavia, Japanese engineers found telecommunications facilities on the island to be quite advanced. One of them even cautioned his fellow Japanese at home not to associate Java with "naked dark-skinned children dancing under the palm trees."[20] More important were the oil fields. In July 1939, as the United States began applying pressure on Japan because of its actions in China, it announced the abrogation of the 1911 Treaty of Commerce and Navigation between the two countries. In response, Japan sought to lessen its reliance on energy imports from the United States by strengthening its relationship with the Dutch East Indies. Timing seemed to favor Japan at first. Following the outbreak of war in Europe, Japan's Ministry of Foreign Affairs briefly sought to prevent a British occupation of the Dutch East Indies and proposed a guarantee of its territorial integrity. Despite the fact that the Netherlands had been invaded by Germany on May 10, 1940, the Dutch became less accommodating with Japan because of Anglo-American support. Rejected by the Dutch government, Japan then concentrated on negotiating on economic matters. On May 20, 1940, Japan presented the Dutch government a list of thirteen vital commodities and asked for assurance of annual quotas.[21]

---

19. For the general military and diplomatic background, see Hata Ikuhiko, "The Army's Move into Northern Indochina"; and Nagaoka, "The Drive into Southern Indochina and Thailand," both in Morley, ed., *The Fateful Choice*, 155–208, 209–41.

20. Takenaka Satomi, "Ran'in ni okeru denwa no gaikyō," *DT* 5.19 (April 1942): 52–59.

21. For a general discussion of Japan-Dutch relations, see Nagaoka, "Economic Demands on the Dutch East Indies."

Faced with a rapidly deteriorating situation in Europe and mounting pressure from the Japanese, the Dutch colonial government in the East Indies agreed to open talks with Japan on a number of issues. The Japanese government was initially divided over priorities to present to the Dutch: the Ministry of Foreign Affairs wanted removal of limitations on Japanese entries (up to 1,633 Japanese were admitted each year), but the Ministry of Commerce and Industry and the Navy considered it paramount to secure supplies of petroleum, tin, and natural rubber. In January 1941, the Japanese government presented Dutch authorities with an extensive list of demands, including relaxation of restrictions on Japanese entering and working in the area and increased Japanese exploitation of natural resources. Eventually the Dutch agreed to short-term contracts for amounts far below Japanese demands.[22]

Upgrading telecommunications and transportation links constituted a separate category in the Japanese-Dutch negotiations. To improve communication links between the two countries, Japan requested permission to "lay a submarine cable using the latest technology, under Japanese control, in order to build secure and efficient communication between Japan and the Dutch East Indies." Since the only existing cable had to go through British-controlled Singapore, the two countries relied heavily on the direct wireless connections established in 1937: some 97 percent of the telegrams exchanged between Japan and the Dutch East Indies went by the wireless route. By setting up a new cable connection, Japan hoped not only to meet the expected increases in traffic but also to protect the secrecy of such communications from a potentially hostile power such as Great Britain or the United States. Japan also requested that the Japanese language be allowed in communications between the two destinations. Later, it added another request: that a Japanese company be allowed to handle traffic with Japan and also communication within the Dutch East Indies as much as possible.[23]

---

22. For studies in Japanese, see Adachi Yasuaki, "Tai-Ran'in kōshō seisaku no keisei," *Rekishigaku kenkyū* 643 (March 1993): 1–16; "Kaisenzen no keizai kōshō: Tai-Ran'in-Futsuin kōshō," in Hikita, ed., *Nanpō kyōeiken*, 101–34. On the Dutch response to Japan's economic expansion, see Peter Post, "Tai-Ran'in keizai kakuchō to Oranda no taiou," in Kobayashi Hideo, ed., *Kindai Nihon to shokuminchi* (3): *shokuminchika to sangyōka.*

23. The first set of demands and the Dutch replies are in *NGNB* II, 474–79. See also Hanaoka Kaoru, "Ranryō Indo haken keizai tokushidan zuiin Hanaoka jimukan hōkokusho," MOC Records II, 663.

Hanaoka Kaoru, a young MOC official specializing in foreign communication, was dispatched to Batavia as a member of the Japanese delegation. Once there, Hanaoka found the atmosphere so hostile to Japan that he could not even carry out planned investigations of telecommunications facilities and services in the Dutch East Indies. To make matters worse, the Japanese delegation was so preoccupied with the issue of importing petroleum that telecommunications issues were pushed aside. In desperation, Hanaoka sought to persuade his colleagues from the Ministry of Colonial Affairs, the Navy, and Japan's consul general to pay some attention to his mission. Because of the apparent political sensitivity of Japan's requests for building telecommunications links to the Dutch East Indies, Hanaoka cautioned his colleagues that they should build their case by emphasizing Japan's technological superiority as well as the economic benefits of Japanese submarine cables.[24]

Even this hope turned out to be unrealistic. Hanaoka had to content himself with presenting more modest requests. In November 1940, he told his Dutch counterpart what Japan wanted to achieve: (1) reduction of tariffs for both telegram and telephone service between the two countries; (2) addition of one circuit each for wireless telegraph and telephone; and (3) relaxation of the ban on using the Japanese language for telephone calls. The demand for a new submarine cable was dropped. To Hanaoka's surprise, the Dutch representative was cool toward these seemingly moderate requests. The Dutch official cited government financial difficulties as a reason for the inability to open new wireless circuits with Japan and attributed the failure to extend hours of Japanese-language telephone service to the lack of qualified Japanese-language specialists in the Dutch East Indies. In the end, the Dutch authorities made a minor concession by permitting the use of three commercial codes by Japanese businessmen.[25] As with other Japanese demands, very little progress was made regarding telecommu-

---

24. Hanaoka Kaoru, "Ranryō Indo haken keizai tokushidan zuiin Hanaoka jimukan hōkokusho," MOC Records II, 663. Hanaoka also published impressions of telecommunications in the Dutch East Indies as "Ran'in no denkitsūshin jigyō," *DT* 4.13 (1941): 64–69; see also Nakamura Karona, "Ran'in no denkitsūshin gijutsu," *DT* 4.13 (1941): 69–72.

25. Hanaoka, "Ranryō Indo haken keizai tokushidan zuiin Hanaoka jimukan hōkokusho," MOC Records II, 663.

nications linkages with the Dutch East Indies. The MOC's wish list for expanding Japan's telecommunications influence in the Dutch colony would remain just that for some time to come.

French Indochina and Thailand (1)

While the Dutch East Indies was important for its energy resources, both French Indochina and Thailand occupied strategic locations and were to serve as stepping-stones to the bigger prize in the region. At the beginning of 1941 Japan's Imperial Liaison Conference adopted the "Outline for Policy Toward French Indochina and Thailand," which called for "formation of inseparable military, political, and economic ties" with the two countries.[26] Although each had different internal and external conditions, both countries seemed vulnerable enough.

Thailand, the only Southeast Asian country to escape the fate of becoming a European colony, maintained a special relationship with Japan long before the war. Political relations between Thailand and Japan grew close throughout the 1930s. The Thai Navy, for instance, ordered ships from Japanese shipyards and sent its cadets for training in Japan.[27] The country occupied an important position in regional communications, including aviation wireless. Bangkok was linked to Saigon and Singapore via submarine cables and also maintained wireless telegraphic links to Europe (Berlin, Paris, London) and to Hong Kong, Calcutta, Batavia, Saigon, Manila, Osaka, and Shanghai. In addition, it boasted overseas wireless telephone communications to Bandung, Manila, Colombo, Saigon, and Tokyo. When Matsumae Shigeyoshi visited Thailand in early 1937 as part of his Southeast Asian tour, he noted that Thailand had established wireless telephone links with foreign countries even ahead of Japan. He concluded, however, that it had paid too much for the equipment imported from Germany's Telefuken Company and was unimpressed with the underdeveloped state of telecommunications inside the country, all of which were under government control. Since, unlike the rest of Southeast Asia, Thailand was an independent country,

---

26. "Tai Futsuin-Tai shisaku yōkō" (January 30, 1941), in Bōeichō, Bōei kenkyūjo, Senshibu, comp., *Shiryōshū Nanpō no gunsei*, 36.

27. For an authoritative overview of Thai-Japanese relations during the Pacific War, see Reynolds, *Thailand and Japan's Southern Advance, 1941–1945*.

Matsumae reasoned, Japan should be able make inroads commercially as long as it "displayed sincerity."[28]

Whether by sincerity or not, Japan sought to become a major supplier of telecommunications equipment to Thailand. In late 1940, when the Thai director of communications visited Tokyo, the MOC presented him with a list of Japanese demands—among them reduction of telegram rates between the two countries and the beginning of telegram service in Japanese. Japan also suggested that Thailand accept a Japanese technical advisor and that Thai students study telecommunications in Japan. Finally, it proposed laying a submarine cable between the two countries as well as an exchange of radio broadcasts and suggested "dispatching several Japanese to Thailand to assist in the undertaking." As it turned out, when it comes to the sensitive issue of communications, even the usually friendly Thai were less than accommodating. The visiting Thai official said a polite "yes" to several requests, such as reduction of telegram rates, telegrams in Japanese characters, and the submarine cable, but in each case wanted further discussion. He gave a categorical "no" to hiring Japanese advisors, explaining that his government had just sent away European advisors, and all technical matters were now handled by the Thai themselves. Although he responded favorably to the suggestion of exchanging radio programs, he quickly added that facilities on his side were not ready. Moreover, he wanted no Japanese sent to Thailand to set up telecommunications facilities.[29]

The two countries had begun to implement telecommunications agreements before the war broke out. For example, the Telecommunications Association in Japan, through the Ministry of Foreign Affairs, had negotiated with the Thai government for the recruitment of Thai students, and the government had agreed. Over 70 applicants took the examinations, but the outbreak of war in the Pacific and Southeast Asia complicated the matter. It was not until June 1942 that five Thai officials arrived in Japan for telecommunications training, remaining there until early 1944.[30]

---

28. Matsumae, "Nan'yō shoppō ni okeru denki tsūshin jigyō," *TKZ* 348 (August 1937): 17.

29. "Nichi-Tai kan denki tsūshin gyōmu ni okeru yūkō kankei sokushin ni kansuru kaidan gaiyō no ken" (October 28, 1940), MOC Records II-467.

30. See Nakayama Ryūji's remarks at a roundtable discussion on February 1, 1942. "Daitōa kensetsu to denki tsūshin kizai o kataru," *DT* 5.19 (1942): 33; "Taikoku ryū-

Japanese like Matsumae who expected relatively easy access to the Thai market in telecommunications would be disappointed as well. Since 1938, member-corporations of Japan's Telecommunications Association had participated in the annual international exposition during the Thai Constitution Day celebration. In 1941, they prepared an extensive list of new telecommunications products for the event to be held on December 8. The golden opportunity of showcasing Japanese technology in Southeast Asia slipped away when the exposition was canceled due to outbreak of the Pacific War on the very day it was to open. Only Japanese manufacturers were present to stage a smaller exhibition demonstrating Japan's new technological progress.[31] Economic diplomacy in Thailand seemed to have reached its limits.

To the east of Thailand, French Indochina occupied a vast area that had come under French rule in 1887 as the Indochina Commonwealth. Its internal telecommunications were the least developed of all colonial possessions in Southeast Asia, however, with only one long-distance telephone circuit linking Saigon and Hanoi, the seat of the French governor-general. Prewar Japanese surveys noted with some contempt that the capital of Hanoi had no international connections. Saigon, the largest city, was the center of overseas communications, with wireless telegraphic links to France, China, Japan, the United States, other parts of Southeast Asia, and French possessions in the South Pacific. Saigon was also the only city with overseas telephone links to Paris, Batavia, Bangkok, Manila, Kuala Lumpur, and Osaka. Prewar Japan's link with French Indochina relied on two circuits: a wireless connection between Saigon and Osaka, and a cable connection via Hong Kong, owned by the Great Eastern Telegraph Company (later Cable and Wireless). The former accounted for 90 percent of traffic, and the latter was used largely for incoming telegrams.[32] Although by no means well developed, these earlier communications links certainly served Japan's interests well.

Unlike Thailand, French Indochina became particularly vulnerable after the start of the European War. After German forces occupied

---

gakusei raichō irai kenkyū keika gaiyō" and "Taikoku ryūgakisei sōbetsukai," *DT* 7.32 (February–March 1944): 30–31.

31. Denki tsūshin kyōkai, *Shadan hōjin Denki tsūshin kyōkai 20-nen shi*, 21, 48–49; the catalogue of Japanese products was printed as "Taikoku kenpōsai kinenhakurankai Nihon kan shuppin denki tsūshin kiki mokuroku," *DT* 5.18 (February 1942): 69.

32. "Honpō-Futsuin kan denpō riyō jōkyō," MOC Records II, 461.

Paris and established the Vichy government in southern France, the position of French Indochina became even more precarious. When the Japanese government pressured Hanoi to cut off all aid routes to the Chiang Kai-shek regime, French authorities gave in to Japanese demands to station troops in northern Indochina.[33] Following Japan's military advance into Indochina, telegrams exchanged between the two areas increased nearly 200 percent in 1941, and the total number of words sent increased nearly sixfold.[34]

Located between southern China and Thailand, French Indochina was a natural link to the Southern Region. Existing land lines linked Hanoi with southern China, and Saigon with Bangkok. Increasingly, French Indochina also came to occupy an important place in Japan's effort to break the so-called ABCD encirclement of Japan, a catchterm in Japan referring to the imagined alliance of America, Britain, China, and the Dutch. A MOC study pointed out that of the three cable routes linking Hong Kong and Singapore—two important British outposts—the cable that passed through Saigon was the shortest and most advanced. Because Saigon was the nexus in the "ABCD clandestine/strategic communication route," Japan should demand from French authorities the right to monitor communication traffic there.[35]

Moreover, the French Indochina Broadcasting Corporation, founded in 1929, operated out of Saigon and boasted at least three 10-kW transmitters and nearly 3,000 subscribers.[36] Shortly before the Pacific War, various agencies in Japan agreed that Saigon would become Japan's news center in Southeast Asia and that direct wireless connection with Tokyo would be established as well. On the eve of Japan's military offensive in the Pacific, the Japanese embassy in Hanoi proposed a number of emergency measures in French Indochina. Given its position as the base for military operations in the Southern Region, Japan must exercise control over intelligence, propaganda, and communications, all requiring the highest degree of secrecy. In particular, Japan should ask French authorities to place postal and telegraph communications with third countries

---

33. Hata, "The Army's Move into Northern Indochina."

34. "Honpō-Futsuin kan denpō riyō jōkyō," MOC Records II, 461.

35. "Tōa ni okeru teisei tsūshin torishimari no kenchi yori mitaru Saigon no ichi," MOC Records II, 461.

36. "Futsuin no tsūshin shisetsu" (May 1941), MOC Records II, 461.

under joint control. Japan must also assume supervision of radio broadcasting. In the meantime, communications between Japan and French Indochina must be strengthened.[37]

Of the many fronts of Japan's Southern Advance before the Pacific War, the failure of the entire Japanese-Dutch negotiations perhaps produced the gravest political consequences. By late 1941, diplomatic pressure seemed to have run its course. Unable to secure much-needed oil from the south and failing to reach a diplomatic breakthrough with the other supplier, the United States, Japan approached its final decision to pursue strategic resources in the region by force.

PHANTOM OF A GREATER EAST ASIAN TELECOMMUNICATIONS COMPANY

On the morning of December 8, 1941, shortly before its bombing of Pearl Harbor, Japanese forces began sweeping across Southeast Asia. Attacks launched simultaneously over the entire Western Pacific scored impressive victories. Within two months, the Philippines, Malaya, Borneo, the Dutch East Indies, and part of Burma had all fallen. Japan had replaced Britain, the United States, and the Netherlands to become the single paramount force in the vast region.

In February 1942, shortly after the fall of Singapore, Japan's Imperial Liaison Conference, the powerful body that combined the General Headquarters and the Cabinet, issued a terse statement spelling out what Japan had set out to accomplish in the new phase of the conflict:

> In order to achieve the objectives of the Greater East Asia War and cope with conditions in the near future, we seek to establish a New Order under the guidance of the [Japanese] Empire, based on the Japan-Manchukuo-China axis, in such fields as the military, politics, economy, and culture. The area of Greater East Asia is defined as follows: Japan, Manchukuo, and China, plus the Southern Region east of 90 degrees East Longitude, west of 180 degrees East Longitude, and north of 10 degrees South Latitude. Decisions concerning other areas shall be made according to future developments.[38]

---

37. "Kōtsū, tsūshin kankei kinkyū taisaku" (December 1, 1941), proposed by Consul Ono in Saigon; MOC Records II, 461.

38. "Teikoku ryōdoka ni shinjitsujō o kensetsusubeki Daitōa no chiiki" (February 28, 1942), in Bōeicho, Bōei kenkuūjo, Senshibu, comp., *Shiryōshū Nanpō no gunsei*, 40–41.

In the months before launching the attack in the Pacific and Southeast Asia, the Japanese Army, Navy, and civilian agencies already deliberated over occupation policies for the Southern Region. The entire area targeted for occupation, some 3 million sq km with a population of 120 million would be placed under military administration but divided between the Army and the Navy.[39] In general, the Army was to control the core areas considered "dense in population and complicated in administration." These included Hong Kong, the Philippines, Malaya, Sumatra, Java, British Borneo, and Burma. The Navy would be in charge of areas described as "sparsely populated and virgin areas reserved for the empire in the future," such as Dutch Borneo, the Celebes, the Molka Islands, Small Sunda, New Guinea, and the Bismarck Islands.[40] To make a smooth transition to Japanese occupation, both services shared the same policy of "using existing administrative structure as much as possible" and "rapidly acquiring important national defense resources."[41]

Restoring and expanding communications links was the major concern for the Japanese. The questions remained what routes would receive priority and what forms of operation would be allowed. In early July 1942, communications links between Java and Tokyo were reopened for public service, followed by the Singapore–Japan service a few days later. By the beginning of 1943, the Philippines, Malaya, Sumatra, the Celebes, North Borneo, and Burma were added. The Japanese government introduced a special category of telegraphic service known as "southbound telegrams." In terms of telegram rates, businessmen's calls for cheaper, flat rates could not easily be satisfied. The entire region was divided into three zones on the basis of geographical distance, each with a separate rate for telegrams in Japanese characters and for European-language telegrams. Although the use of the Japanese language was no doubt a boost to business, the use of codes—always fa-

39. Bōeichō, Bōei kenkyūjo, Senshibu, comp., *Shiryōshū Nanpō no gunsei*, 17. The best English-language introduction to the Japanese military bureaucracy in Southeast Asia is Akashi, "Bureaucracy and the Japanese Military Administration," 46–82.

40. "Senryōchi gunsei jisshi ni kansuru riku kaigun chūō kyōtei" (November 26, 1941), Bōeichō, Bōei kenkyūjo, Senshibu, comp., *Shiryōshū Nanpō no gunsei*, 96–97.

41. "Principles for Administration of Southern Areas Adopted by the Liaison Conference of 20 November 1941," in Lebra, *Japan's Greater East Asia Co-Prosperity Sphere*, 113–17; "Teikoku no shigenken o ikani subeki ya" (February 28, 1942), Bōeichō, Bōei kenkyūjo, Senshibu, comp., *Shiryōshū Nanpō no gunsei*, 226–27.

vored by businesses for secrecy and savings—was initially forbidden for security reasons.[42]

As telecommunications routes with Southeast Asia were reopened, discussions over organizational schemes continued. In early 1942, business newspapers in Japan reported plans to establish large telecommunications companies in Malaya and Luzon. In March, it was reported, both the military and the MOC had announced the principle of private management of telecommunications (excepting broadcasting) in the Co-Prosperity Sphere. According to one plan, a telecommunications enterprise would be established in Japan as the parent company, while companies responsible for operating or constructing telecommunications in various regions within the sphere would become its subsidiaries.[43]

There seemed to be good rationale for unifying telecommunications operations in Southeast Asia. To ensure smooth and efficient operations, the Southern Economic Council, a business group, suggested that instead of setting up telecommunications separately in each area as in China, all operations should be placed under unified management and unified government supervision: "If there were more than two supervising government agencies, even if there was an institutional body, telecommunications might not be able to fully realize its functions."[44] Japanese employees at the Manchukuo-based MTT had offered their version of an Asian communications corporation encompassing all of Greater East Asia. The new entity, capitalized at 2.3 billion yen and set to consolidate all telecommunications enterprises in East Asia, would be headquartered in Mukden or Shinkyō. As the first telecommunications "national policy company" founded outside Japan, the MTT apparently had leaders with enough confidence to think they could pull off such a scheme. Although the plan might appear outlandish from the MOC's standpoint, MTT executives even considered changing the Manchurian climate to make it more comfortable for Japanese who would come from Japan proper for such an endeavor. Reflecting confidence in Japan's engineering prowess, the scheme proposed filling in

42. Denmukyoku, Gaishinka, "Tai Nanpō denki tsūshin no gendankai," *TKZ* 414 (February 1943): 61–62.

43. *Nikkan kōgyō shinbun*, January 14 and March 22, 1942.

44. "Nanpō tsūshin ni taisuru minkan dantai no koe," *TDTZ* 2 (April 1942): 58–61.

the Bering Strait so that warm currents would raise the temperature of Manchuria![45]

To cope with the greatly expanded sphere of Japan's control, MOC officials proposed setting up a new telecommunications company of mammoth proportions under Japanese control for the entire East Asian region. Tentatively named the Greater East Asian Telecommunications Public Corporation, the vertically integrated company would absorb the ITC and the JTTCC. The new company would manage all external telecommunications in Japan and Greater East Asia, as well as internal telecommunications in the Southern Region, through direct management or through investment. In addition, this new company would manufacture and sell telecommunications equipment and engage in construction and maintenance on behalf of the Japanese government at home and throughout the empire.[46] Such a company would not only exercise strong control over joint-stock companies to be established in recently occupied Southeast Asia but also gradually assume control over existing enterprises in Manchukuo and China as well. The company would be placed under the supervision of the minister of communications, although the Army and Navy ministers also would be responsible for military-related matters. Moreover, an advisory committee consisting of concerned government agencies would discuss important issues about operation of the company. Unlike earlier telecommunications enterprises set up by the Japanese in Manchukuo and China, however, the new company was named a "public corporation" (*kōsha*), considered "a kind of extension of the state apparatus" due to the highly "public" nature of its mission. As such, it would operate not on the basis of civil law or commercial law used in Japan but enjoy the status of an *eidan*, a wholly government-owned and -managed legal entity.[47] The government would appoint all managers unilaterally and would separate capital from management. In a word, the company would be able to operate swiftly and smartly in business because of private management, but in its planning operations it would consider highly public interests.

45. See Hase Kiyoshi's speech at a roundtable discussion in Tokyo on April 17, 1941, *DT* 4.14 (1941): 44.

46. Teishinshō, "Daitōa denki tsūshin kōsha setsuritsu yōkō dai-1-ji shian" (May 1942), MOC Records II, 86.

47. For a brief discussion of *eidan*, see Johnson, *Japan's Public Policy Companies*, 74.

The MOC's new proposal was its latest effort to take advantage of the new imperial expansion to "achieve unified control and guidance" of telecommunications enterprises throughout the imperium. The military in the south had different ideas, however. In early August 1942, the newly established superintendent of military administration issued a comprehensive directive on military administration in Southeast Asia. The directive recognized that the restoration of communications "is necessary for industrial development and enlightenment and propaganda. It is also helpful for stabilization of public life." It went on to state that "telecommunications in the Southern Region shall gradually move toward private management after its operation and finances become stabilized. Worth special attention is utilization of existing facilities and confiscated equipment. Needless to say, as long as it does not contradict the basic spirit of Japanese control, locals should be used as widely as possible to ensure self-support [*jikatsu*]." It also proposed a Southern Communications Network, where current military communications facilities would also be transferred to, or used along with, the regular military administrative communications network, circumstances permitting. This would reduce the burden on the military and prove economical as far as the consolidation was concerned. As before, existing facilities would be counted as "contributions" to the privately operated enterprises, and measures would be taken to evaluate the facilities and to plan other companies.[48]

Given the military's strong preference to keep operations local, it is perhaps not surprising that little came of the calls for "unified management and unified government supervision." As it turned out, the MOC had little influence on overall policies in the Southern Region, which were placed under Japanese military administration. The MOC had already been excluded from the powerful Sixth Committee on the Cabinet Planning Board, which was largely responsible for formulating basic economic and financial policies for that area. Moreover, regardless of the policy guidelines set in Tokyo, once the military occupation began, military authorities in Southeast Asia took many things into their own hands. Although they needed technical and administrative personnel to help run the military occupation in Southeast Asia, they would

48. Gunsei sōkanbu, "Gunsei sōkan shiji" (August 7, 1942), in Bōeichō, Bōei kenkyūjo, Senshibu, comp., *Shiryōshū Nanpō no gunsei*, 303.

have been the last to welcome a super-company designed and controlled by civilian bureaucrats in Tokyo.

The government in Tokyo and the military eventually found a compromise solution and once again turned to the ITC to perform most duties under orders from the military. As we have seen, the ITC was created as a convenient tool for the MOC and had proved a valuable asset. In July 1942, the Cabinet Planning Board decided that the ITC was to take over construction and maintenance of telecommunications facilities in the newly occupied areas on behalf of the military. It was to operate interregional communications service between occupied areas in the south, on one hand, and Japan, Manchukuo, China, and other countries, on the other. In addition, the ITC would manage enemy properties such as telecommunications equipment factories, and provide maintenance and construction of communications facilities for such special purposes as broadcasting and aviation. Operational costs would be met by the revenues from its service, but if there was a deficit, the government would make it up.[49] In a directive addressed to ITC President Ōhashi Hachirō, Vice Army Minister Kimura Heitarō emphasized that this was a temporary measure, to be eventually replaced by state-run or other types of operations.[50]

To supervise ITC operations in the South, the Army, Navy, and Communications ministries agreed to coordinate their activities.[51] As called for in previous plans, the ITC would reroute existing telecommunications facilities from the "Euro-American system" to a new Greater East Asian network centered on Japan. This would include both cable and wireless. Within the region, the ITC must guarantee adequate communications facilities for the administration and exploitation of urgently needed resources. In addition, it must prepare communications facilities to ensure the safety of major shipping and air routes. The government proposal called for expansion of the ITC to set up necessary offices in the area to ensure smooth operations.[52] For

49. Cabinet Planning Board, "Nanpō chiiki ni okeru denki tsūshin jigyō itaku keiei ni kansuru ken" (approved by the Sixth Committee on August 12, 1942), *KSSS* 8: 583.

50. Vice War Minister (Kimura Heitarō) to ITC President (Ōhashi Hachirō) (October 9, 1942), MOC Records II, 86.

51. "Kokusai denki tsūshin kabushiki kaisha kantoku ni kansuru Riku Kai Tei sanshōkan oboegaki an" (June 8, 1942), MOC Records II, 86.

52. "Nanpō denki tsūshin jigyō jutaku keiei tōryō an," MOC Records II, 86.

that purpose, the ITC established a Southern Headquarters in Singapore in August 1942, initially to repair, construct, and operate telecommunications facilities in the Malaya area on behalf of the Amy. A branch was set up in the Philippines at the same time. Later, the ITC took over the same task for Burma. At the end of 1942, it moved into Java, again acting on orders from the Army. Finally, Sumatra was transferred to ITC operation at the beginning of 1944.[53] The ITC also set up an East Indies Headquarters in Makassar on the island of Celebes for areas under Navy occupation. Branches were established in Borneo, Amboina, Small Sunda, and New Guinea.

The ITC proved to be the right solution. It comfortably settled into its new role of operating telecommunications in the Southern Region on behalf of the Army and the Navy, adapting its annual expansion plans to meet the military's urgent needs. In early 1943, for instance, the ITC was to set up wireless facilities for local and long-distance telephone and telegraph service in Singapore, Rangoon, and Manila, all under the Army administration, as well as in Navy-controlled areas. In Army-administered areas, plans were drawn up to privatize telegraph and telephone service. The ITC was also to provide capital ranging from 15 to 35 million yen to set up local telecommunications companies in Java, Sumatra, Burma, and Malaya, although this plan never materialized due to Japan's defeat.[54]

Consolidation of telecommunications operations through ITC proceeded, although at a slower pace than many MOC officials would have liked. Within the GGT, opinion was divided over whether to turn over completed cable to the ITC. In 1942, supporters of the transfer finally prevailed.[55] In 1943, as the colonial government transferred long-distance cable facilities in Taiwan to the ITC, the company issued new shares, increasing its total capital to 85.8 million yen. In May 1944, the ITC took over telecommunications operations in the South Sea islands from the Nan'yō Agency as the islands themselves came under direct administration of the Navy. Two months later, Hong Kong was added

53. *KDTKKS*, 53–54.

54. Tsuda Ryūzō, "Taiheiyō sensō toki no gaichi tsūshin," in Teishin gaishi kankōkai, *Teishin shiwa* II: 588. For ITC facilities in Southeast Asia, see *KDTKKS*, 112–20. For ITC plans of construction in both Army and Navy areas in Southeast Asia in 1943, see NHK Records S-4-175.

55. *KDTKKS*, 39.

to the ITC's sphere of operations. The ITC established a local branch office (*jimusho*) in 1943, which was then upgraded to a branch office. In Hong Kong, the ITC constructed a microwave telephone link between Hong Kong and Canton, the only one in the region.[56] By 1945, the ITC was the largest shareholder of the JTTCC. In fact, by then, in addition to being a de facto long-distance telecommunications monopoly in newly occupied Southeast Asia, the ITC had become the largest integrated telecommunications operator in the Japanese imperium (see Fig. 3).

PHANTOM OF INDEPENDENCE

Japan's justification for its declaration of war against the United States and the European powers in Asia was to break the ABCD encirclement and to remove the yoke of Western colonialism and to liberate Asia. Prewar Japanese planning had already suggested granting some form of independence to Burma as a political strategy to encourage anti-British sentiment in India, and to the Philippines as long as it cooperated with Japan in a conflict against the United States. Given Japan's early victories in its military operations in the region and the strong opposition from its military, talks of independence were dropped in favor of direct Japanese military administration, at least for the time being. It was only after the tide of war in the Pacific began to turn against Japan that it belatedly moved to promise independence or autonomy to selected regions in Southeast Asia in order to secure their cooperation. As Japan's resources were spread thin, it also had to retreat to a less direct form of control.[57]

From early 1943, Japan deliberated the future of occupied areas in Southeast Asia. On May 31, Japan's Imperial Liaison Conference adopted an "Outline on Directing the Politics of Greater East Asia." It stipulated

---

56. *DDJS* 6: 483. Before the war, Chinese authorities had decided to install ultra-shortwave in Canton to connect with nearby towns; see W. H. Tan, "Telephonic Communications in China," *FER* 32 (November 1936): 509.

57. For leading works on this subject, see Iriye, *Power and Culture*; and Hatano, *Taiheiyō sensō to Nihon gaikō*.

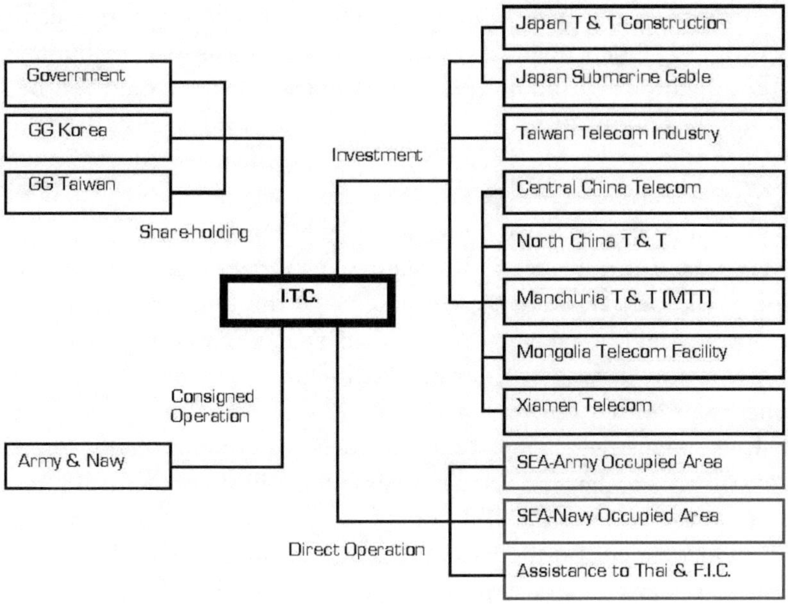

Fig. 3 ITC operations in Greater East Asia, 1941–45
(SOURCE: *KDTKKS*, 47).
GG: Government General; T & T: Telegraph and Telephone; SEA: Southeast Asia;
F.I.C.: French Indochina

that Malaya and most of the Dutch East Indies were to be incorporated into Japanese territory. Together making up of over 60 percent of the population of Southeast Asia, these areas would supply Japan with important natural resources and remain under Japanese military administrations. Burma (except Shan and Karen territories) and the Philippines (except Mindanao) were to become independent shortly. Japan would then convene an assembly of leaders of various countries in Great East Asia "to demonstrate to the world the completion of the Greater East Asia Co-Prosperity Sphere as well as its firm determination to carry the war to its conclusion."[58] On August 1, Burmese nationalist Ba Maw proclaimed the independence of Burma and immediately declared Burma would join the Greater East Asia War on Japan's side.

---

58. "Senryōchi kizoku fukuan" (January 14, 1943); "Senryōchi kizoku fukuan no setsumei" (January 14, 1943); "Daitōa senryaku shido daikō" (May 31, 1943), all in Bōeichō, Bōei kenkyūjo, Senshibu, comp., *Shiryōshū Nanpō no gunsei*, 44–46, 49.

A closer look at the operational arrangements, however, reveals that the promised independence was largely on paper only. As for the nerve system of the would-be independent countries, Japan would still be in a position of domination. In the case of Burma, for instance, the Treaty of Alliance as well as the Secret Military Agreement signed on the same day gave Japan de facto control over the newly independent country. Japan would place "a small number of highly capable Japanese inside its government for guidance." Although transportation and communications would be put under the sovereignty of the Burmese government, measures would be taken to ensure that Japan's special requests would be met, particularly when it came to military operations.[59] In reality, even the promised transfer of communications facilities failed to materialize, as Japan's military authorities deemed communications operations too vital to Japan's defense against the Allied counteroffensive to be left to the natives. The Telecommunications Bureau, nominally part of the independent Burmese government, was immediately placed under Japanese military supervision. In May 1944, the bureau was absorbed by Japan's Burma Army.

On September 26, 1943, the Republic of Philippines declared independence. Speaking via international telephone, Japanese reporters stationed in Manila gave detailed descriptions of the celebrations and did their best to convey a sense of excitement to readers at home. Responding to the comment that "the wait for independence has finally ended," the Japanese reporter in Manila agreed, adding "all thanks to Japan."[60] The post-independence Philippines had an identical arrangement to ensure Japanese control as with Burma.[61]

Having driven away the Europeans and the Americans, the Japanese in occupied Southeast Asia had to embrace the local population in other ways as well. It was only because of Japanese manpower shortages, more

---

59. *DDJS* 6: 501–5; "Daitōa sensō kansui no tame Biruma dokuritsu shisaku ni kansuru ken" (January 14, 1943) and "Biruma dokuritsu shidō yōkō" (March 10, 1943), both in Bōeichō, Bōei kenkyūjo, Senshibu, comp., *Shiryōshū Nanpō no gunsei*, 41–49. For English versions of these documents, see Trager, ed., *Burma: Japanese Military Administration, Selected Documents, 1941–1945*, 144–89.

60. *Yomiuri*, September 26, 1943.

61. "Hitō dokuritsu shidō yōkō" (June 26, 1943), in Bōeichō, Bōei kenkyūjo, Senshibu, comp., *Shiryōshū Nanpō no gunsei*, 52–53. For a comparison of the Burma and Philippines arrangements, see Hatano, *Taiheiyō sensō to Nihon gaikō*, 103–27.

than anything else, that the native populations were involved in the lower levels of telecommunications operations. The Japanese simply replaced Westerners in positions of importance. In Burma, for instance, as many as 2,000 Burmese and Indians were employed in postal and telecommunications, after 200 Japanese were dispatched from the MOC. In the Philippines, locals outnumbered Japanese in the ITC by 2 to 1. As many as 150 locals received training at the ITC in British Malaya.[62] The Japanese seemed generally satisfied with the cooperation from the natives. In New Guinea, for instance, the Japanese used native youth in telephone exchanges. Since they did not understand Japanese except a few short phrases, the Japanese could feel safe even with their military communication. While they were not nearly as intelligent as the Indonesians, one Japanese later claimed, these Papuan youth were very honest. In Java, natives made up the vast majority of some 600 employees of ITC. In total, nearly 10,000 Indonesians were employed in the company's East Indies Bureau in mid-1944 as compared to over 300 Japanese.[63] Some Japanese, like MOC engineer Amishima Takeshi when he traveled to Java in mid-1942, felt a particular "friendliness" among native Indonesians toward Japan. Curiously, Amishima singled out the Philippine people for their lack of national consciousness.[64]

Decisions on granting independence seemed to be quite unrelated to "friendliness" or "national consciousness," however. It was not until August 1944 that the Imperial Liaison Conference agreed to announce future plans of independence for the Dutch East Indies. Remarkably it was less than a month before its own surrender that Japan's Supreme War Council finally decided to grant independence to the Dutch East Indies.[65]

Ironically, just as Japanese officials were ironing out internal differences and outlining policies toward "post-independence" Southeast Asian states, they were still at work with the unfinished task of absorbing Thailand and French Indochina fully into Japan's telecommunications network. These two countries, which were not under direct Japanese

62. *DDJS* 6: 504, 490, 512, 533.

63. *GKDTS* 12: 348–49, 478.

64. Amishima Takeshi, "Nanpō chiiki o tabishite," *TKZ* 410 (October 1942): 23.

65. See "Tōindo dokuritsu sochi ni kansuru ken" (July 17, 1945), Bōeichō, Bōei kenkyūjo, Senshibu, comp., *Shiryōshū Nanpō no gunsei*, 76.

occupation for the most part and were thus designated as the Second (*Otsu*) Region in the wartime southern Sphere, continued to pose headaches for Japan.

Incorporation of Thailand into the Greater East Asia Co-Prosperity Sphere remained a policy objective for Japan during the Pacific War. When Japan launched its military operation in the Pacific in December 1941, it demanded the right of passage for its troops in Thailand, which was granted grudgingly. Japan's initial military victories in Southeast Asia gave Japanese officials confidence that the Thais would give in on other demands soon. Indeed, under Prime Minister Phibun Songkhram, the Thai government signed a mutual defense treaty with Japan in early 1942, and went on to declare war on Britain and the United States. In September 1942, Japan's Imperial Liaison Conference spelled out guidelines regarding finances, natural resources, industry, and trade. In shipping, for instance, a Japanese-Thai joint venture was to become the sole shipping company, enabling Japan to exercise "guidance" over its operations.[66]

About the same time, through the Japanese embassy in Bangkok, Japan began negotiating in earnest with the Thai government to start telegram service in Japanese characters between the two countries. Two MOC officials were dispatched to Bangkok for the purpose. Time and again, Thai officials cited technical and financial difficulties. Even a Thai who had studied in Japan over three years, the Thai official insisted, could not satisfactorily write *katakana*; the result would be not just more communication errors but a decrease in communication capability. The Japanese side was perfectly aware of the fact that this was a delaying tactic and that the real problem was the Thais' concern for their communications sovereignty. On the other hand, the Thais were eager to start service with Rangoon and Singapore, a desire Japanese negotiators hoped to exploit to achieve their own goals. Still, the Thai side showed little enthusiasm for making preparations to send telegrams in Japanese characters, a demand they believed to be a temporary wartime measure. In their opinion, Japan would have to continue to use European-language telegrams after the war was over.[67] The negotiations

---

66. "Tai-Tai keizai shisaku yōkō" (September 29, 1942), in Bōeichō, Bōei kenkyūjo, Senshibu, comp., *Shiryōshū Nanpō no gunsei*, 154–56.

67. Ambassador (Tsubokami) to Foreign Minister (Tōgo), August 14 and October 28, 1942; "Nichi-Tai kan Wabun denpō gyōmu toriatsuki ni kansuru Taigawa to kōshō

began to show promise only after a more conciliatory minister of communications had taken office in Thailand, but they were again delayed, by natural intervention this time—flooding problems in Thailand.

As the negotiations stalled, some Japanese back in Tokyo demanded tougher action. At a meeting convened by the new Greater East Asia Ministry in May 1943, Army representatives expressed impatience with the government's "extremely lukewarm" approach in dealing with Thailand. In contrast, the Greater East Asia Ministry, which had taken over diplomatic relations with Asian countries from the Ministry of Foreign Affairs, emphasized the need to "respect Thai independence as a matter of principle." The prevailing view at the meeting was that the semi-private ITC should facilitate the "permeation of Japanese influence" by supplying telecommunications equipment and by constructing facilities in Thailand.[68] As its military fortunes began to wane, Japan's diplomats could ill afford to alienate Thailand, which was a nominal, albeit often difficult, ally in its Asian diplomacy.

The Japanese exhausted "all sorts of tactics of persuasion" that included promised delivery of free telecommunications equipment. Ultimately, as one MOC official reported from Thailand, "the silent pressure of the Japanese military might" proved effective in breaking the deadlock in the negotiation. Acting under the order of Japan's Southern Army, the Japanese military attaché in Bangkok approached the Thai military authorities with the request to open direct communications with the Southeast Asia region under Japanese military control. The Thai military agreed.[69] Soon afterward, the Thai government finally gave way and signed the new agreement with Japan in June 1943, revising the Wireless Communications Agreement signed between the two countries in 1932. According to the new agreement, wireless for direct bilateral

---

keika," forwarded to Director, MOC Telecommunications Bureau (October 12, 1942), MOC Records II, 466. In mid-1942, Thailand reopened trade with Japanese-occupied areas such as Singapore, China, Java, Malaya, and the Philippines. Although Japan was Thailand's largest export destination, it was replaced by Singapore after 1943; Numnonda, *Thailand and the Japanese Presence*, 87–88.

68. "Tai, Futsuin tsūshin kankei uchiawasekai no ken" (May 18, 1943), MOC Records II, 466.

69. "Taikoku ni okeru Wabun denpō toriatsukau kaishi nado ni kansuru ken" (March 18, 1943), containing the letter from Hasegawa in Bangkok to Head, MOC Foreign Communication Section, MOC Records II, 466.

communications would be operated by stations near Osaka and outside Bangkok, and both governments promised to maintain optimum conditions. Certain time slots during the day and night would be set aside for Thailand to transmit all direct telegrams to Japan (including Korea, Taiwan, and the mandate islands). Thailand would now accept and send telegrams in Japanese characters between Bangkok and Japan, Manchukuo, and China. Since Thailand insisted that such service be operated by Thai hands, Japan offered to train 22 Thai telegraph operators. The basic currency would be the Japanese yen, since Thailand had already signed on to Japan's "yen bloc" in May 1942. In return, Thailand was allowed to keep the existing international treaties as well as its own domestic laws intact; thus the appearance of sovereignty was maintained. The agreement would take effect immediately for three years, six months' prior notice being required for its abrogation.[70]

With this agreement, Thailand became the only country, apart from Manchukuo and the puppet regime in China, to accept telegrams in Japanese characters as part of its regular international service. Understandably jubilant, Japanese newspapers heralded the agreement as a major milestone in the expansion of Japan's communications sphere in Asia. In an editorial entitled "East Asian Communication Co-Prosperity Sphere," the *Mainichi shinbun* opined that to really solidify the East Asian communications network, "it is necessary to build a cultural and intellectual basis by way of common language and common script. Considering this, we must say that handling telegrams in Japanese characters in Thailand is great progress." The editorial ended with the "expectation that French Indochina would follow such wisdom."[71] The new agreement handed the MOC a much-needed victory, but the euphoria in the Japanese press almost derailed the agreement. The Privy Council in Tokyo complained that it had been left in the dark about the treaty, and the Cabinet Legal Affairs Bureau also became apprehensive about possible legal oversights. It took considerable effort by the MOC to persuade skeptics that telegrams in Japanese characters

70. MOC, "Nichi-Tai kan Wabun tsūshin gyōmu kaishi ni atari," MOC Records II-467. A text of the agreement in English can be found in MOC Records II-466.

71. *Mainichi shinbun*, June 12, 1943.

were only one part of the entire treaty and, as such, could be left to the agency concerned—the MOC.[72]

Understandably, Thailand was less happy about the agreement than Japan had expected. According to a report from Japan's embassy in Bangkok, almost immediately after the agreement was signed, Thai-language broadcasting based in Delhi voiced criticism that the agreement was harmful to Thailand because Thai telegraph offices had to handle telegrams in Japanese characters with no benefit to their own country. It even alleged that Japanese technicians stationed in the Bangkok post office could be privy to Thai state secrets through their censorship of telegrams.[73] Thailand's attitude on telecommunications matters reflected its domestic politics as well as ambiguity in its wartime diplomacy. In June 1943, Prime Minister Phibun still expressed confidence about an Axis victory, yet he steadfastly refused Tokyo's invitation to attend the Assembly of Greater East Asian Nations. A month after the agreement, Prime Minister Tōjō visited Bangkok and transferred to Thailand four Malay states and two Shan states claimed by Thailand, ostensibly in gratitude for Thailand's aid to Japan. During the Pacific War, as many as 150,000 Japanese troops were on Thai soil. When the tide of war turned decisively against Japan, however, the pro-Japanese cabinet under Phibun was quietly replaced in Thailand, costing Japan a major ally in the government.[74]

French Indochina, although theoretically under the control of a European ally, the Vichy government, continued to pose difficulties for Japan during the war. In February 1942, as Japan's military operations swept across Southeast Asia, eighteen Japanese employees of the ITC arrived in Saigon. Ostensibly reporters for Japan's Domei News Agency, they began constructing a wireless facility for Japan's military. Completed three month later, this wireless station worked for the headquarters of Japan's Southern Army in Saigon until it moved to Singapore.[75] Negotiations with French authorities in French Indochina

---

72. Mokuji Teizō, "Wabun tsūshin no kokusai shinshutsu," Teishin gaishi kankōkai, *Teishin shiwa* II: 501.

73. Ambassador to Thailand (Tsubokami) to Minister of Greater East Asia (Aoki) (sent June 16, 1943), MOC Records II, 467.

74. Reynolds, *Thailand and Japan's Southern Advance, 1941–1945*.

75. *KDTKKS*, 53.

over terms of operation proved no easier than negotiations in Thailand. Japan pressed the French authorities to allow telegrams in Japanese characters to be sent and received in the colony, but to little avail. In mid-1942, Kimura Yōji, the MOC representative in the area, sounded desperate in his report to Tokyo. Since the issue of telegrams in Japanese characters was an entirely "political" one, he noted, he had reached the limit as far as operational matters were concerned. Apparently, French officials also resorted to various stalling tactics, claiming that they had to discuss matters with their home government. They added that since the existing international telecommunications treaty would be affected under such circumstances, a new treaty should be negotiated in Tokyo between the French embassy and the Japanese Ministry of Foreign Affairs. Another difficulty for Japan was the French insistence that the gold franc remain the currency of calculation.[76]

It was not until March 21, 1944, more than a year after the agreement with Thailand, that Japanese and French officials affixed their signatures to a similar agreement in Saigon. Japanese-language service would begin with Tokyo, followed by Singapore, Manila, and Hong Kong. Although Japan had demanded the same arrangement as the one it had reached with Thailand in 1943, it had to accept a less satisfactory agreement, in part due to its own dwindling fortunes in war. Actual telegram service in Japanese characters between Japan and French Indochina did not start until August 1944. The French opposed any changes in European-language telegrams. Although Japan demanded that Indochina be included in the so-called East Asian telegram system, it acceded to the French position and decided to consider such issues later on. Again, due to French suspicion of possible espionage, telegrams in Japanese characters were to be handled by a separate telegraph office. Japan dispatched five Japanese to help train Japanese-language telegraph operators.[77] This less-than-satisfactory arrangement remained in place until March 1945, when the Japanese deposed the French colonial government in a pre-emptive strike.

---

76. Kimura to MOC Director of Telecommunications (July 6, 1942), MOC Records II, 460.

77. Mokuji Fujizō, "Futsuin Daitōa tsūshinken no ichiyoku tantō," *TDTZ* 4 (July 1944): 3–5; *DDJS*, 6: 496–97.

Despite its initial military success, Japan had only limited success in a speedy and full incorporation of Thailand and French Indochina into its imperial telecommunications network. As Japan began to lose the initiative in the war, however, officials like Shigemitsu Mamoru of the Greater East Asia Ministry emphasized the need for diplomatic offensives in Asia to prevent a total defeat in the war. Though militarily no match for the Japanese, both Thailand and French Indochina were adept at exploiting the diplomatic advantages afforded them as Japan sought to keep them as token allies. The fact that different Japanese agencies often worked at cross-purposes helped these two besieged governments. The Japanese military dealt with French Indochina and Thailand regarding communications with other parts of Southeast Asia, whereas communications with Japan proper remained in the hands of the MOC. Although the military preferred leaving communications matters to local representatives of the Greater East Asian Ministry as a way to assert greater control, the MOC expressed reservations about doing so, on the grounds that the MOC had always been involved in that task. By resorting to delaying tactics or playing the Japanese bureaucracies against each other, both Thailand and French Indochina managed to give up as little as possible.

Japan's expansion into Southeast Asia did not simply follow the same trajectory as its own continental expansion or that of the Western imperialist powers. As this chapter shows, Japan's prewar Southern Advance was more cautious and opportunistic than its expansion into Manchuria and China proper. MOC officials and Japanese industrialists hoped that exports of telecommunications equipment would not only balance increasingly large imports of raw materials from the region but also build Japanese influence there, creating a de facto economic dependence on Japan. It was not until the decade before the Pacific War that Japan's attempt to extend direct telecommunications links into Southeast Asia began in earnest. An examination of Japan's ambitious plans for expanding its submarine cable telecommunications network into Southeast Asia after the outbreak of the European War reveals a larger, evolving geostrategy that sought to integrate the region firmly into Japan's sphere of influence. Given their unmistakable political implications, it is hardly surprising that these demands met with dogged resistance and had little chance of success before December 1941.

Japanese power in Southeast Asia was never as omnipotent as Japanese planners wished to believe, even after Japan's occupation of much of the region. Japan's expansion into Southeast Asia and the South Pacific after December 1941 was a spectacular accomplishment, but its control over the vast area did not ultimately move far beyond the phase of military operations. Technological overstretch certainly had much to do with it. Japan's wartime dealings with Thailand and French Indochina, as with its promise of independence to the Philippines and Burma, also illustrate the complexities of international politics even within the Co-Prosperity Sphere. To some extent, Japan had become a captive to its own rhetoric of "co-existence" when dealing with such countries as Thailand.

Compounding the problem of diplomatic resistance from governments in Southeast Asia were the internal differences among various Japanese groups operating in the region, each with its own priorities and goals. It is not surprising that the MOC did not give up its long-cherished dream of creating a single corporation to manage all trunk cables and circuits in the empire, leaving local communications to its subsidiaries. By mid-1942, however, it was clear that the Greater East Asian Telecommunications Public Corporation would not work due to resistance from the military, which was running the show in Southeast Asia to a much greater degree than elsewhere in the new imperium. The MOC had to settle for a more moderate solution—an expanded ITC. Such a solution worked well for Southeast Asia and the South Seas. For building a unified telecommunications system in Greater East Asia as a whole, however, bureaucrats in Tokyo had to pin their hopes on a regional organization rather than on such a company. It is the rise and fall of Japan's overall imperial telecommunication system that we shall turn to in Part IV.

PART IV

Network, 1939–1945

As is well-known, electricity travels at the same speed as light. Clearly, just from a technological point of view, something that transmits at such a high speed must be handled by a single entity. As the most important issue for the construction of Greater East Asia at the hands of Japan, it is a matter of course that [all telecommunications companies] must be absorbed into the nerve system of Japan.

As such, it goes without saying that all communications in the Greater East Asia Co-Prosperity Sphere must become unified. It is also time to reconsider all the existing communications in China, Manchuria, and colonies such as Korea, Taiwan, Karafuto, and Nan'yō.

—Matsumae Shigeyoshi, 1943

Briefly stated, imperialism is a system that splits up collectivities and relates some of the parts to each other in relations of *harmony of interest*, and other parts in relations of *disharmony of interest or conflict of interest.*

—Johan Galtung, 1971

CHAPTER 9

Systemic Integration

The belief in the connection between communications and imperial unity is not new. Charles Bright, a British electrical engineer at the end of the nineteenth century, was acutely aware of the great distance that lay between Britain and its overseas colonies. His solution was an imperial telegraphic network. "Being a highly individualistic people both by temperament and by tradition," Bright reasoned, "we have, unfortunately, no *bushido* here to fall back upon."[1] Bright's observation was typical of many of his contemporaries in the West who believed that Japanese society was tightly knit thanks to its traditional values. Thus, to Bright, the modern technology of instant communications was the only effective way to compensate for the lack of a collective ethos on the part of the individualistic Anglo-Saxons.

Nearly four decades later, as Japan launched its own imperial telecommunications network in Asia, few Japanese would take comfort in the naturally endowed traits commonly attributed to their society. To meet the needs of an imperium extending over a vast area in East Asia, Japan created the largest physical network spanning the region. By the end of the 1930s, Japan had added numerous cable and wireless links to telecommunications facilities under its control in Japan, Manchukuo, occupied China, and the South Sea islands. In the meantime, this rapidly expanding imperial communications system, like the imperium itself, became increasingly complex in terms of internal structure. In real life, complexity is not simply a function of multiplication of links; nor was the problem confined to technology. It was largely organizational

1. Bright, *Imperial Telegraphic Communication*, 99.

and political, since telecommunications service continued to operate under separate chains of command in Japan's home islands, its colonies of Korea and Taiwan, and Manchukuo and China. As we saw in Chapter 7, even in the same area, such as Manchukuo, facilities for serving the public, military, police, and railway were often managed by different agencies. As advocates of integrated operations repeatedly warned, this situation threatened to undermine the effectiveness of the entire imperial telecommunications network.

The solution seemed obvious: if multiple agencies created greater risk of conflict and miscommunication, better organization and coordination thus became the key to establishing greater control. As one MOC official aptly put it, "To fully realize the potential of telecommunications, it is necessary not only to have facilities that embody superior technology but also to create an equally superior communications system to regulate the use of communications."[2] The creation of the ITC in 1938 and its subsequent reorganization (Chapter 6) represented a major step in that direction, but it stopped short of unifying all telecommunications in East Asia. Since it was charged primarily with construction and maintenance of facilities for interregional communications, business operations (*gyōmu*) involving interregional traffic continued to be decentralized. Integrating separate Japanese-controlled telecommunications operations throughout East Asia into a cohesive single system, or an "East Asian telecommunications bloc" as it was sometimes called, remained one of the greatest challenges for the Japanese.

ORGANIZING THE IMPERIAL NETWORK

The East Asian Telecommunications Conference

As Japan began to extend its telecommunications network into its newly controlled territories after 1931—first Manchuria, then China, and later Southeast Asia—organizational complexity as much as distance became an increasingly obvious problem. Following the example of coordination for military transportation by railway between Manchuria

2. Mokuji Fujizō, "Daitōa denwa tsūshinken no kakudai," *TDTZ* 3 (March 1943): 14.

and Korea, the Kwantung Army proposed in 1934 to convene a similar meeting in Manchukuo to coordinate communications matters between Manchukuo, Korea, and Japan proper. The MOC's response was predictably cool; it suggested consultation among the MOC and other proper authorities before deciding whether such a meeting was necessary.[3] From Tokyo's perspective, devolution of authority was as worrisome as disruption of communication.

For those who were actually operating telecommunications in Japan's empire, lack of coordination among different operators had serious consequences. The NCTT discovered this problem the hard way. As Japan's long-distance NLC to Manchukuo neared completion, the NCTT proposed extending the cable to North China so as to open direct telephone service with Japan. In its ambitious five-year business plan, the NCTT was to build telecommunications cables connecting Tianjin, the largest commercial city in North China, with Shanhaiguan, an important city on the border with Manchukuo, in anticipation that the MTT would extend its own cable to the same location to form a critical interregional link. In this way, the two regional telecommunications systems would be connected with secure and powerful cables. After completing the first phase of construction, the NCTT was understandably proud of its own contribution to the cause of "advancing the frontier of East Asian telecommunications." However, the NCTT discovered that the MTT's priorities had shifted toward development of northern Manchuria in response to demands from the Kwantung Army. Lack of coordination left the NCTT's grand design only partially fulfilled.[4]

The NCTT attempted to salvage the situation but soon encountered a different problem, even after the cable construction was completed. As NCTT officials discovered, agreeing on a formula for tariff and payment allocations among several telecommunications operators was no easy task. At the end of 1939, the NCTT entered into discussions with both the MTT and the MOC for leasing one circuit on the Japan–Korea–Manchukuo long-distance cable. The MTT agreed to lease a circuit but demanded a hefty fee—over 6,000 yen per month—to compensate for its loss of revenue on the circuit. By the NCTT's calculation,

---

3. "Man-Sen-Naichi kan no tsūshin tōsei kaigi no ken" (February 1934), Riku Man Dainikki 1934-3/32, Imperial Japanese Army and Navy Archives (1868–1945).

4. Hokudenkai, *Kahoku denden jigyō shi*, 91.

some 670 units of telephone calls per day—an impossible target—would be needed to generate enough income simply to pay the MTT's fee under this arrangement. This made the leasing arrangement "ridiculous," as one disappointed NCTT executive put it. Stressing the public utility nature of this interregional connection, the NCTT's president, who had been an MTT director, asked the MTT to reconsider, but with little success. The MOC, while sympathetic to the NCTT, added a condition that further complicated the problem: it demanded that the tariff for calls over the proposed North China–Tokyo telephone circuit be equal to the MOC's rate for Shanghai–Tokyo calls. Although it made sense for the MOC to standardize all telephone rates between China and Japan, at 7.50 yen, the rate was too low for the NCTT to meet its operating costs. As a result, even after all cable connections had been tested and found in working order, interregional service was delayed due to disagreement over the rate. The negotiations dragged on for several months before the three parties finally ironed out their differences.[5] Instead of such ad hoc negotiations, something more regular and permanent was apparently needed.

Not surprisingly, Watanabe Otojirō, the energetic ex-MOC official who headed business operations for the NCTT in Beijing, became an early advocate of an interregional telecommunications forum. To cope with the new developments in East Asia, Watanabe noted, Japanese business groups were already calling for greater organizational integration in the region. For instance, the heads of Japan's chambers of commerce in Osaka, Kobe, Keijō, Dalian, Tianjin, and Qingdao were discussing a pan–Yellow Sea economic federation. In early 1939, Watanabe proposed to convene a regional coordination conference at which representatives from telecommunications companies in Manchukuo, China, and Mongolia would discuss common principles for a Japan-Manchukuo-China communications network as well as joint efforts to build it. According to his proposal, the four regional operators on the continent—the NCTT, CCTC, MTT, and MTFC—would meet twice a year to discuss their annual operation plans as well as to make arrangements to facilitate interregional communications. As Japan gradually consolidated its control on the Asian continent, Watanabe argued, there

5. NCTT, Eigyōbu, Gyōmuka, "Ka-Nichi tsūwa no kaitsū ni itaru made" (July 1939), NCTT Records 2028/1493.

was a need for telecommunications operators there to formulate a common position based on their shared circumstances and common interests vis-à-vis Japan's East Asian communications policy; in the meantime, they also needed to resolve their conflicting interests through mutual coordination and deeper understanding. An ex-MOC official himself, Watanabe did not leave the MOC completely in the dark. In sending his proposal to Tokyo, Watanabe invited the MOC to send an observer to the first meeting, which was to be convened shortly in Beijing.[6]

After some discussion within the ministry, Tamura Kenjirō, director of the MOC's Telecommunications Bureau, responded to Watanabe's invitation by emphasizing that "the relationship among Japan, Manchukuo, and China in communications requires an inseparable, unified system." To strengthen the East Asian communications bloc, Tamura suggested, it was imperative to set up "an East Asian communications forum consisting of *all* concerned agencies [in these countries]." In other words, important issues about telecommunications on the continent should be discussed not only among the newly established companies there but also with concerned parties in Japan and elsewhere. The "continental forum" proposed by the NCTT, Tamura pointed out, appeared to conflict with such a goal by marginalizing Japan. But since preparations were already under way, the MOC would allow a one-time-only gathering for the sole purpose of "creating an atmosphere for the future East Asian communications forum."[7] In addition to such lukewarm response from Tokyo, Watanabe found that support from telecommunications operators in Manchukuo was not forthcoming.[8] In the end, the NCTT abandoned the proposed forum altogether.

The MOC's objection to Watanabe's proposal was understandable because it had its own plans for such a forum. As the MOC began planning the East Asian telecommunications network, systemic coordination assumed greater urgency. Always considering itself the center of

---

6. "Man-Ka-Mō tsūshin renraku kaigi an" (1939), appended to NCTT Head of Business Department (Watanabe) to MOC Directors of Telecommunications and Engineering (March 11, 1939), MOC Records I, 200.

7. Tamura to Watanabe (March 39, 1939), MOC Records I, 200.

8. MTT Director of Telecommunications to NCTT Director of Operations (April 8, 1939), NCTT Records 2028(2)/49.

the East Asian telecommunications network, the MOC in Tokyo was not pleased with local initiatives that seemed to undermine its role in empirewide telecommunications matters. As we have seen, the MOC had tried to influence the new telecommunications operators in Manchukuo and China. Although its explicit concern was China, the MOC's Committee on Administration of Communications in China was a major organizational step toward coordinating communications policies throughout the empire.[9] In an internal memo dated June 1938, the MOC went a step further and proposed an "East Asian telecommunications conference under MOC guidance" to "secure cooperation from other concerned agencies to carry out the East Asian communications policies adopted by the MOC."[10] To underscore the importance of its leadership, the MOC noted Britain's leading role in the European bloc and America's influence at international conferences and in the formation of a pan-American bloc. To realize Japan's communications policy toward the East Asian region, including Manchukuo, North China, and Central China, the MOC argued, Japan should establish a permanent council on East Asian communications and serve as its leader in order to "maintain inseparable connections with countries on the continent and to form an Asian bloc occupying one third of the world."[11]

Under MOC auspices, the East Asian Telecommunications Conference finally met for the first time in Tokyo in late November 1939. In addition to MOC officials, there were representatives from Japanese government agencies in the colonies and in the mandate territories, as well as from the "joint venture" telecommunications companies in Manchukuo and occupied areas in China. Representatives from facilities companies, such as the JTTCC and the ITC, were also present, as were broadcasting operators from Japan and its colonies, for consultations on technical matters. Despite its claim to be an East Asian gathering, almost all the delegates were Japanese, a phenomenon that sparked mixed reactions even among the attendees. Some delegates

9. Kubo Odō, "Tai-Shi teishin gyōsei shori no shin kikō," *TKZ* 354 (February 1938): 2–4.

10. "Hokushi oyobi Chūshi ni okeru tsūshin kaisha ni taisuru tōsei yōryō" (draft, June 1, 1938), MOC Records I, 264.

11. "Tōa ni okeru denki tsūshin kaigi jochi ni kansuru ken" (n.d.), MOC Records I, 200.

welcomed this as an unmistakable sign of Japan's unchallenged strength; others felt that Japan's mission in Asia would be better served by the inclusion of some non-Japanese delegates. This can be seen as an early indication of the lack of agreement over the definition of the East Asian community.[12]

There seemed to be ready agreement, however, on the mission of the meeting. Participants signed a prospectus (*shūisho*) that declared:

> Responding to the new situations in East Asia, it is of utmost urgency at this moment to strengthen business and technical ties in telecommunications between Japan, Manchukuo, and China, and to fulfill most thoroughly its comprehensive functions, in order to fully accomplish telecommunications' mission in building the East Asian New Order. It is essential to set up an appropriate agency of liaison and consultation and to maintain mutual organic ties among telecommunications operators in Japan, Manchukuo, and China.[13]

An annual East Asian Telecommunications Conference, with the explicit purpose of promoting coordination among all telecommunications operators in areas under Japanese control, hence would become the main vehicle for ensuring the smooth operation of a single telecommunications system in East Asia. Membership in the conference eventually included four types of participants: (1) the MOC and other Japanese government agencies in charge of telecommunications matters; (2) telecommunications companies from Manchukuo and China; (3) the ITC and JTTCC, which were involved in facilities construction and maintenance in the empire; and (4) the NHK and other agencies in charge of broadcasting in the colonies and China. Not all of them had an equal presence. Judging from the number of proposals put forward by the participating entities, the MOC was predominant, with nearly one third of all proposals. The MTT, the oldest national policy company in imperial telecommunications, came next; but its two regional cousins in China, the NCTT and CCTC were also active, as was the ITC (see Table 12).

12. "Tōa ni okeru denki tsūshin kaigi jochi ni kansuru ken" (n.d.), MOC Records I, 200.

13. "Tōa denki tsūshin kyōgikai shūisho," in Tōa denki tsūshin jimukyoku, *Tōa denki tsūshin kyōgikai dai-1-kai kaigi gijiroku*, appendix, 1.

Table 12
Participants and Proposals at the
East Asian Telecommunications Conferences, 1939–43

Participants	Total proposals
Ministry of Communications (MOC)	64
Manchurian Telegraph and Telephone Co. (MTT)	38
North China Telegraph and Telephone Co. (NCTT)	20
International Telecommunications Co. (ITC)	17
Central China Telecommunications Co. (CCTC)	15
Korea Government-General (GGK) Communications Bureau	12
Inner Mongolia Telecommunications Facility Co. (IMTFC)	10
Japan Telegraph and Telephone Construction Co. (JTTCC)	7
Inner Mongolia Postal and Telecommunications Administration	6
Korea Broadcasting Society	5
Taiwan Government-General (GGT) Communications Bureau	3
Japanese Broadcasting Corporation (NHK)	3
South Seas (Nan'yō) Agency	2
China Broadcasting Corporation	2
Karafuto Communications Bureau	0
Taiwan Broadcasting Corporation	0
North China Broadcasting Corporation	0

SOURCE: Tōa denki tsūshin jimukyoku, *Tōa denki tsūshin kyōgikai kaigi gijiroku* (various years).

The East Asian Communication Bloc in Action

Early indications suggested considerable progress with the organization. Beginning with the second conference in 1940, for instance, the MOC initiated annual consultation among participants concerning their proposed expansion of connecting circuits. In order to ensure smooth communications within Japan's newly expanded sphere of influence, conference participants drafted an "East Asian Telecommunications Agreement" in late 1940. An elaborate document containing 49 articles, the agreement dealt with matters ranging from the composition and operation of both cable and wireless communications circuits to tariff setting and tariff payments. It also included measures for rerouting traffic in case of interruptions in service due to accidents. The agreement was supposed to replace all existing bilateral agreements among its members, including the broadcasting associations and facilities companies. A permanent secretariat was set up in Tokyo; its expenses were to be shared by all members. The secretariat was charged with coordinating correspondence and publishing a journal devoted to East Asian telecommunications. It was also responsible for organizing general assemblies for future revisions of the agreement.[14]

A significant, if somewhat symbolic, development was the designation of a new category of telecommunication traffic in the 1940 agreement. All telephone and telegraph exchanges among the participating operators in East Asia would henceforth be called "East Asian communication" (*Tōa tsūshin*), as distinct from "international communication" (*kokusai tsūshin*), which would refer only to communication between signatories of the agreement and countries outside the East Asia region. In other words, in terms of telecommunication, East Asia was to become a distinctive region separate from the rest of the world. The official scripts for all telegrams were Japanese *katakana* as well as the Latin-Arabic alphabet; this allowed both Chinese- and European-language telegrams to continue to be sent under separate designation codes. All payments for East Asian communication—both telegrams and telephone calls—were to be calculated in Japanese currency.[15]

Signed on January 22, 1941, the East Asian Telecommunications Agreement was a major step toward realizing Japan's plan to create an independent telecommunications network in East Asia. Although initially the agreement existed only on paper and its terms were deliberately vague, it marked the beginning of a separate sphere of telecommunications in East Asia under Japanese control. More than simply a forum devoted to free discussion and exchange of views, the agreement would have binding authority over its signatories, except in those matters requiring the approval of supervising governmental agencies in each area.[16] At the special meeting convened for signing the agreement, Minister of Communications Murata Shōzō spoke of the "world-historic" changes

---

14. Tōa denki tsūshin jimukyoku, *Tōa denki tsūshin kyōgikai dai-2-kai kaigi gijiroku* (1940).

15. "Tōa denki tsūshin ni kansuru gyōmu kyōtei," signed on January 22, 1941. However, as an internal MOC opinion memo indicated later, Chinese-language telegrams became an independent category out of consideration for China's "appearance." See Gyōmuka, Denshin-kakari, "Tōa denki tsūshin ni kansuru gyōmu kyōtei kaisei iken," MOC Records I-319.

16. Watanabe Otojirō, "Tōa denki tsūshin kyōgikai to Tōa denki tsūshin gyōmu kyōtei to no hikaku" (September 1942), MOC Records I-309.

taking place in Europe and Asia. To establish the Greater East Asia Co-Prosperity Sphere, Murata pointed out, Japan, China, and Manchukuo must cooperate closely in order to form a strong community (*kyōdōtai*), which would, in turn, become the core of the Co-Prosperity Sphere and driving force behind it. As a basic instrument for political, economic, and cultural contacts between Japan, Manchukuo, and China, Murata reiterated, telecommunications in East Asia must be further strengthened.[17] In early 1941, the members also adopted a set of East Asian Telegraph and Telephone Regulations, in which they pledged to cooperate in telecommunications operational matters such as improving the business system and technology, as well as network consolidation.[18]

Apart from the practical need for coordination among telecommunications operators in areas under Japan's control, there was an ideological dimension to this emerging regional system in East Asia. In short, Japan was attempting to create an alternative to the International Telecommunications Union (ITU), which had been founded in Madrid in 1932 through the merger of the International Telegraphic Union and the International Radio Conference. As one MOC memo argued, since the ITU was created for the convenience of communications by Western countries, its regulations were intended to meet the communication needs of Western countries. As a latecomer, Japan had adopted Western methods at the dawn of its international communications, and it had to abide by ITU regulations. At present, with all the great changes taking place in East Asia and with Japan becoming the leading country in the region, the MOC memo pointed out, it "goes without saying that Japan must establish a new telecommunications union as well as regulations, with an emphasis on the convenience of international telecommunications among East Asian countries."[19]

As the annual meeting became regularized and its scope broadened, many participants called for turning the forum into a more permanent organization along the lines of the ITU. The MOC also saw the need for a more formal East Asian telecommunications union, on a par with

---

17. Teishinshō denmukyoku, *Tōa denki tsūshin ni kansuru gyōmu kyōtei teiketsu kaigi gijiroku* (January 1941), 3.

18. Denmukyoku, Gaishinka, "Tōa denshin denwa kisoku no seitei," *TKZ* 392 (April 1941): 100–111.

19. MOC, "Tōa denki tsūshin seisaku" (n.d.). I thank Professor Hikita Yasuyuki for making a copy available to me.

regional organizations elsewhere. In an internal document written in 1940, the MOC outlined three main reasons for the formation of such a union:

1. From an operational point of view, the fact that the majority of telegrams in East Asia are written either in Japanese *kana* or in Chinese ideographs calls for a special East Asian regional telegram system. Moreover, both long-distance telephones and allocation of the wireless spectrum call for regional arrangements similar to those existing in Europe but separate from the general international agreements.
2. Technically, a research organ made up of specialists is needed to study and promote standardization of telecommunications equipment, as is a consultative body of experts to unify maintenance and operation in the region.
3. From the perspective of international politics, solidarity of telecommunications operators in East Asia is necessary to drive out well-entrenched third-country interests as well as to defend "East Asian interests" at international conferences.[20]

The outbreak of the Pacific War drastically changed the scope and orientation of Japan's external telecommunications. On December 8, 1941, eleven of its international circuits signed off. Altogether, 21 of Japan's international communications links with foreign countries ceased to operate (thirteen wireless telegraph, three submarine telegraph, two phototelegraphy, and three wireless telephone) within a few weeks. Japan's international communications partners now consisted of its two allies in Europe—Germany and Italy—as well as neutral countries such as the Soviet Union, Sweden, and Switzerland. In Latin America, after links with Mexico and Brazil stopped in early 1942, Japan could communicate only with Argentina, Peru, and Chile. For intelligence and propaganda purposes, the Japanese government managed to open new direct routes to Lisbon, Hanoi, Vichy France, Turkey, and Spain.[21] In terms of the sheer number of destinations, the onset of war in Decem-

20. Teishinshō, "Tōa denki tsūshin rengō no kessei o hitsuyō to suru riyū" (1940), MOC Records I, 310.

21. Denmukyoku, Gaishinka, "Daitōa sensō boppatsu go no taigai denshin renraku," *TKZ* 405 (May 1942): 37–41. Interestingly, the wireless contact between Tokyo and RCA's station in San Francisco continued to operate until December 13, 1941, six days after the attack on Pearl Harbor. The MOC attributed this delay in cutoff to inertia on both sides and cited the presence of many Japanese citizens in the United States as an additional factor.

ber 1941 thus isolated Japan from the world outside East Asia to an extent unknown since the beginning of its international communications.

While losing many communications links with Europe and the United States, Japan gained quite a few in Asia. Above all, it was now in control of a communications network of unprecedented scale in Asia. When communications links with Southeast Asia were gradually restored and operated by the Japanese, Japanese-language telegrams could reach nearly all parts of East and Southeast Asia for the first time. In terms of sheer space, the coverage of Japanese-language telegrams had reached its peak. The presence of "purely Japanese-style telegrams" in all parts of East Asia, as MOC officials put it, was "a landmark historic event" for East Asian telecommunications.[22] In June 1942, half a year into the Pacific War, the signatories of the East Asian Telecommunications Agreement met in Tokyo to adjust to the new geopolitical conditions as well as to simplify procedures. Seeing new opportunities on the horizon, Watanabe Otojirō of the NCTT made a series of proposals calling for a "thoroughly new concept of the international telegram."[23]

The outbreak of the Pacific War not only expanded Japan's sphere of telecommunications operations in Asia but also created new needs and organizational challenges. Unlike Manchukuo and occupied areas in China, the Southern Region was under direct military occupation; this created yet another type of telecommunications operation. The MOC had far fewer resources in Southeast Asia than in other areas of the Co-Prosperity Sphere, and the military exercised direct control over telecommunications matters there. Unlike the "national policy companies" operating in Manchukuo or in occupied areas in China, a company like the ITC was reduced to a "robot," as its employees recalled later, that simply took orders from the Japanese military authorities in charge of the occupation.[24] Moreover, Japan's operations in Southeast Asia experienced problems similar to those encountered elsewhere, such as lack of coordination between local military authorities and Tokyo and rivalry between the military services (see Chapter 8).

---

22. Denmukyoku, Gaishinka, "Tai-Nanpō denki tsūshin no gendankai," *TKZ* 414 (February 1943): 61.

23. Teishinshō denmukyoku, *Tōa denki tsūshin gyōmu kyōtei dai-2-kai kaigi yobi kaigi gijiroku* (June 1942), 35–36.

24. *KDTKKS*, 53–55.

In the meantime, developments in Europe provided another impetus for further organizational integration. In September 1942, the German minister of communications notified the Japanese government of Germany's plan to convene a European Postal Union conference in Vienna in October 1942. MOC officials noted with satisfaction that Germany had expressed much interest in the formation of a Greater East Asian telecommunications sphere and indicated a desire to build a "new world order" on the basis of partnership with Japan. Taking a clue from this, Nakamura Jun'ichi, MOC director of telecommunications, urged that Japan should not be left behind by its ally in Europe. The time had come, he reasoned, for Japan to finally launch a Greater East Asian telecommunications union, comparable to the new bloc emerging in Europe under German aegis and equal to the ITU under the control of Britain and the United States. Indeed, as Nakamura saw it, a Greater East Asian telecommunications "co-prosperity sphere" had already emerged, in that all members of the telecommunications conference were being treated equally and the operating costs of the secretariat were shared by all.[25] To further promote solidarity among the employees of the various companies and government agencies and to improve operational skills, the telecommunications conference sponsored the First East Asian Telecommunications Skills Contest in September 1942. Held in the Manchukuo capital Shinkyō, a total of 73 telegraph operators representing thirteen entities in different parts of the imperium competed in twelve fields such as typing and transcribing telegrams.[26]

In October 1943, when the annual East Asian telecommunications conference met in Tokyo for the fifth time, it was renamed the First Greater East Asian Telecommunications Conference, a somewhat belated acknowledgment of the newly expanded imperium. In addition to signatories of the earlier agreement, the Cabinet Planning Board as well as Army and Navy officers were present as observers. The Japanese had invited a high-ranking official of the Thai Ministry of Communications to attend, but Bangkok shrewdly declined the invitation on the

---

25. Tōa denki tsūshin jimukyoku, *Tōa denki tsūshin kyōgikai dai-4-kai kaigi gijiroku*, 156–58. See also the handwritten comments at the end of a MOC document, "Tōa denki tsūshin kaigi kiyaku oyobi Tōa denki tsūshin gyōmu kyōtei teiketsu ni kansuru ken" (October 6, 1943), MOC Records II-591.

26. Nagatani Etsurō, "Tōa denki tsūshin kyogi daikai," *TKZ* 411 (November 1942): 56–59.

ground that "too many agencies are consulted." As a result, several Thai government technicians who happened to be in Tokyo on official business attended instead. The Japanese had better luck with French Indochina; a French official was flown to the conference on a Japanese military airplane.[27] In appearance, at least, organizational evolution was complete, with the inauguration of a Greater East Asian Telecommunications Bloc.

The East Asian telecommunications network can be seen as an extension of Japan's earlier experience with system-building in the empire when its transportation links grew beyond the home islands. Beginning in 1924, the Japanese held annual meetings to coordinate railway transportation between Japan, Korea, and Japanese-operated railways in Manchuria. By the late 1930s, as both railway and shipping routes operated by Japan greatly expanded, the Japanese also sought to coordinate all modes of transportation within East Asia. By 1940, the Japan-Manchukuo-China Consultative Committee on Transportation had developed into an impressive umbrella organization. Headed by Japan's Railway Ministry, it included the colonial governments of Taiwan and Korea, three railway companies, five shipping companies, and three aviation companies.[28] Moreover, the formation of the East Asian telecommunications bloc paralleled other concurrent schemes for constructing imperial networks in the imperium. One of the most important was an East Asian broadcasting network, also known as the East Asia Radio Corporation Chain. Through similar conferences and agreements, radio operators made arrangements to exchange broadcast programs and relay the so-called East Asian Broadcasting from Tokyo. It operated largely by wireless facilities throughout the imperium and hence was closely tied to the electrical communications network that used both wireless and cables.[29] These various imperial networks in East Asia represented Japan's effort to better integrate the imperium. In other words, the Japanese imperium in East Asia after the late 1930s had indeed become truly an empire of networks.

---

27. "Daitōa denki tsūshin kaigi kankei ikken: jinji kankei," in JMFA Archives.

28. For a list of participants, see Nichi-Man-Shi kōtsū kondankai dai-ichi chōsakai jimukyoku, comp., *Tōa kyōeiken nai kōtsuryō tōkei (1936–1940)* (Tokyo, 1942).

29. For a comprehensive study of Japan's wartime overseas broadcasting, see Kitayama Setsurō, *Rajio Tōkyō*. The East Asian broadcasting network is briefly discussed in English in Namikawa Ryō, "Japanese Overseas Broadcasting."

TECHNICAL AND OPERATIONAL CHALLENGES

Setting up organizational structures was only the first step in building a fully functioning system. From the beginning, as the records of the annual East Asian Telecommunications Conference make clear, the Japanese viewed their task as essentially two-dimensional—requiring both technical and business (*gyōmu*) network-building. Accordingly, the MOC structured the conferences into two separate section meetings (*bukai*): a Business Section and a Technical Section. In this way, the conferences also mirrored the division of labor in the MOC itself. But because many issues required the skills of both technical and administrative experts, a Joint Business and Technical Section was soon added to address them.

Coordinating Technologies

The East Asian Telecommunications Conference tackled a wide range of technical matters. At the second conference, in 1940, Shinohara Noboru, head of the MOC's Investigation Section, outlined six technical areas that were of particular relevance to building an East Asian long-distance network: (1) toll-phone dialing systems, (2) relay and exchange in long-distance lines, (3) standards for testing the coordination of circuits, (4) transmission of standard frequencies, (5) aviation wireless, and (6) wireless phototelegraphy.[30] Accordingly, the Technical Section was divided into seven subcommittees, responsible for investigation, lines, mechanics, transmission, wireless, architecture, and coordination, respectively. Altogether, they reflected particular concerns with standardization of telecommunications equipment and maintenance, cooperation in developing technical solutions, and the perennial problem of wireless frequencies.

One of the major tasks facing the East Asian Telecommunications Conference was standardizing telecommunications equipment and technical practices throughout the whole system. Many different types of equipment and practices existed in different areas, and sometimes even within the same area, creating enormous complexity. Moreover, the

30. Tōa denki tsūshin jimukyoku, *Tōa denki tsūshin kyōgikai dai-2-kai kaigi gijiroku*, 145.

increasingly widespread shortage of spare parts necessitated an exchange system among the various operators. Setting technical standards, as Shinohara put it, constituted the third and final stage of technological development, following research and implementation. At the second conference, Shinohara, who had collaborated closely with Matsumae Shigeyoshi in development of the NLC, took special care to propose a uniform technical standard of NLC transmission, to be used in all trunk cables in East Asia.[31] To keep communications facilities in working order, a maintenance conference devoted primarily to the Japan–Manchukuo cable was set up at the third conference.[32] In March 1940, operators adopted "Basic Guidelines Concerning the Establishment of Telecommunications and Technical Standards of Construction and Maintenance in China." The purpose was threefold: (1) to facilitate technological control of construction and maintenance of telecommunications in East Asia; (2) to save material and reduce imports from outside the yen bloc; and (3) to take into consideration the special circumstances in China.[33] Considerable attention was paid to measures to cope with interruptions in trunk circuits, as well as other emergencies.

As the ITC gradually assumed responsibility for much of the construction and maintenance of trunk routes in East Asia, it offered a solution to some of the problems. However, other technical problems continued to consume the attention of conference participants. Of all the technical problems, wireless frequencies were by far the most complex and most urgent.

Allocation of the wireless spectrum was not new or unique to Japan. The relatively low cost and ease with which wireless communication facilities could be set up, compared to cable operations, created their own problems, as did the fact that wireless was carried through open space and extremely vulnerable to interference. Since the 1910s, there had been two separate systems of wireless communications in Japan: the civilian wireless under the control of MOC and various military wireless communications used by the Army and Navy. To deal with the interna-

---

31. Ibid., 146–47.

32. Ibid., 175–78.

33. "Shina ni okeru denki tsūshin yo kiki no kikaku settei yōkō" (February 12, 1940); "Shina ni okeru denki tsūshin hōshiki aruiwa kensetsu, hoshu no gijutsu hōshiki nado no settei ni kansuru kihon yōkō" (March 15, 1940), NHK Records.

tional conference on frequency allocation, the three ministries set up a meeting in 1921 to allocate radio frequencies within Japan. As wireless use in Japan became more widespread, especially after the start of the radio broadcasting, the issue of coordination and regulation became more important. In 1929, the meeting became the permanent Three-Ministry Frequency Regulation Council. After the founding of Manchukuo, the need to coordinate frequency use between Manchukuo and Japan led to a Japan-Manchukuo Frequency Regulatory Forum in 1936. Although the idea of setting up a single regulatory agency for all Japanese wireless operations in East Asia was discussed, it came to naught. After 1939, wireless frequency allocation became a constant theme at the annual East Asian Telecommunications Conference.[34] Almost half of the 25 proposals submitted to the Technical Section at the 1941 conference, for instance, concerned the problem of wireless.

By then, much of Japan's telecommunications traffic within the Co-Prosperity Sphere had come to rely on wireless, the official emphasis given to long-distance cable communications in the imperium notwithstanding. As Japan expanded its wireless communication routes, the need to coordinate the use of radio frequencies became a great concern. This became even more pronounced after the outbreak of the Pacific War because Japan's communication with the Southern Region was carried almost entirely by wireless facilities (see Map 5). In addition, newly expanded broadcasting programs, using many of the same facilities as wireless telegraphy and telephony, compounded the problem. Differences in equipment, electrical currents, order and protocol, and especially the huge number of frequencies used throughout Japanese-controlled areas in East Asia meant nothing short of chaos in wireless communications. Poor coordination made the situation worse, as even a relatively minor problem could cause a blackout of crucial communications. For example, between 14:20 on September 16 and 9:00 on September 19, 1942, the wireless connection between Tianjin and Shanghai was shut down for a total of 48 hours. As a result, some 1,400 telegrams were delayed, for the first time since the founding of the NCTT. Only after the company dispatched an employee to Shanghai did it discover the source of the problem: wireless operators in Shanghai had confused

34. *Zoku teishin jigyō shi*, vol. 6, reprinted in *KSS*, 2: 776–89.

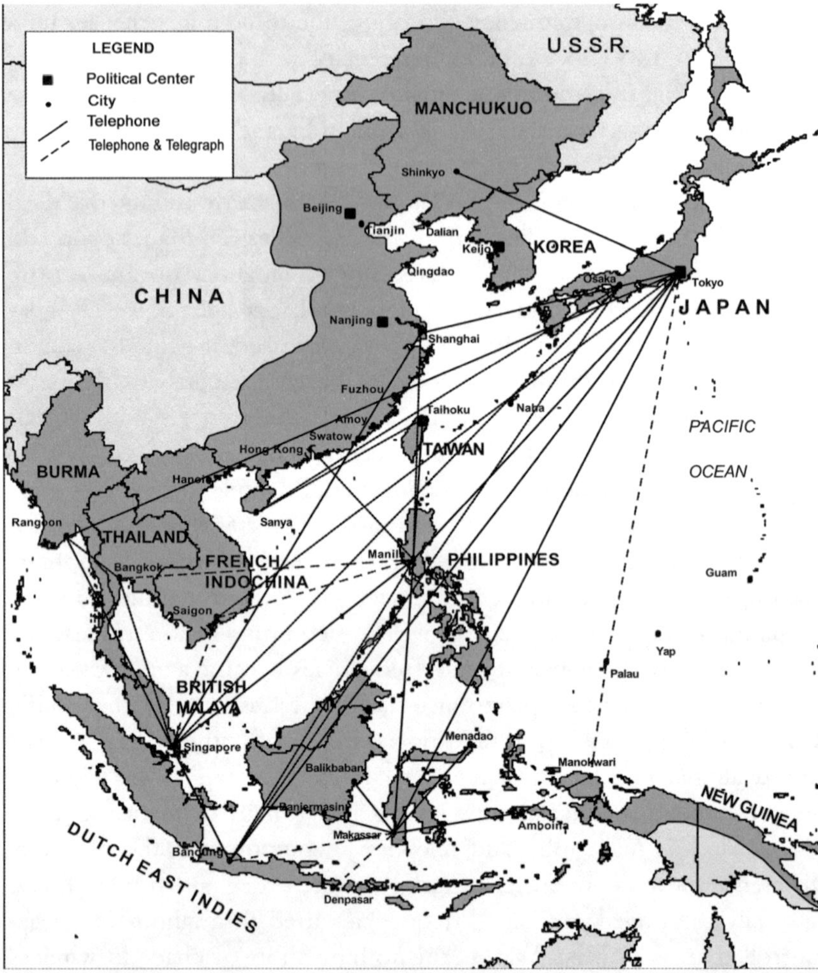

Map 5 Major wireless routes in Greater East Asia, 1943 (SOURCE: MOC Records II, 804).

frequencies designated for Tianjin, Osaka, and relay broadcasting.[35] The wireless problem got worse as more and more of Japan's submarine cables were put out of commission. As an MOC official noted in late 1943, since communications between important bases in East Asia and Japan proper were carried out mostly through wireless, their importance to the war effort increased in proportion to construction in East Asia, es-

35. Daitōa denki tsūshin jimukyoku, "Daitōa denki tsūshin kyōgikai dai-1-kai kaigi gijiroku (1943)," handwritten copy, MOC Records I-330.

pecially in the southern region. The state of wireless communications between Manchuria and Japan and particularly between Japan and the South Seas, he admitted, was not satisfactory.[36]

As a remedy, the NCTT proposed the immediate establishment of a headquarters for frequency control throughout East Asia. To eliminate the problem of generating wrong frequencies by mistake, some suggested, it was necessary for all wireless transmitters to adjust to a standard frequency (*hyōjun denpa*). The MTT proposed transmitting this standard frequency over wires throughout East Asia as the most secure method. After devoting much of its discussion to the wireless issue, the First Greater East Asian Telecommunications Conference in 1943 concluded that it was necessary to establish a central authority to regulate frequencies and asked all members to strive to create the right "atmosphere," both at the center and in the field.[37]

Given the fact that wireless was used heavily by both the Army and the Navy—the Navy alone would claim almost 10,000 different frequencies for its own use by ships, aircraft, and land stations[38]—it is not surprising that the East Asian Telecommunications Conference was unable to make substantive decisions that would affect the military. Immediately following the outbreak of the Pacific War, a Headquarters for Regulating Frequencies was established by the MOC, Army and Navy, and the Cabinet Information Bureau. It was not until 1944, when the war turned decisively against Japan, that further progress was made on the problem of frequency control. In April of that year, a Frequency Bureau (Denpakyoku) was established within the new Communications Board specifically for the purpose of controlling frequencies throughout Japan.

Wrestling with Tariffs

Of all the operational issues to be ironed out among various telecommunications operators in Japan's new empire, telegraph and telephone tariffs were one of the most important and most difficult. The charge for

36. Ibid.

37. "Daitōa denki tsūshin kyōgikai dai-1-kai kaigi teian kettei yōryō" (October 1943), MOC Records I, 328.

38. Samejima Sunao, *Moto Gunreibu tsūshin kachō no kaisō*, 12.

telegrams and telephone calls affected not only hundreds of thousands of users but also the revenue needed to sustain such operations. Under the Japanese government monopoly, tariffs had been a relatively simple matter in Japan and its formal colonies. But operators in different parts of the empire were of widely varying characters: there were government monopolies in Japan proper as well as its colonies; quasi-business operations in Manchukuo and China; and, after the outbreak of the Pacific War, new rates set largely by the military administration in Southeast Asia. To make things even more complicated, because Japan still maintained "international communications" links with countries outside the East Asia Sphere, it had to calculate those charges in francs first and then figure out the proper exchange rate with the yen, since the yen was the basis of telegram rates within the sphere.

Beginning in April 1938, Britain's Imperial Communication Committee adopted the Empire Flat Rate, set at 1s. 3d. per word for all plain-language telegrams to and from the United Kingdom and between its dominions and colonies.[39] Such a revolutionary move had far-reaching implications. "What is the objective of such a tariff policy?" asked Watanabe Otojirō in one of his essays on communications policy. After comparing Japan's rate system with the new British imperial rates, he concluded that the British flat-rate policy was based on "an exclusive and long-term plan to complete the all-British imperialist integration through an independent and powerful network of submarine cables and beam wireless that connect British territories throughout the world."[40]

When Watanabe looked at Japan's rate structure in the emerging East Asian telecommunications system, he had plenty of reasons to worry. As Fig. 4 shows, there was anything but a uniform rate in East Asia under Japanese control. Telegram rates in Japan—both domestic and with other regions—were relatively low, ranging from under 0.10 yen to 0.20 yen per word. Primarily because of the large volume of traffic, rates in Manchukuo and North China had been set relatively low as well, not exceeding 0.20 yen. Inner Mongolia and Central China, in part due to their higher operating costs, charged higher rates, often 0.30 yen. As a result, it was cheaper to send a telegram from Japan to Inner Mongolia (0.20 yen per word) than vice versa (0.30 yen).

39. Barty-King, *Girdle Round the Earth*, 252–53.

40. Watanabe, *Denki tsūshin kokusaku to denki tsūshin jigyō*, 377–82.

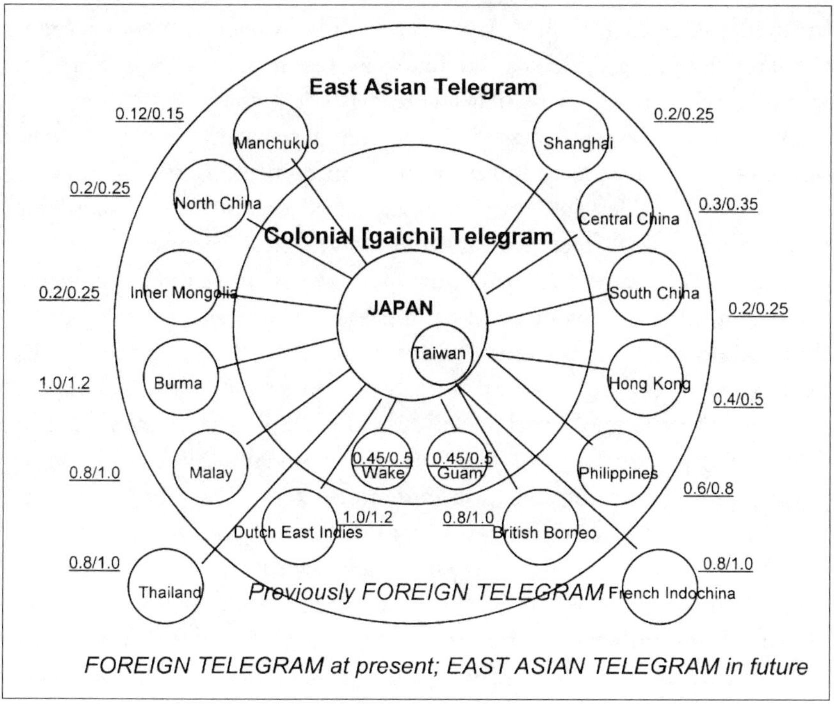

Fig. 4 Outbound telegram rates in East Asia, 1942. NOTE: Rates (yen) are per each five characters in Japanese telegrams and one word in European-language telegrams. However, rates for Guam- and Wake-bound telegrams are minimum charges per telegram. Minimum charge for all other telegrams are five times the base rates indicated above (SOURCE: MOC Records I, 297).

Although simplification of tariffs was often suggested, it was by no means an easy task. The key to standardizing telegram rates was revenue allocation, since the services on two ends of any given route needed to work out a satisfactory formula between them. Upon receiving complaints about the MOC's refusal to adjust proportions of revenue allocation, Watanabe called for "Greater East Asianism" (*Daitōashugi*) as a new principle to be applied in tariff policy. However, as one MOC official explained, adjusting tariffs was extremely difficult because it was intrinsically tied to purchasing power and hence to the currency systems of various areas in East Asia.[41] Progress was slow and limited in scale. For instance, when the MOC announced a revision of the East Asian Telegraph and Telephone Regulations in 1942, it

41. *Tōa denki tsūshin kyōgikai dai-4-kai kaigi gijiroku*, 53–56.

brought the rates for the "East Asian wireless telegrams"—telegrams exchanged between wireless facilities in Japan, China, and Manchukuo—closer to rates for regular telegrams.[42]

As with the wireless, tariff policy was a constant theme of many discussions at the telecommunications conferences. A special meeting on tariffs was called in February 1943, based on the East Asian Telecommunications Agreement signed in 1941. The meeting, which was devoted to telegram tariffs only, dragged on for more than two weeks. Nine proposals were submitted on the basic principles of tariff calculation and on methods of tariff collection. A Tariff Subcommittee of experts was set up and reached agreement on some of the issues. But when all the delegates met to discuss the guiding principles of East Asian telegram tariffs, which would affect revenue allocation among participating telecommunication operators, there was little common ground. From the start, delegates from different parts of the imperium simply took the opportunity to state their own preferences.[43] The biggest dilemma was how to balance the overall political imperative for uniform rates with very different local economic realities. For instance, government telegrams and press telegrams served specific political objectives and had to be placed at reasonable rates, yet cost also had to be considered. Again, no solution was forthcoming.

In October 1943, in conjunction with the First Greater East Asian Telecommunications Conference, the Tariff Subcommittee met for the second time. Delegates agreed in principle to begin investigating per-unit operating costs, using 1942 as the basis for eventual revenue allocation. Although this was sound in theory, delegates soon realized that the different economic conditions within the East Asian region made it impractical. In particular, runaway inflation in Central China since early 1943 compelled CCTC executives to demand reconsideration of the basic method of setting tariffs. The three-day discussion that ensued appeared to some participants to consist of constant lecturing by Japanese delegates based in China on the impact of economic problems there. Not surprisingly, delegates postponed any decision on tariff allocation

42. Denmukyoku, Gaishinka, "Tōa denshin denwa kisoku nado no kaisei," *TKZ* 410 (October 1942): 82–84.

43. "Ryōkin iinkai o orite," *TDTZ* 4.3 (February 1943): 47–51.

until further studies of operating costs were made. Once again, the tariff meeting had failed to produce concrete solutions.[44]

Tariff was one of the few subjects discussed at the Second (and last) Greater East Asian Telecommunications Conference, held in Tokyo in late 1944. The focus of these discussions was the unilateral "tariff surcharge" (*fuka ryōkin*) adopted by China-based telecommunications companies to cope with rising operating costs due to the rampant inflation there. In January 1944, the CCTC had to raise telegram and telephone rates by almost 100 percent. Six months later, they were raised again. Delegates from elsewhere wanted to see this surcharge replaced by a new "tariff by agreement" (*kyōtei ryōkin*) between two or more operators, but met with stubborn resistance. Representatives from China-based companies argued that the different base prices of each company had been compounded by the unique economic conditions in China. In the end, the meeting had to settle with a compromise: a tariff surcharge was now recognized as part of the "tariff by agreement." Thus, as late as 1944, the goal of a standardized East Asian Tariff System seemed as remote as ever. One Communications Board official tried to put a positive spin on it by claiming that the recognition of the surcharge did provide a temporary solution to the perennial problem of proper ratios of revenue distribution among various operators. As he put it, this might well be a noteworthy step by creating "tariff by agreement" for the first time in East Asia.[45]

Ultimately, efforts to build a unified tariff structure ended in dismal failure. Nagatani Takeo, a Japanese employee in the Business Department of the NCTT, candidly admitted the inherent contradiction:

> The fundamental reason the Tariff Committee could not reach a conclusion that could be put into effect was that telecommunications in East Asia, like mining in iron and coal, have been based on separate management geared toward self-sufficiency. To improve their operations in each area, [telecommunications enterprises] must adopt an autonomous tariff policy in their own areas. Ultimately, [such a practice] opposes the very essence of telecommunications embodied in the East Asian tariff policy. [Thus] if these enterprises must

---

44. "Ryōkin iinkai dai-2-ji kaigi no gaikyō," *TDTZ* 3 (November 1943): 52–54.

45. Tanaka Shizuo, "Daitōa denkitsūshin kaigi dai-2-kai kaigi no gaikyō," *TDTZ* 4 (November 1944): 3–10.

conform to the Greater East Asian tariff policy at any cost, they will end up bankrupting themselves.[46]

The 1944 conference made some progress on other tariff-related matters, however. Delegates agreed to count the addressee's name in Japanese-language telegrams as three words (up from two, as previously), while keeping the minimum length the same, at five words. Although this encouraged shorter telegrams, the goal of unifying the tariff on addressee names in European-language telegrams failed to reach consensus. Perhaps the greatest accomplishment was the unification of rates between Japanese and Chinese telegrams, a demand insisted on by the telecommunications operators in occupied China. Not only would each word in Chinese-language telegrams be charged the same tariff as in Japanese-language telegrams, but the addressee's name in the former was also set to count as three words.[47] This was a symbolic victory, albeit a belated one. Chinese-language telegrams, which had been similar to European-language telegrams in tariff structure, could now be considered a step closer to Japanese-language telegrams.

ORGANIZATIONAL AND STRUCTURAL CHALLENGES

The MOC Versus Colonial Korea (Act II)

The persistent difficulties in systemic integration of the telecommunications business in Japan's imperium reflected deeper structural problems. The MOC had foreseen operational and technical problems when setting up the annual telecommunications conferences back in 1939, but there were unexpected organizational problems as well. Although MOC efforts to build an integrated telecommunications network through the East Asian Telecommunications Conference seemed to have some success, this arrangement brought about further complications. And despite the fact that the conferences were almost always all-Japanese affairs, delegates could not avoid major conflicts over organizational matters.

---

46. Nagatani Takeo, "Daitōa denki tsūshin no airo o daikaiseyo!" *TDTZ* 4 (May 1944): 25.

47. Ibid.

The first and perhaps most dramatic disagreement over the conference was again between the MOC and the colonial administration in Korea. In previous international telecommunications treaties involving Japan and its colonies, the GGK had signed under the umbrella category of "Japanese Agencies in Charge of Telecommunications," a category that included the MOC as well as other colonial administrations. Essentially, such an arrangement relegated all authority to the MOC. But at the Second East Asian Telecommunications Conference in October 1940, the GGK representatives insisted on being treated as an independent signatory in the proposed East Asian Telecommunications Agreement. Korea should play a more prominent role in the East Asian telecommunications network, they explained, now that the agreement included China and Manchukuo and thus differed from previous international treaties. After the conference, Japanese officials from Korea continued to argue their case. Since Japan's communications sphere now included operators in Manchukuo and China, they insisted, Korea's legal autonomy under its governor-general must be preserved by accepting it as an independent signatory.[48]

MOC officials viewed the matter quite differently, of course, feeling that a unified administration would give Japan a stronger international position and maintain consistency with previous practice as well. In memos sent to the GGK before the January 1941 conference for signing the East Asian Telecommunications Agreement, the MOC harped on the old theme of the "harmonious union between Japan and Korea" as the "core of East Asia," which in turn formed the axis of a broader East Asia together with Manchukuo and China. Ultimately, it pointed out, these concentric circles would include Southeast Asia to make the Greater East Asia Co-Prosperity Sphere. The MOC assured the GGK that under the new agreement Korea's authority in dealing with Manchukuo and China would not be diminished, and Korea's special demands could be met via separate bilateral agreements. If Korea insisted on becoming independent of the Tokyo government in external telecommunications matters, the MOC implied, the "Fundamentals of National Policy" would be endangered, with grave consequences. As a practical matter, the MOC pointed out, none of the institutional

48. Director of GGK Bureau of Communications (Yamada) to Director of MOC Bureau of Telecommunications (Yasuda) (December 26, 1940), MOC Records I, 311.

adjustments that the GGK demanded could be accomplished before the new agreement was to take effect on April 1, 1941.[49]

The MOC's worst nightmare came true at the general conference for signing the East Asian Telecommunications Agreement in January 1941. After the last round of discussion of minor amendments to the agreement, Korea's representatives consulted with Keijō (Seoul) one more time and then refused to sign. By then, the situation had reached crisis proportions. GGK officials again insisted that Korea's governor-general enjoyed complete administrative authority in Korea, and he exercised communications administrative authority completely independently as far as Korea was concerned. Although the MOC had been commissioned to build various projects in Korea in the past, they noted, this was not the same as transfer of administrative authority:

> Within the Japanese empire, there is not a single government agency that is not based on *kansei* (rules of government organization). Even the smallest post office is always based on *kansei*. However, the term "responsible government agency" [in the proposed agreement] has no basis in *kansei*, and therefore, even if the agreement is passed unanimously here, it cannot be considered part of the Japanese government.[50]

With the entire agreement threatening to collapse before their eyes, MOC officials were furious and called GGK's refusal to participate as a member of the imperial government "suspicious." It was embarrassing, to say the least, that the major obstacle to unity in East Asian telecommunications should come from Japan's own colonial officials in Korea. To avoid a greater disaster, the MOC had to accede to GGK demands by entering into a separate agreement with Korea outside the conference and by promising to treat Korea as a separate entity in East Asian telecommunications matters instead of simply part of the Japanese government.

The GGK's abstention was not just for theatrical effect. Shortly after this episode, a Japanese official from the GGK Communications

---

49. Nakayama to Fukuda (December 18, 1940), and Teishinshō, "Tōa denki tsūshin gyōmu kyōtei ni okeru Chōsen teishin no tōjisha chii ni kansuru iken" (January 7, 1941), both in MOC Records I, 311.

50. Teishinshō denmukyoku, *Tōa denki tsūshin ni kansuru gyōmu kyōtei teiketsu kaigi gijiroku* (January 1941), 32–33, 38–41. For the agreement reached, see "Oboegakisho" (January 22, 1941), MOC Records I, 311.

Bureau explained the rationale to his counterparts representing other areas in the empire in terms of the special political character of Korea:

> In Korea, we must pay close attention to governing an alien people (*iminzoku*) in a way that is more than can be imagined in Japan proper, and [must] adopt special measures appropriate for the situation. . . . Things not in accordance with this principle, even if they are extremely rational and perfectly legitimate, are considered inappropriate and even illegitimate in Korea. We public servants in Korea cannot forget, even for a single minute, this principle of governing an alien people.[51]

Statements like this seemed incongruous with, if not contradictory to, the official policy of "Japan and Korea as one" (*naisen ittai*), which had been promoted since the late 1930s. Indeed, just a few months after the meeting, both Korea and Taiwan would be given the same *naichi* status as the home islands and transferred to the jurisdiction of the Home Ministry. That resistance to the perceived eradication of a separate identity would come from the colonial bureaucracy itself was ironic but not beyond comprehension.[52] Obliterating the differences between Korea and Japan would inevitably jeopardize the separate sphere of authority that the Japanese bureaucracy in Korea had enjoyed for decades. This tendency toward autonomy had been reinforced, at least in part, by the relative stability of personnel in the GGK bureaucracy. Although occasional lateral transfers to other colonies or even to Japan proper did occur—GGK Communications Bureau Director Yamada Tadatsugu had served in the Kwantung Leased Territory, for example—the majority of Japanese serving in the GGK tended to have grown up and stayed in Korea.[53] This lack of personnel transfer nurtured the growth of a separate identity for Japanese serving in Korea vis-à-vis Japanese in Japan. In this context, even relatively mundane issues such as different pay scales could become a source of friction. Referring to the prospect of the MOC operating inside Korea, one GGK official hinted that GGK employees would be displeased to work side by side

---

51. Fukakawa Toshio, speech at a panel discussion; in *TDTZ* 2 (May 1942): 4.

52. For a discussion of the policy implications of *naisen ittai*, see Eckert, *Offspring of Empire*, 236–238; on the new status of Korea and Taiwan, see Hatano, *Taiheiyō sensō to Ajia gaikō*, 106–8.

53. "Chōsen no kansei, gyōsei mondai o Hagiwara shi ni kiku," in Chōsen shiryō kenkyūkai, *Chōsen sōtofuku kansei to sono gyōsei kikō*, 25–27.

with MOC employees because the latter had better chances for promotion. As a bureaucrat, he was voicing a real concern. Despite high-sounding rhetoric, the governments in both Tokyo and Keijō were often dominated by bureaucratic self-interest.

Visions and Interests in Conflict

The discord between the GGK and the MOC was the tip of the proverbial iceberg of conflicting interests that plagued Japan's system-building efforts. As one Japanese participant lamented after attending the conference, the proposals often contradicted one another, and perfectly amicable people engaged in futile argument due to their different perspectives.[54] Indeed, the organizational problems were inseparable from conflicting visions of the Greater East Asia Sphere itself. As result, even deeper dissension existed as to the structure of the imperial telecommunications system.

Early in 1942, as Japanese forces were sweeping through Southeast Asia, Tokyo set out to unveil new blueprints for the Greater East Asia Co-Prosperity Sphere. On February 16, the day after the fall of Singapore, Prime Minister Tōjō Hideki delivered a speech to the 79th Imperial Diet calling for the "establishment of an order of co-existence and co-prosperity that is based on the principle of Japan at its core and all countries and peoples in Greater East Asia in their appropriate places." To reach such a goal, in March the Japanese government formally launched a Council for the Construction of Greater East Asia. Headed by Prime Minster Tōjō himself, the council included government ministers and vice ministers serving as directors of its commit-tees. Some 50 businessmen, scholars, and former government officials were selected as council members. The council was the most ambitious program the government had established thus far to gather advice on all policy issues concerning East Asia except military and diplomatic affairs. It was organized into a general committee and four other committees on general policy, education and culture, population and ethnicity, and economic policy, respectively. To formulate specific economic policies, the council set up four additional committees to study

54. Sugitani Hidenosuke, "Nihon shinkō tsūshingaku taibō ron," *DT* 5.21 (July–August 1942): 12.

mining, industry, electric power, agriculture, trade and finance, and so on. In short, the council was Japan's most concerted effort to date to plan its new imperium in Greater East Asia.[55]

The Eighth Committee of the Council for the Construction of Greater East Asia was charged with formulating transportation and communications policies in the Co-Prosperity Sphere. Headed by Minister of Communications Terashima Ken, its members included ITC President Ōhashi Hachirō and Kajii Takeshi, the MOC engineer turned president of the NEC. Although shipping consumed most of the committee's time, aviation and communications were addressed as well. Nagakawa Sōnosuke, vice minister of the Railway Ministry, called for policy guidelines for a Greater East Asian aviation network and "unified management" of aviation in the entire area. "Aviation is a rapid means of transportation that literally obliterates thousands of miles," noted Nagakawa. "It therefore plays the most appropriate role in the construction of Greater East Asian Economic Order based on the great principle of 'Eight Corners Under One Roof,' with Japan at the core."[56]

Addressing the issue of telecommunications, Ōhashi Hachirō compared the Greater East Asia Co-Prosperity Sphere with other similar spheres in Europe and America. Tellingly, Ōhashi again evoked the familiar example of the British empire, which derived its power from the control of submarine telegraph cables, although he was quick to point out that "Japan's principle of building a Greater East Asia Co-Prosperity Sphere was, of course, fundamentally different from the British-style colonial policy." He proposed considering two aspects of future telecommunications expansion in Greater East Asia: trunk lines connecting each region, and local communications. While affirming the need to place Japan at the center of the Co-Prosperity Sphere, Ōhashi suggested that some circular lines might be useful to connect major, central cities in each region. Underlining the importance of Japan's manufacturing sector, Ōhashi cautioned that "even after peace in the future, the sphere would depend on Japan's technology and equipment

55. *DKSKS*, 1: 1–3. *DKSKS* is the most complete record of the deliberations by the Council for the Construction of Greater East Asia, although many sessions were conducted with such secrecy that apparently no minutes were kept. For a digest, see *KSSS* 4: 1257–341. For a background discussion, see Kawahara Hiroshi, *Shōwa seiji shisōshi kenkyū*; and Furukawa Takahisa, *Shōwa senchūki no sōgō kokusaku kikan*, 270–76.

56. *DKSKS*, 2: 132–36.

instead of those from Britain and the United States."[57] The "Deliberation of Fundamental Policies of Transportation and Communication in Greater East Asia," adopted by the subcommittee, essentially reaffirmed the earlier vision of Japan's telecommunications hegemony:

> With Imperial Japan (*kōkoku*) at its core, [we should] strengthen the Greater East Asian trunk communications circuits that connect Imperial Japan and various regions in the Sphere as well as between various regions. In the meantime, [we shall] plan the expansion of an international communications network that centers on Imperial Japan and ensure the advantageous communications rights of Imperial Japan in the world.
>
> To achieve this goal, [we must] strengthen communications enterprises, control frequency, expand communications equipment manufacturing as well as communications research institutions, and ensure the supply of communications personnel.
>
> [We must] establish an appropriate organization consisting of concerned agencies for the purpose of building a unified communications system appropriate for the character of the Greater East Asian Sphere, ensure Imperial Japan's guiding power over communications, and set up a Greater East Asian Communications Sphere.[58]

This proposal reflected the basic position of the MOC: since Japan was the indisputable nexus of all communications in East Asia, the MOC insisted that communications routes always radiate from Japan in the center. For officials in Tokyo, there was never any doubt as to the benefit as well as necessity of Japan serving as the hub of regional and international communications for all of East Asia. In October 1938, for instance, the MOC proposed that all international telephone calls placed in East Asia go through Japan. This would apply not only to Japan's own colonies but to occupied areas in China as well. Only in areas such as Taiwan or Central China, where secure telephone links with Japan proper by cable were not yet available, would direct international telephone calls be allowed with adjacent areas such as Hong Kong and Manila. With such measures, the MOC could accomplish greater control over telephone communications in East Asia.[59] Japan's military vic-

---

57. Ibid., 142–52.

58. *KSSS* 4: 1219–324. The document also called for strengthening of broadcasting and meteorology facilities.

59. Kōmukyoku, "Gaichi oyobi Manshūkoku o kokusai denwa tsūwa kuiki hen'nyū ni kansuru ken" (October 28, 1938), MOC Records II, 159.

tories in the Pacific War no doubt strengthened Japan's claim to predominance in the telecommunications sphere in Asia, now fortified with new geostrategic considerations. Japanese leaders envisioned a postwar world that would eventually be divided into three or four spheres, controlled by Germany, the Soviet Union, Japan, and the United States, respectively. Japan would become the undisputed hegemonic power in Greater East Asia.

A highly centralized communications network also had many supporters outside the ministry. As Uchino Tadao, a Japanese working for the NCTT, argued during the Pacific War, planning for Greater East Asian telecommunications must ultimately be based on geopolitical theory, taking into consideration national defense, population, geographical conditions, and trends in communication use. The most urgent task, he pointed out, was the completion of a circular long-distance cable connecting Japan, Manchukuo, and China, which would turn the Sea of Japan, Yellow Sea, and East China Sea into "inland seas." Uchino emphasized that the precondition of a Greater East Asia Sphere was that Japan be its center. As a result, the first step in telecommunications was to ensure communication with Japan under all circumstances. Due to the variety of political conditions in the Southern Region, each component there should communicate directly with Japan, rather than through a regional center. Moreover, communication between the Continental Sphere and the Southern Sphere should in principle go through Japan (see Fig. 5).[60]

In early 1943, Komatsu Saburō, head of the ITC's Business Division, painted his vision of Japan's communications hegemony in Asia. Organization of the Greater East Asian Telecommunications System, Komatsu emphasized, must "thoroughly reject the exploitative, imperialist control system used by Britain and the United States," replacing it with a system with "Japanese characteristics." By this he meant that all wireless telegraph and telephone connections, as well as cable routes within the system, must radiate from Japan to various parts of Greater East Asia. All international communications circuits must also extend from Japan to countries outside the Co-Prosperity Sphere. Komatsu was explicit about its purpose:

---

60. Uchino Tadao, "Daitōa chiseigakuteki kanten yori suru denki tsūshin ritchi no kenkyū" (Part II), [NCTT] *Kenkyū zasshi* 3.2 (December 1943): 51.

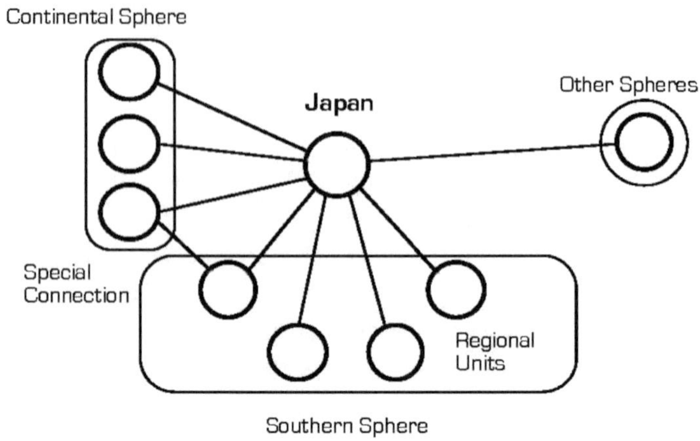

Fig. 5 One scheme of Greater East Asian telecommunications, 1943 (SOURCE: Uchino Tadao, "Daitōa chiseigakuteki kanten yori suru denki tsūshin ritchi no kenkyū," pt. 2, [NCTT] *Kenkyū zasshi* 3.2 [December 1943]: 51).

That all external communication must be relayed through Japan is not for the purpose of suppressing and exploiting other peoples [in the Sphere]. Rather, making our country's cultural, and especially political, power absolutely the strongest is the absolute precondition to shower other peoples in their rightful places with Imperial Glory. To accomplish this task requires the concentration of communication power, which controls thought.[61]

Although blissfully ignorant of the contradictions in his own logic, Komatsu was apparently mindful of the political implications of such a Japan-centered system. His disclaimer was hardly persuasive, however.

Not everyone in the East Asian Telecommunications Bloc readily agreed on the merits of a telecommunications system that eliminated all other regional centers in favor of Japan. Some regional telecommunications operators in China had their own calculations. In the pages of a leading Japanese journal on telecommunications, for instance, NCTT President Inoue Otsuhiko presented a different vision. He emphasized that the new East Asian communications network should consist of circular as well as radial trunk lines. The radial lines from the Japanese home islands would reflect the continental countries' dependence on Japan for financial and material assistance in their construction. But as

61. Komatsu Saburō, "Kokusai denki tsūshin no dōkō," *TKZ* 416 (April 1943): 30.

the countries on the continent moved toward a self-sufficient economy based on their local natural resources, Inoue reasoned, they would develop from "infancy to adulthood in relationship with Japan" and gradually strengthen economic relationships among themselves. Following this logic, the Japan–Manchukuo cable would extend to North China and, by way of Xuzhou, connect with Nanjing and Shanghai in Central China, thus forming a trunk cable link throughout the continent that then returned to Japan by sea. In Inoue's view, this circular cable route would supplement the radial lines as a second route, which was necessary since part of the radial network could be cut off.[62]

Inoue and his supporters emphasized that the Greater East Asia Communications Sphere had to be based on the solid financial foundation of each operator. Consequently, the highly lucrative international wireless bases ought to be dispersed within the sphere instead of concentrated in Japan. To justify establishing itself as a center for international telecommunications, the NCTT argued that it was already handling a high volume of international telegrams, and that direct wireless routes with foreign destinations would not only save foreign currency but also bring the company an extra 500,000 yen each year. Moreover, although North China's foreign trade was depressed after the China War had begun, as resources became developed in the area, its foreign trade was bound to increase. The NCTT was working on its own plans for future expansion of communications links with Europe through Central Asia, along a planned railway link.[63] Likewise, the CCTC argued that its international wireless telegraphic traffic with areas outside the East Asia Sphere brought in revenue vital for its survival, since it had to keep internal tariffs at an artificially low level for political reasons.[64]

Differences also persisted over the question of decision-making authority in Japan's imperial telecommunications network. The slow pace of deliberations and implementations at the East Asian telecommunications conferences was a constant source of frustration for their participants. This sentiment was even evident at a roundtable discussion

62. Inoue Otsuhiko, "Denki tsūshin no Daitōa kyōeiken kakuho," *DT* 4.13 (1941): 30–34.

63. "O-A renraku tsūshin kensetsu keikaku yōryō" (n.d.), NCTT Records 2028(2)/44; "Kokusai tsūshin no kichi kettei narabini keika rosen ni kansuru teian riyūsho" (n.d.), NCTT Records 2028(2)/49.

64. Tōa denki tsūshin jimukyoku, *Tōa denki tsūshin kyōgikai dai-3-kai kaigi gijiroku*, 35.

immediately after the January 1941 signing of the agreement. Several participants called for greater unity in policymaking but admitted that it was impossible to reduce several axes of authority into a single one. The MOC, they concluded, did not have the full power to decide telecommunications policy. Honda Shizuo, a former MOC official then serving on the Cabinet Technology Board, acidly remarked that although there had been expansion plans in the MOC, there was no real domestic telecommunications policy, let alone a telecommunications policy for East Asia. Shinohara Noboru of the MOC tried to clarify the issue of unification. While telecommunications operations could be left to different parties, he insisted, construction and maintenance had to be entrusted to the most skilled experts. However many axes there were, what was most important was where the communications axis was located and how it worked with axes of authority in other fields. Shinohara expressed frustration that, despite the MOC's strenuous efforts, the importance of telecommunications had not been fully appreciated by others in the center. As a result, many recommendations by the Telecommunications Committee had not been implemented, and proposals put forward by the MOC concerning the Southern Advance had been shelved. Yasuda Takesuke, director of the Telecommunications Bureau, echoed Shinohara's disappointment.[65]

The authority of the East Asian telecommunications conference was curtailed from the beginning. Although telecommunications operators on the continent seemed to have submitted to the MOC's will, the establishment of such a forum became a bone of contention in Tokyo when it was discussed at the directors' meeting of the inter-ministerial Telecommunications Committee. The Navy and the Asia Development Board agreed in principle that governmental supervising agencies, such as the GGK and GGT, should be included in the conference as originally planned. The Army did not oppose the idea of a conference per se, but insisted that it remain separate from the inter-ministerial Telecommunications Committee and that there be no permanent secretariat that could lend it the appearance of a decision-making body. Apprehensive that the MOC intended to use the conference to formulate East Asian communications policy, the Army opposed inclusion of the GGK and GGT, and rejected any decision-making role for the con-

65. "Kōa no denki tsūshin o kataru zadankai," *DT* 4.13 (January 26, 1941): 69.

ference. Even terms such as "comprehensive function" and "organic ties" in the conference's platform raised the suspicion of the Army, which insisted that the platform refer only to specific operational matters, not to general policies. Therefore, even before its founding, the mission of the East Asian telecommunications conference was limited to little more than an exchange of opinions. After considerable consultation among the ministries, it was agreed that membership would be initially confined to telecommunications operators in Manchukuo and China, with the understanding that, at an appropriate time in the future, governmental supervising agencies might also be included.[66]

Over the years, Watanabe Otojirō made several elaborate pleas for increasing the authority of the Telecommunications Conference and transforming it into an administrative organ, so as to realize the vision of a unified telecommunications network in East Asia under Japan's control. Writing in 1942, he commented:

> Based on the character of telecommunications, random establishment of facilities without coordinated plans in the Greater East Asian Co-Prosperity Sphere is impermissible. Matters such as selection of communications centers, design of communications circuits, selection of relay routes, method and speed of communication, standards of communications skills, selections of frequencies and regulations of waves, and measures for breakdowns must be solved on the basis of a unified whole.[67]

Watanabe called for a proper balance among local operational autonomy, dependence on Japan, and a sound financial basis for operations. He asked the MOC and the Cabinet Planning Board to consult operators in the field when drawing up new plans. None of his petitions, however, seemed to have much effect.[68]

All in all, only a few of the numerous proposals were actually acted upon, and even then, they were usually delegated to a new committee.

66. "Tōa denshin kanjikai no iken matomari taiyō" (September 6, 1939) and "Tōa denki tsūshin kyōgikai ni kansuru gun oyobi Kōain no ryōkai ni kansuru ken," both in NCTT Records, 2028(2)/49; Arisue Seizō (Chief, Military Affairs Section, Military Affairs Bureau, Army Ministry) to Takeuchi Tokuji (Chief, General Affairs, Bureau of Manchurian Affairs) (October 26, 1939), MOC Records I, 301.

67. Watanabe, *Denki tsūshin kokusaku to denki tsūshin jigyō*, 399.

68. "Daitōa denki tsūshin kaigi dai-1-kai kaigi gijiroku" (Part 1, October 1943), unpublished manuscript, MOC Records I, 328.

The First Greater East Asian Telecommunications Conference in late 1943 decided to set up a Permanent Committee to study long-term plans to strengthen telegraph and telephone operations. Another committee was set up at the same meeting to investigate business systems. Neither results. Other proposals were deemed too difficult. For instance, the NCTT proposed a systematic study of an ideal form of telecommunications operations in Greater East Asia in late 1943, only to be told that "as its impact is extremely wide, it is difficult to reach a decision." The MOC was to study how to study the subject.[69]

Writing in the official journal published by the East Asian Telecommunications Secretariat in 1944, Nagatani Takeo summed up the fundamental problems in Japan's endeavor of integrating the Greater East Asian telecommunications network. First, he blamed the lack of recognition of the importance of telecommunications among the general public. "Although one often hears the mantra 'Telecommunications is power,'" he lamented, it was nothing more than self-consolation. Second, Nagatani cited lack of unification of telecommunications enterprises as an obstacle to implementation of telecommunications policy. The failure to reach a common "tariff by agreement," he felt, had resulted from the localism of telecommunications operators in various areas. In his view, the telecommunications conferences were discussions without much real authority, and even when agreements were reached, they were nothing more than temporary measures.[70] Nagatani thus highlighted the wide gap between vision and policy as well as the reality of entrenched interests that resisted fundamental change. By then, however, time was about to run out.

At the height of the Pacific War, a Japanese telecommunications specialist based in occupied China published an essay calling for a "new discipline of Japanese communication studies." As its author, Sugitani Hidenosuke, saw it then, "Japan's strength of leadership in East Asia must lie, above all, in its effort to control its vast space, and then in

69. Daitōa denki tsūshin jimukyoku, "Daitōa denki tsūshin kyōgikai dai-1-kai kaigi teian kettei yōryō" (October 1943), MOC Records I, 328.

70. Nagatani Takeo, "Daitōa denki tsūshin no airo o daikai seyo!" *TDTZ* 4.3 (May 1944): 23–32.

directing various people in the sphere toward a Greater East Asian consciousness."[71] In other words, the success of Japan's empire-building in East Asia would have to rest on these twin pillars of a communications network and a Japan-led community.

Today, Sugitani's name is long forgotten but his conceptualization of communication is echoed in what British communications scholar G. J. Mulgan calls "double nature of control." In his important book *Communication and Control*, Mulgan distinguishes between two types of control—control as exogenous, imposed, abstracted, and rationalized, and control as endogenous, communicative, and shared. "Their twin histories, the one that of tools, weapons, techniques and structures, the second that of language, of communality, of self-regulation and nurture," he argues, "run in parallel."[72] Communications networks are designed to foster community, yet the proper functioning of communications also depends precisely on a sense of community.

As this chapter illustrates, Japan's telecommunications network for Greater East Asia was one of the large technical systems (LTS) in the imperium. Like railway, aviation, shipping, and broadcasting, such systems played an important role in the last decades of Japan's imperial expansion, in creating an East Asian bloc under Japan's leadership. Indeed, the ambitious nature of Japan's plan for an East Asian long-distance cable network should not be underestimated. The East Asian telecommunications network was a system that included technical, economic, political, and cultural components.

As Japan's imperium grew larger and more complex, so did the challenge of systemic control. As frequently seen with other system-builders, the Japanese found themselves facing nearly insurmountable political and organizational problems as often as they encountered technical ones.[73] The system that Japan sought to build in East Asia was fraught with fundamental contradictions. There were contradictions between the need to create an integrated, rational, and efficient economy on the one hand, and the political need to maintain the façade of a community of voluntary nations under Japan's leadership on the

---

71. Sugitani Hidenosuke, "Nihon shinkō tsūshingaku taibō ron," *DT* 5.21 (1942): 12–15; also published in *TDTZ* 2.5 (1942): 19–23.

72. Mulgan, *Communication and Control*, 4, 54.

73. Hughes, Foreword to Summerton, ed., *Changing Large Technical Systems*, ix.

other. There were also tensions between the center and the periphery, each of which competed to exert control over the system or part of it. Japan's difficulties in organizing an integrated telecommunications system would compound the technological troubles that came to haunt the network in its last months of existence.

CHAPTER 10

Operation, Meltdown, and Aftermath

On the night of August 9, 1945, at an Imperial Liaison Conference attended by Emperor Hirohito that lasted till the early hours of the next morning, the Japanese government reached the difficult decision to accept the Potsdam Declaration on the condition that the emperor's authority be maintained. A few hours later, telegrams were sent to the Japanese embassies in Switzerland and Sweden, with instructions to pass them on to the Allied Powers. That evening, Japan also broadcast its acceptance via wireless news telegrams and overseas broadcasting. Mutō Tomio, a former high-ranking Japanese official in the Manchukuo government, learned of the momentous decision on the same night. He immediately rushed to the office of the Manchukuo embassy in Tokyo, from which he sent a coded telegram to an old Japanese friend who was a senior official in Manchukuo, urging him to prepare the Japanese there for a ceasefire with Soviet forces, which had declared war on Japan on the same day. For reasons beyond Mutō's control, the telegram apparently never reached the intended party in Manchukuo. It was not until August 14—four days later—that the Kwantung Army commander learned from the telecommunications company MTT about the emperor's upcoming broadcast the next day and the impending decision to end the war. Mutō would regret for the rest of his life that he had not tried to place a long-distance telephone call to his friend in Manchukuo to convey the crucial message. The lack of prior notification, he thought, exacerbated the chaos and increased Japanese casualties in Manchukuo during its last days.[1]

1. Mutō Tomio, *Watashi to Manshūkoku*, 446–47. The Kwantung Army received the formal order through official communication channels on August 16.

What happened to Mutō's telegram may never be known. Nor is it clear that he could have placed an immediate long-distance telephone call to Manchukuo. By then Japan's imperial telecommunications network in Asia was severely battered by the war and was deeply mired in both technical and organizational problems. It was no small wonder that when advance notice about Emperor Hirohito's broadcast as well as the post-broadcast orders to lay down arms were transmitted over the same network, imperial communications worked as well as it did.

A LIVING NETWORK

Given the enormous importance Japan placed on its telecommunications network at home and in East Asia, it is only appropriate to ask how it actually worked and what kind of information went through it. Surveys conducted by the Japanese government or colonial agencies as well as by the Japanese-controlled telecommunications companies provide valuable glimpses into the flow of information through the imperial network. The composition of traffic, changes in volumes, as well as the composition of users reveal an important dimension of the daily life of Japan's empire.

A nationwide survey of all paid telegrams at nearly all 7,000 telegraph offices throughout Japan proper, conducted in mid-1933, measured all government, civilian, and press traffic that went through the Japanese domestic telecommunications system with the exception of those related to its operations, which were free. Although official telegrams (*kanpō*)—both civilian and military—were most important for governing the empire and enjoyed discounted rates as well as priority in using telecommunications media, their share in the total traffic was relatively small (1.6 percent). The large share (98.4 percent) of nongovernment use is noteworthy. This was also true with telegraph traffic between the home islands and major colonies (see Fig. 6).[2] Within Japan's colonies, the share of government telegrams was slightly higher. A survey in Taiwan found that paid telegrams for governmental business occupied 3 percent in 1934;[3] according to surveys conducted by MTT

2. Teishinshō denmukyoku, *Denpō kōryū jōkyō ni kansuru chōsa* (Tokyo, 1935).

3. Ishida Minoru, "Taiwan ni okeru denshin gyōmu no gaikyō," *Denshin kyōkai kaishi* 320 (May 25, 1937): 13.

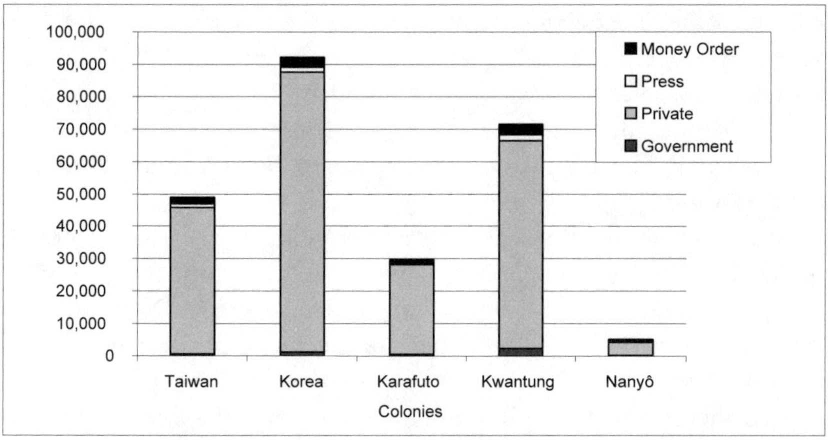

Fig. 6 Composition of imperial telegraphic traffic, 1933 (SOURCE: Teishinshō denmukyoku, *Denpō kōryū jōkyō ni kansuru chōsa* [Tokyo: 1935]). Data collected over an 11-day period.

in 1938 and 1939, among all paid telegrams sent and received in Manchukuo, the share of government telegrams stood at 5–6 percent.[4] In terms of traffic volume and revenue, therefore, Japan's telecommunications network was largely used and sustained by nongovernmental users.

By far the largest share of telegrams in the 1933 Japanese survey fell under the category of private telegrams (95.1 percent); press telegrams and money-order telegrams (*kawase denpō*) counted for 0.6 percent and 2.9 percent, respectively. In the colony of Taiwan, as much as 63 percent of all paid telegrams was related to business transactions or the stock exchange, followed by 24 percent related to social functions, according to the survey in 1937.[5] The results were similar in Manchukuo. Surveys conducted by MTT in 1938 and 1939 show that among all paid telegrams sent and received in Manchukuo, business-related telegrams (59 percent and 46 percent, respectively) constituted the largest category, followed by personal use (between 35 to 46 percent).[6] It should not be surprising that business-related telegrams were the largest category of paid telegrams. Telecommunications services have played a

4. Kishimoto Hajime, "Wagasha no kokumin no kojinteki seikatsu ni taisuru kōken," (MTT) *Gyōmu shiryō* 10.3 (March 1943), 269.

5. Ishida, "Taiwan ni okeru denshin gyōmu no gaikyō," 13.

6. Kishimoto, "Wagasha no kokumin no kojinteki seikatsu ni taisuru kōken," 269.

crucial role in expanding the geographical sphere of economic activities since the Meiji era. Since Japan's endeavor to build the Greater East Asia Co-Prosperity Sphere was supported by Japanese business interests, large and small, the overwhelming majority of business branches established in the post-1937 period were to be found in China and Southeast Asia.

The relatively small shares of press telegrams in the surveys can be misleading. For one thing, since they had enjoyed a discount rate since 1906, they tended to be much longer than average. New agencies spearheaded the use of some of the latest technologies. When the completion of the Japan–Manchukuo trunk cable in late 1939 greatly improved the prospects of phototelegraphic service, the number of photographic telegrams exchanged between Manchukuo and Japan jumped fourfold from 157 in 1939 to 632 in 1940, with the majority (74 percent) sent from Japan. Despite its prohibitive cost for ordinary customers, the phototelegraphy circuit had become a crucial component of Japan's imperial media network.[7]

In a fundamental sense, the construction of Japan's empire in Asia was made possible by an outward movement of the Japanese population. In addition to the ambitious plans for agricultural migration, travel in the imperium can be attributed largely to administrative and business activities in the wake of Japan's military occupations. It is estimated that by August 1945, some seven million Japanese were residing outside Japan proper in various colonies or occupied areas. Roughly half were civilians. Meeting the communication needs of these millions of Japanese residents and travelers, many with family and relatives in the home islands, was a formidable task. Telegrams related to a variety of personal matters therefore occupied a large share of communication traffic. As noted above, surveys conducted by MTT in 1938 and 1939 show that between 35 and 46 percent of all paid telegrams sent and received in Manchukuo fell in the category of personal use. According to a single-day study in November 1939, congratulatory and condolence messages constituted 10 percent of personal telegrams, roughly another 10 percent was related to transfers of money, and 33 percent for the purpose

7. *Dentsū tsūshin shi*, 7–8. *Manshū denshin denwa kabushiki kaisha 10-nen shi* (1943), 955–56. The time required for transmitting a photograph between Mukden and Tokyo, which could be sent in three different sizes, was about one hour.

of announcing arrivals of passengers or shipment of goods.[8] Telegraphic use was closely related to personal travel, typically for announcing arrivals or departures. A survey by NCTT in 1939, for instance, found that 26 percent of all telegrams exchanged between Japan and North China were related to personal travel.[9] In October 1940, NCTT surveyed a total of 14,474 telegrams sent and received between North China and Japan (including Korea). Of all the outgoing telegrams, 30.7 percent were concerned with business; the rest dealt with personal appointments (6.8 percent), travel (25.7 percent), and greetings (6.8 percent), among others. Of incoming telegrams, as many as half of the nonbusiness telegrams dealt with personal travel.[10] A similar survey of telegraphic traffic between North China and Manchukuo around the same time reveals that noncommercial telegrams (personal, travel, greetings, money, gifts) outstripped business telegrams fourfold (4.7 times in incoming telegrams), a fact the author attributed to the active "exchange of personnel due to the current situation."[11]

Telegraphic money orders played an increasingly important role for individuals as well as institutions in Japan's imperium. In addition to postal money orders, money-order telegrams partially made up for the lack of regular bank transfers and became indispensable to the increased economic activity. In 1940, for instance, money-transfer telegrams accounted for slightly over 10 percent of all telegrams exchanged between North China and Japan (including Korea). Interestingly, some 65.5 percent of the incoming ones (93.5 percent of the total amount) came to North China from Korea, reflecting increasing economic ties between the two areas.[12] In 1943, a total of 345,935 money-transfer telegrams were sent to Japan from North China, as compared to 14,345 sent within North China and 17,662 from North China to Manchukuo.[13] The extremely high volume of money-transfer telegrams sent from China to Japan (including Korea) shows that the telegram had

---

8. Kishimoto Hajime, "Wagasha no kokumin no kojinteki seikatsu ni taisuru kōken," (MTT) *Gyōmu shiryō* 10.3 (March 1943), 269–71.

9. (NCTT) *Eigyō geppō* 25 (August 1940), 39.

10. "Denpō kōryū ni kansuru chōsa" (Ka-Nichi denpō), NCTT Papers, 2028/1352.

11. Ibid.

12. Ibid.

13. "Shōwa 18-nen kawase denpō hasshin jōkyō chōsho," (NCTT) *Eigyō geppō* 68 (January 1944), 15.

become an important channel of monetary flow between Japan and the occupied areas.

It is true that throughout the imperium, Japanese access to and use of telecommunications was disproportionately high. When available, survey data show a disparity along ethnic lines among those who used the service. According to one survey, 67 percent of all telegrams handled by the MTT in the mid-1930s were sent by Japanese. The Chinese, who made up more than 90 percent of the population, sent only 13 percent of all telegrams. Of the telegrams exchanged between Manchukuo and China proper, the Chinese sent only 23 percent, compared with 67 percent sent by the Japanese.[14] A one-day survey in Manchukuo in 1939 showed that over 80 percent of private-use telegrams were sent by Japanese, with Chinese and Koreans making up the remaining 20 percent.[15]

Compared with Manchukuo, the Chinese use of telegraphic service in North China was considerably higher. A three-day survey conducted by NCTT in 1940, for instance, showed that 41.6 percent of telegrams sent and 43 percent of telegrams received in North China were in Chinese. On certain routes, Chinese-language telegrams even surpassed those in Japanese. For instance, Chinese-language telegrams exchanged with Central China made up 55.8 percent (outgoing) and 60 percent (incoming) of the totals.[16] But Japanese dominated the external traffic in occupied China as a whole. In October 1940, the NCTT surveyed 14,474 telegrams sent and received between North China and Japan (including Korea). All but 2 percent were in Japanese.[17] According to a survey conducted in Central China by the CCTC in mid-1939, nearly 60

14. Maeda Naozō, "Manshū ni okeru denshin denwa jigyō no keiei," *TKZ* 330 (February 1936): 151–61; Manshū denshin denwa kabushiki kaisha, *Manshū denshin denwa kabushiki kaisha 10-nen shi*, 951–52.

15. Kishimoto Hajime, "Wagasha no kokumin no kojinteki seikatsu ni," (MTT) *Gyōmu shiryō* 10.3 (March 1943): 270–71. In the largest category of "miscellaneous telegrams," which included announcing arrivals and sending money, close to 30 percent were sent by Koreans and Chinese.

16. NCTT, Eigyōbu, Eigyōka, Eigyō chōsa kakari, "Denpō kōryū ni kansuru chōsa" (1. Naikoku denpō) (May 1940), NCTT Papers 2028/1352.

17. NCTT, Eigyōbu, Eigyōka, Eigyō chōsa kakari, "Denpō kōryū ni kansuru chōsa" (Nikka) (May 1941), NCTT Papers 2028/1352. Strictly speaking, of course, some Chinese may have sent telegrams in Japanese to take advantage of the better rates.

percent of all telegrams sent and received in that region were in Japanese. Chinese-language telegrams made up only one quarter of the total traffic; only in telegrams sent to and from North China were Chinese-language telegrams a majority.[18]

In telephone use a great disparity also existed between the Japanese and non-Japanese population. In Beijing alone, as one 1943 NCTT study acknowledged, some 110,000 Japanese residents accounted for 86 percent of all telephone usage; the approximately two million Chinese accounted for the remaining 14 percent. Only in long-distance phone use was the gap less glaring, although the Japanese still accounted for 58 percent of all usage. NCTT justified this disparity by "the fact that most Japanese are very much involved in active production." It noted that the Chinese in fact tended to use the telephone more than the telegraph because of the latter's cumbersome nature. Moreover, the study pointed to the relatively widespread use of the telephone in some parts of North China, especially in the East Hebei area, which alone accounted for 26 percent of all telephone usage in the region.[19]

Even allowing for local variations, it often served Japanese purposes for the non-Japanese population in the imperium to make use of Japanese-controlled telecommunications service as individual customers (as opposed to institutional ones). Such measures not only helped expand the revenue base of Japan's telecommunications operations but also enabled the Japanese to demonstrate their superior technology and management to the population they sought to control.

Available telecommunication traffic data show a steady increase in communications within Japan's sphere of influence in the 1930s after a decline in the years of the Great Depression. In contrast, international traffic remained largely flat and then declined significantly after the outbreak of the Pacific War. In 1940 alone, Japan exchanged some 12 million telegrams with its colonies, Manchukuo, and occupied China, more than ten times Japan's total telegraphic traffic with the rest of the world.[20] The somewhat incomplete record of both telegraph and

18. Kōain Kachū renrakubu, *Chūshi ni okeru denpō kōryū jōkyō, denwa tsūwa jōkyō chōsa setsumei shiryō* (November 1939), esp. 1–3.

19. "Denki tsūshin no riyō," (NCTT) *Kenkyū zasshi* 3.1 (1943), 110.

20. *Denmu nenkan* 1942, 297–99; *Denmu nenkan* 1943, 336–37. The accounting year covers twelve months from March of each year.

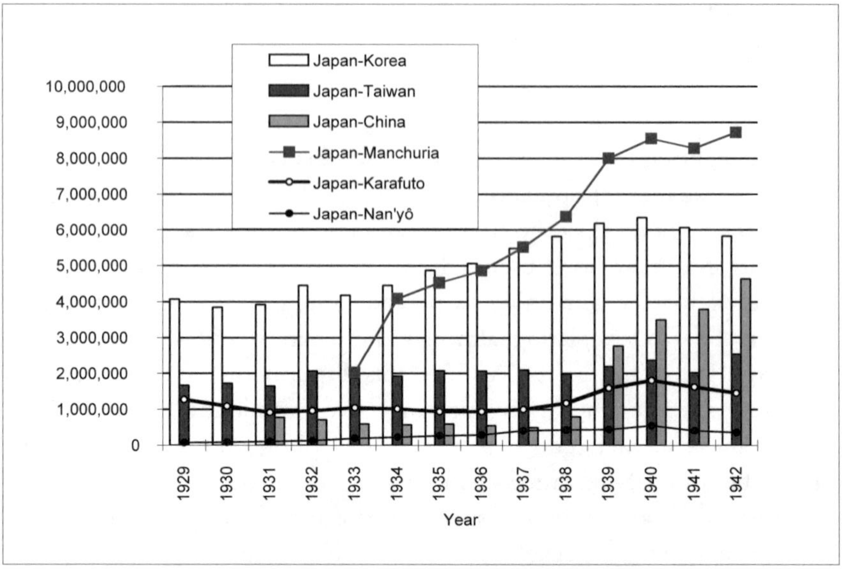

Fig. 7 Telegraph traffic within the imperium, 1929–1942 (SOURCE: *Denmu nenkan* 1943, 336–37).

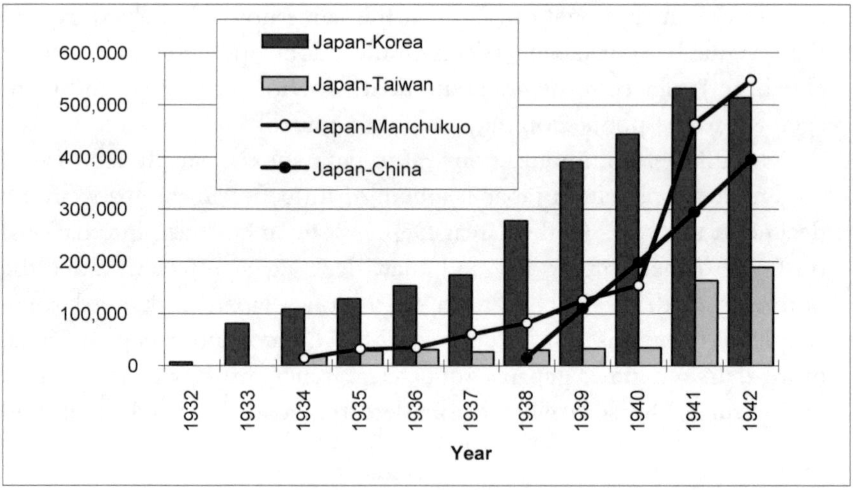

Fig. 8 Telephone traffic within the imperium, 1932–42 (SOURCE: *Denmu nenkan* 1943, 338–39). Unit system was in use in Taiwan and Manchukuo before 1941 and 1940 respectively.

telephone traffic between Japan and the major areas within the formal and the informal empire during the 1930s reveals that Korea led other colonies in generating and receiving telecommunications traffic with Japan proper, and the sharpest increase in aggregate telegraphic and telephone traffic was between Japan proper and newly incorporated areas such as Manchukuo and China (see Figs. 7 and 8).

As this brief discussion shows, Japan's imperial telecommunications network was not simply an essential government tool; rather, it had become indispensable to the daily life of a living imperium. The ever-increasing communications traffic soon began to spell trouble for Japan, due to the stagnation of network expansion, the chronic problem of coordination, and intensification of war-related damages.

PITFALLS OF JAPANESE TECHNOLOGY

Even before Japan launched the war in the Pacific, when the MOC was continuing to produce ambitious expansion plans, Japan's telecommunications network in East Asia had already experienced setbacks and was beginning to show signs of serious strain.

The setback in network expansion is best illustrated by the fate of the Nagasaki–Shanghai submarine telephone cable, one of the key components of the East Asian telecommunications network and one of the most ambitious submarine cable projects ever launched by the MOC. Together with the Japan–Manchukuo long-distance cable, which was being extended southward on land to Beijing and Tianjin, the planned Nagasaki–Shanghai submarine cable would complete the core of Japan's telecommunications cable network in Northeast Asia. It was also to serve as the first stage of network expansion in Southeast Asia. By then, Japan had already built a 3,760-ton cable ship, the *Tōyō maru*, to replace the aging *Okinawa maru*. The age of submarine cable expansion seemed ready to begin.[21]

---

21. Nippon denshin denwa kōsha, Kaiteisen shisetsu jimusho, *Kaitensen hyakunen no ayumi*, 284–304. When the *Tōyō maru* was designed in 1936, some had proposed building a larger one, at 6,000–7,000 tons, in anticipation of the future expansion of Japan's submarine cable network. The tonnage was eventually reduced, largely because of the maintenance cost but also to enable the ship to operate in the shallow coastal waters of Japan. However, the fuel capacity (coal instead of oil, in case of wartime shortages) was raised from 500 tons per month to 700 tons to allow a wider range of operations.

The Nagasaki–Shanghai telephone cable posed an unprecedented technological challenge in submarine telephony. Unlike submarine telegraphy, long-distance submarine telephony had a higher relay requirement. Preliminary exploration began in July 1938. Given the 454-km stretch of ocean between Cheju Island and Shanghai, the idea of floating relay stations was considered. In the end it was decided to construct a relay station on an uninhabited island halfway in between, which would also serve as a beacon in the busy shipping lane between Japan and Central China. The budget for investigation and construction of this relay station alone was set at 1 million yen. Since coaxial cable was still a new technology, the MOC decided to use lead-shielded submarine cables that had a normal depth limit of 300 meters. To provide adequate amplification in such a long submarine cable, some Japanese engineers proposed inserting amplifiers inside the cable, using small but extremely stable vacuum tubes, with electricity in direct current provided from both ends of the cable. Actual construction proved even more difficult than expected, and cables produced by the three leading cable manufacturers in Japan failed to withstand the high water pressures and support high-quality voice transmission, as required.

By early 1940, the MOC had to revise the design for the Shanghai–Nagasaki cable once again. As Japan moved to a war economy under national mobilization, continuing with the costly research and securing the required material—1,200 tons of copper and 5,000 tons of lead for this cable alone—became increasingly unrealistic. Ironically, when the Cabinet approved the plan to build a submarine telephone cable from Taiwan to Batavia (the South China Sea western circular cable) in late 1940, the Nagasaki–Shanghai cable project was already in serious trouble. It was highly doubtful that, even if the Dutch authorities had agreed to Japan's demand to land the new cable in the Dutch East Indies, Japan would have been able to do so, at least in the short run. The Shanghai–Nagasaki telephone cable project was finally abandoned after the outbreak of the Pacific War.[22]

22. Nippon denshin denwa kōsha, Kaiteisen shisetsu jimusho, *Kaitensen hyakunen no ayumi*, 304–30. After the war, Shinohara Noboru regretted not having obtained a patent on his idea of inserting amplifiers inside the cable. The same technology was used by the United States on the Key West–Havana cable as well as the transatlantic telephone cable. See Shinohara, "Sōka keburu kara musōka keburu e," in Teishin gaishi kankōkai, *Teishin shiwa* II: 245; also in *DTJGKS*, 42, 101–2, 261–63.

Slowing expansion after 1940 was another sign that the vast system as a whole was in trouble. Much of the original East Asian cable telecommunications network existed only on paper, as Japan was badly behind schedule in adding new communications circuits. In telegraphy alone, ten circuits between Japan and Korea, eight between Japan and Manchukuo, and another two between Japan and North China were thought necessary but were never built.[23]

Japan's ambitious plans of expanding the telecommunications cables southward did not materialize, either. In October 1940, recommendations to strengthen telecommunications at home were issued before the ink had barely dried on the blueprint for the massive cable expansion into Southeast Asia, indicating a conflict over priorities. After the outbreak of war in the Pacific, Japan did gain control of 15,000 nautical miles of existing submarine cables in the region, but even this turned out to be a hollow victory. Most of the cables in Southeast Asia turned out to be non-operational or to require enormous resources for repair and maintenance. Japan acquired half-finished cables in a British depot in Singapore, but all that Japan was able to produce was external shielding on 300 nautical miles of cable core that was already manufactured. Although Japan's occupation of Southeast Asia gave it direct access to gutta-percha (GP)—the raw material indispensable to making submarine cables—for the first time, its efforts at refining GP in Singapore failed. As the Japanese discovered, although there was an overabundance of such material as GP, aluminum, and tin in Southeast Asia, Japan remained short of copper and nickel, which were indispensable to producing telecommunications equipment. Despite requests from the military, even attempts to produce gum-coated wires locally did not succeed.

Japan did increase its domestic production of GP submarine cables; indeed, the ebb and flow of Japan's manufacturing capabilities were clearly reflected in such production (see Fig. 9). GP cable produced for the MOC, to be used for its planned expansion of the East Asian network, reached a peak in 1940, after which cable for the military—the Navy, for the most part—increased sharply and occupied the largest share. After 1943, however, production began to drop as Japan's

23. *DDJS*, 6.

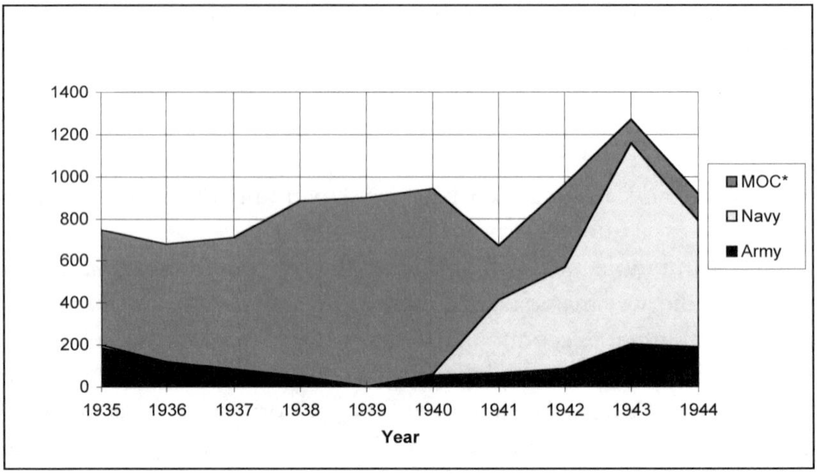

Fig. 9 Gutta-percha submarine cable production in Japan, 1935–44 (SOURCE: *Nihon Taiyō kaitei densen kabushiki kaisha shi*, 62).

economy suffered increasing shortages of material.[24] In the meantime, Japan also experienced problems with conventional submarine cables.

As the number of submarine cables still in operation proved to be insufficient, Japan had to rely primarily on wireless communications with Southeast Asia. Despite the fact that Allied forces destroyed some equipment before their retreat, the Japanese did inherit a large number of wireless facilities. In the Philippines alone, according to one Japanese account, more than 600 wireless sets, large and small, were seized.[25] In addition, Japan made progress in wireless-related research. Japanese scientists conducted extensive research on the ionosphere, which is critical for stable wireless communications. By the end of the war, Japan had become a world leader in this area, having set up numerous ionosphere ob-

24. On raw material in Southeast Asia needed for telecommunications manufacturing, see "Denki kizai no shigen taisaku o kataru," and Arai Hiroshi, "Denki kizai no jūyō fuzoku shizai no taisaku," both in *DT* 5.21 (August 1942): 40–61, 61–67; on GP, see Shimabara Sadakichi, "Marai no gutta percha ni tsuite," *DT* 6.28 (September 1943): 21–28. See also *Nihon taiyō kaitei densen kabushiki kaisha shi*, 72–73.

25. Niwa Osamu, "Shōnantō, Java ni okeru rokatsu musen kizai," *DT* 5.22 (October 1942): 8–10; Asamura Saichirō, "Kōryaku chokugo no Shōnan musen tsūshin," *DT* 34 (May–June 1944): 20–21; Amishima Takeshi, speaking at the Forum on Greater East Asian Construction and Telecommunications Equipment, *DT* 5.19 (March–April 1942): 40.

servation stations in China and Southeast Asia.[26] Reliance on wireless created its own problems, however. A breach in Japan's communications security would prove to be one of the leading causes of Japan's military setbacks in the Pacific. To be sure, the Japanese were concerned with the unprotected nature of wireless communication. To prevent leakage of sensitive information, the Tokyo International Communications Promotion Society (*Tōkyō kokusai tsūshin shinkōkai*) hosted a meeting in mid-1943 on security matters, bringing together business executives in the fields of finance, general trade, shipping, and newspapers.[27] Such efforts were either too little or too late. The Allies succeeded in decoding much of Japan's wartime military and diplomatic wireless communications.

A SLOW MELTDOWN

Toward a "Nervous Breakdown"

In late 1943, ITC President Ōhashi Hachirō made an unsettling admission at the First Greater East Asian Telecommunications Conference: because telecommunication demand within Japan had increased sharply after the war broke out in China, there were severe congestion problems throughout the network. A telegram that had taken 40 to 50 minutes to transmit before the China War could now take as long as six or seven hours to reach its destination. Even an ordinary long-distance telephone call from Osaka to Tokyo could require a wait of eight or nine hours. Comparing the problem to the near-disastrous state of telecommunications service in Japan during World War I, Ōhashi wryly called the situation "a severe nervous breakdown."[28]

This was not the first time an alarm had been raised over the deterioration of Japan's East Asian communications network. At the Third

26. Nakamura Fumio, "Denpa denba," *DJ* 31 (February 1981): 65–70. In 1944, Japan had fifteen such observatories throughout Asia, compared to eight for the United States and four for Britain. After the Japanese surrender, the United States showed much interest and produced a *Report on Japanese Research on Radio Wave Propagation*, 2 vols. (Tokyo: General Headquarters, United States Army Forces, Pacific Office of the Chief Signal Officer, 1946).

27. Tokyo kokusai tsūshin shinkōkai, *Tsūshin ni kansuru gun-kan-min kondaikai kiroku* (June 1943).

28. *DKSKS*, 2: 152.

East Asian Telecommunications Forum in mid-1941, the GGK had reported that the average transit time in Korea for telegrams between Japan and the continent had increased to two hours.[29] Congestion was by no means a new problem to Japan's telecommunications service at home, especially during times of economic prosperity. However, it was quite a serious matter when congestion became a systemic feature of the whole imperial network.[30]

Although strains in the existing telecommunications network were not entirely new, the war changed telecommunications in the Japanese empire in fundamental ways. Enemy destruction and sabotage directly damaged telecommunications stations. In Japanese-occupied areas in Manchukuo and China proper, such facilities were favorite targets of Chinese guerrillas for years. Destruction of communications facilities topped the list of "enemy sabotage activities" recorded by the Japanese authorities. In North China, for instance, the Communist-led Eighth Route Army was active in areas where NCTT operated telecommunications. Throughout the NCTT's seven-year existence, Chinese destruction was a constant problem for the company. After one such attack disrupted the trunk line between Beijing and Tianjin, recovery and repair cost the NCTT more than 160,000 yen.[31] By NCTT's own counting, such incidents increased from 205 in 1938 to 660 in 1939 to 820 in 1940 to 1,991 in 1941.[32] In May 1944, the NCTT president had to take special measures to cope with destruction of communications lines between Baoding and Shijiazhuang.[33] Chinese guerrillas in the south similarly disrupted Japan's communication by severing or stealing subma-

---

29. Tōa denki tsūshin jimukyoku, *Tōa denki tsūshin kyōgikai dai-3-kai gijiroku*, 17–18.

30. The colony of Taiwan seemed to be an exception because telegraphic traffic dropped to an all-time low (180,000) in the first half of 1945, even below the level at the time of the Japanese takeover. The all-time high was reached in 1940, with a total of 3,870,000 (*DDJS*, 6: 300).

31. Hokudenkai, *Kahoku denden jigyō shi*, 178–79.

32. "Kita-Shina kaihatsu KK kankei dantai keibi daichō kaigi jikō hōkoku" (March 2, 1942), NCTT Records 2028/1166. In comparison, in 1941 the Japanese record was 307 instances of damage to the railway, 57 cases of arson, and 55 assassinations, among others.

33. Hokudenkai, *Kahoku denden jigyō shi*, 177–79. On Japan's "communications protection movement," which covered both railway and telecommunications facilities in North China, see Lincoln Li, *The Japanese Army in North China*, 193–99.

rine cables.[34] Even in the colony of Korea, considered to be "politically stable," Japan's telecommunications facilities suffered from sabotage by Koreans.

With the outbreak of the war in the Pacific, Japan's East Asian telecommunications network went from expansion to contraction. The war in the Pacific particularly affected Japan's telecommunications operations in the Southern Region, where nearly all cable communication ceased operation due to destruction and lack of repair. Although Japan controlled a vast cable network in East Asia, the lack of cable ships as well as the gradual loss of supremacy on the high seas prevented the Japanese from repairing the existing cables, some of which had been cut by the Japanese Navy at the onset of the conflict. Even Japan's Nagasaki–Shanghai telegraphy cable, which had served Japanese interests well since 1915, went out of operation in August 1943.[35] Moreover, manufacture of new cable began to decline precipitously after 1943. The only exception was a number of short submarine cables laid by the Navy on its own in late 1943 and 1944 to connect several islands in the Philippines and the Dutch East Indies and totaling less than 300 km.[36] In 1944, the government belatedly drew up plans for building ten cable ships in the next five years, to be deployed in Shanghai, Manila, Hong Kong, Singapore, and Batavia, but none of them materialized. Worse, Japan's small fleet of cable ships was being lost to Allied submarines and bombers, which sank them one after another, crippling Japan's ability even to do maintenance work on the existing cables. In early 1944, after successfully connecting the American-built Shanghai–Manila submarine cable through Taiwan, the *Nan'yō maru* was sunk near Okinawa by an American submarine. The loss of one of Japan's best cable ships was a major blow to its submarine cable communications network.[37] After all three submarine cables connecting Japan and Taiwan

34. Nippon denshin denwa kōsha, Kaiteisen shisetsu jimusho, *Kaitensen hyakunen no ayumi*, 435–37.

35. Ibid., 181.

36. "Kyū kaigun ga fusetsushita kaitei densen chōsho" (October 20, 1948), in "Shūsen mae Nihon rikukaigun fusetsu un'ei no kaitei densen ni kansuru chōsa no ken," Postwar Records, F'2.2.11, Japanese Diplomatic Record Office. This report was based on the recollection of those involved in the work.

37. Nippon denshin denwa kōsha, Kaiteisen shisetsu jimusho, *Kaiteisen hyakunen no ayumi*, 395, 421–23, 477–82.

went out of operation, Japan was unable to conduct repair work and had to rely on wireless.[38]

The year 1943 marked a turning point in Japan's imperial enterprise. At home, increasingly frequent air raids by American forces resulted in heavy damage to Japan's telecommunications infrastructure as well as its production capability. One government engineer estimated it to be ten times worse than the havoc caused by the 1923 Great Kanto Earthquake: 97 percent of telegraph circuits and 39 percent of toll telephone circuits in Japan proper were destroyed by the end of the war. Telephone subscriptions were reduced by over half a million, equivalent to 48 percent of the peak level.[39] In March 1944, for example, American bombing caused severe damage to the Osaka factory of the Japan Submarine Cable Company, Japan's only manufacturer of submarine cables. The GP-refining and -coating facilities, considered the core of the entire factory, as well as finished cables in a huge cable tank, were lost in the fire. Two months later, the Yokohama factory of the same company was largely destroyed in bombing raids.[40]

In addition, shortages of materials, funding, and even personnel made it increasingly difficult to keep telecommunications operating smoothly.[41] The war in the Pacific made it harder and harder to obtain the materials needed to repair the existing telecommunications network. Although the economy was affected across the board, shortages of copper and lead hit telecommunications manufacturing especially hard. In late 1942, Japan admitted for the first time that material shortages due to the outbreak of the Pacific War had created problems for the five-year plan, drawn up in late 1941, for expanding interregional telegraph and telephone circuits in East Asia. At the First Greater East Asian Telecommunications Conference in late 1943, the chief of planning in the MOC's Engineering Bureau also admitted that the five-year plan was not being fulfilled. The GGK representative then complained that the military maintained such tight control over materials that intervention by a

---

38. Taiwan sōtokufu, *Taiwan tōchi gaiyō* (Taihoku, 1945), 200.

39. Yonezawa Shigeru, "Denki tsūshin no fukkō to gijutsu mondai," *TKZ* 444 (January–February 1947): 4. See also Nakamura Fumio, "Shūsen," *DJ* 33 (August 1983): 41.

40. *Nihon taiyō kaitei densen kabushiki kaisha shi*, 66, 80–83.

41. Though written more than half a century ago, Jerome B. Cohen's *Japan's Economy in War and Reconstruction* still provides a good general picture.

third party was required in order to obtain any. As a result, the conference's Subcommittee on Materials was asked to hold biweekly meetings to adjust communications equipment production.[42]

Steady deterioration in the network's performance resulted as it became overburdened. Using essentially the same technology as wireless telephony, radio broadcasting relied heavily on existing telecommunications facilities. As the imperium expanded, broadcasting demands on the telecommunications system also increased. At the end of 1941, for instance, the ITC devoted eighteen transmitters, including three 50-kW and several 20-kW ones, to broadcasting a total of 53.25 hours of radio programs in 16 languages each day.[43] As Japan's war effort became desperate, such overseas broadcasts were increasingly viewed as crucial to eroding the enemy's morale while boosting the spirits of Japan's own troops and civilians. The result was that the hours and reach of radio broadcasts were often increased. After the government ordered the ITC to give higher priority to building new broadcasting facilities, the ITC's 1943 business plan included a "high-power broadcasting facility consisting of three 100-kW transmitters and one 50-kW transmitter." Even in its last annual business budget, drawn up in early 1945, ITC allocated nearly half the funds earmarked for wireless operations to expanding broadcasting facilities.[44] Japan's overseas broadcasting activity reached its peak in early November 1944, when Japanese radio stations broadcast fifteen programs in 21 languages, for a total of 33 hours of programs each day.[45] For telecommunications operators, however, radio broadcasting became a drain on limited materials and personnel.

There were other new demands on the deteriorating communications system as well. Allied air raids added a new burden on Japan's communications system and forced the Japanese government to focus on

42. Daitōa denki tsūshin jimukyoku, "Daitōa denki tsūshin kaigi dai-1-kai kaigi gijiroku" (pt. 2, October 1943), unpublished transcript, MOC Records II-330. For wartime shipping and shipbuilding capabilities, see Cohen, *Japan's Economy in War and Reconstruction*, 250–70.

43. *KDTKKS*, 439.

44. For the ITC's annual business plans, see "Shōwa 18-nendo jigyō keikakusho" (June 21, 1943), "Shōwa 19-nendo jigyō keikakusho" (July 10, 1944), and "Shōwa 20-nendo jigyō keikakusho" (January 24, 1945), all in NHK Records. For technical information, see *KDTKKS*, 220–22.

45. Kitayama, *Rajio Tōkyō*; *KDTKKS*, 438–40.

home defense. After a series of major blunders in the Pacific theater, the Japanese government declared a new stage of "decisive battles" (*kessen*) on all fronts at home. As a result, Japan's telecommunications network became increasingly geared toward coping with defense—and with last-ditch operations in the home islands should Japan face invasion. An unprecedented 146 million yen was allocated to the 1943 telecommunications budget. Of this, 40 million yen was to go to a special PBX network for the purpose of national defense. Another 17.5 million yen was committed to fortifying telegraph and telephone lines against air raids, and 48 million was earmarked for expansion of telephone service and for conversion of all toll lines in Japan to cable. As one MOC official admitted, the plans were heavily tilted toward military and defense matters.[46] Already in September 1943, the General Headquarters had decided to establish an "absolute national defense perimeter" by the middle of 1944. In addition to repairing damaged facilities, the MOC now had to cope with civil defense, which meant that an early warning system had to be given top priority. Moreover, because the military did not possess its own circuits at home, it had to use MOC facilities in its own preparations to fight to the bitter end. Even expansion at home did not go as planned, however. In 1944, communications equipment production in Japan met only 42 percent of its targeted amount, although aviation communications did better, at about 80 percent. According to one postwar study, only 50 to 60 percent of the planned home-defense communications network was completed by August 1945.[47]

Increased communication traffic put a heavy strain on human resources, which were already weakened due to attrition and conscription. The MOC estimated that in 1939, well before the peak period, the per capita workload—both for government communications employees and for messengers—had increased nearly 50 percent since 1932.[48] More serious was the shortage of skilled personnel in both service and production, for many adult males were being conscripted into the military and sent off to fight. At the First Greater East Asian Telecommunications Conference in late 1943, the MOC proposed training female tech-

---

46. Yoshida Tadashi, "Kessen dankai ni okeru denki tsūshin," *DT* 6.24 (February 1943): 8–13.

47. Nakamura Fumio, "Shūsen," *DJ* 33 (August 1983): 41–43.

48. MOC Records I, 305.

nicians and deploying them throughout East Asia, since it had become increasingly difficult to recruit male technicians.[49] Apparently, since many of its technicians were sent to the Southern region, the ITC's technical institute also came to focus on training women recruits. After a vigorous recruitment campaign, the first class enrolled 50 graduates from various women's high schools in 1941.[50] When Kajii Takeshi, then head of Sumitomo Electric Industries, returned to Tokyo from a trip to Sendai in 1944, he noted in his diary that he had seen young conscripts being herded to the front. The scene reminded him of Germany near the end of World War I, which he considered to be an ominous sign for Japan.[51] After 1944, he got a reprieve. Several telecommunication manufacturers, including Sumitomo Electric, were designated "military supply enterprises," a label that exempted their employees from conscription. In addition, temporary workers, members of the female *teishintai* (volunteer corps) and even students were recruited to work in important industries.[52] Poorly trained workers created new problems, however. At the Third East Asian Telecommunications Conference in early 1941, for instance, a number of delegates already complained about the poor quality of vacuum tubes, the heart of wireless communication.[53] Moreover, the shortage of skilled personnel probably had a more detrimental effect on service. In 1943, for instance, the MTT's telegram error rate reached an unprecedented 12 percent. It was rumored that almost all telegrams contained some sort of error. The MTT therefore set a goal of eliminating telegram delays of more than five hours and error rates of more than 10 percent.[54]

---

49. "Daitōa denki tsūshin kaigi dai-1-kai kaigi teian jikō" (October 1943), 12; MOC Records I, 328.

50. *KDTKKS*, 41.

51. Kajii Takeshi, *Kajii Takeshi ikōshū*, diary entry of July 18, 1944.

52. *Nihon taiyō kaitei densen kabushiki kaisha shi*, 78.

53. Tōa denki tsūshin jimukyoku, *Tōa denki tsūshin kyōgikai dai-3-kai kaigi gijiroku*, 156–57. On the role of vacuum tubes in wartime Japanese military communications, see Nakamura Fumio, "Kokuun sayūshita shinkūkan," *DJ* 32 (February 1982): 43–48.

54. *DDJS*, 6: 392.

Countermeasures

To cope with these problems and to prepare for homeland defense, Japan attempted a number of solutions. From the top, it began with a reorganization of the government in late 1943. In the course of this reorganization, the Cabinet Planning Board was abolished and a new Ministry of Munitions was created to oversee military production.[55] The Ministry of Communications became the semi-independent Board of Communications (Tsūshin-in) in the newly established Ministry of Transportation and Communications (Un'yu tsūshinshō). The Telecommunications Bureau and Postal Bureau were amalgamated into the Operations Bureau (Gyōmukyoku). The Engineering Bureau remained intact, and a Frequency Bureau (Denpakyoku) was created to deal with the more serious problem of frequency allocation and control. Two new bureaus were set up—Defense Communications and Communications Supervision—to meet the new demands. New priorities can be seen in such reorganization (see Fig. 10).

In the same year, the MOC called on the general public to contribute "idle telephones" for use in factories and business firms involved in the war effort. A year later, measures were toughened so that punishment would be meted out if infrequently used telephones were not surrendered to the authorities. The government also launched a campaign "for promotion of communication spirit" and, in a bid to obtain cooperation from corporate customers, convened a roundtable conference in May 1944 that brought together 250 representatives of banks and business firms.[56]

The Japanese sought technical solutions as well. Delegates to the telecommunications conferences suggested remote control of relay stations and a new design for overhead wires, largely to cope with shortages of manpower and material. Both the MOC and the MTT worked

55. For a recent study of the organizational changes during the war, particularly in the "comprehensive national policy agencies," see Furukawa, *Shōwa senjichūki no sōgō kokusaku kikan*, especially 291–318.

56. Radio Tokyo broadcasts on April 6, 1943, May 6, 1944; in Office of Strategic Services, Research and Analysis Branch, comp., *Transportation and Communication in Japan*, 200–203.

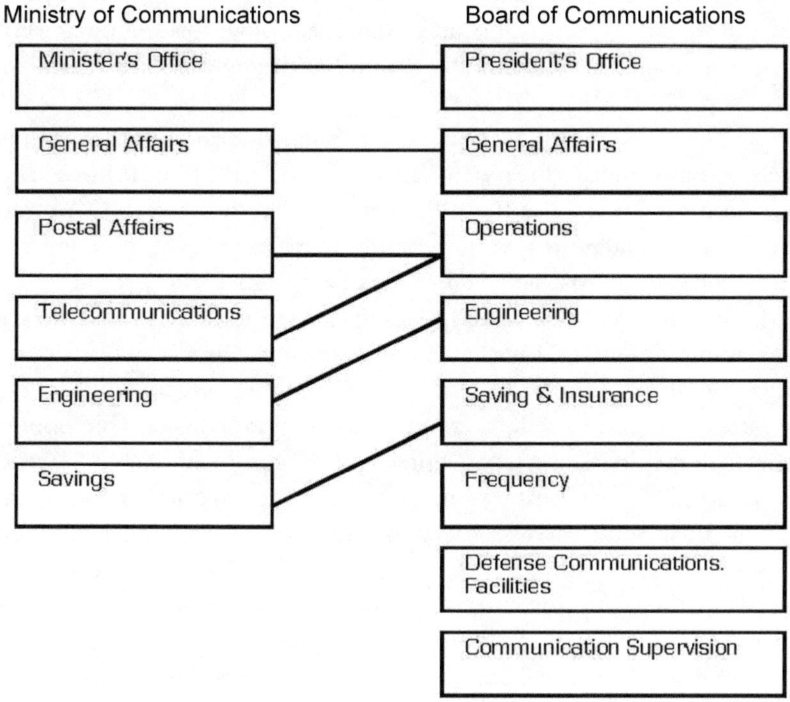

Fig. 10 Changes in Japan's communications administration, 1943 (SOURCE: *Zoku teishin jigyō shi*).

to develop ultra-shortwave and extremely ultra-shortwave, to reduce interference and to offset the shortages of material.[57] The ITC suggested aluminum-covered cable, since aluminum was an "East Asian resource" relatively easy to obtain. Technical solutions could also be "low tech." For example, the Kumamoto Communications Bureau in southwestern Japan announced in 1944 that it would replace metal bells in telephone sets with porcelain bells made in the famed Arita district. According to the bureau, the porcelain bells produced a clear ringing sound and their performance was entirely satisfactory. An estimated 1 million bells were

57. Daitōa denki tsūshin jimukyoku, "Daitōa denki tsūshin kaigi dai-1-kai kaigi teian kettei yōryō" (October 1943), MOC Records I, 331; Tōa denki tsūshin jimukyoku, *Tōa denki tsūshin kyōgikai dai-4-kai kaigi gijiroku*, 111.

requisitioned—enough to refit all the telephones in Japan.[58] Such an innovation had an ironic ring since Matsumae Shigeyoshi, Kumamoto's native son, might have extolled it as another example of "Japanese-style technology."

As telecommunications congestion reached nearly unprecedented levels, additional measures were taken to cope with the problem. Simplification of service was the most common option. The MOC terminated several categories of telegraphic service, such as greeting and condolence telegrams, and launched a campaign to limit unnecessary telegrams and telephone calls. In late 1943, the First Greater East Asian Telecommunications Conference adopted a resolution calling for simplification throughout East Asia so as to meet wartime needs. But measures designed to reduce congestion met with considerable opposition, since they reversed Japan's previous telecommunications practices and affected regular business revenues. Having invested much in improving their services over the years, the NCTT and MTT were reluctant to simplify for fear of losing customers. The 1943 conference also passed a resolution encouraging the use of shorter telegrams in order to ease congestion. Domestic Japanese-language telegrams and those sent to other areas in East Asia and the Southern Region were shortened to ten characters per message at the basic rate, a reduction from the previous fifteen characters. The premium for "urgent domestic telegrams" (*naikoku kinkyū denpō*) was increased from twice that of ordinary telegrams to three times.[59]

These measures notwithstanding, congestion became so severe that, at the Second Greater East Asian Telecommunications Conference in 1944, the Board of Communications decided to set up a Telegraph Decongestion Command Center (Denshin sotsū shirei honbu). To ease communication flows, it designated "command circuits" (defined as "telegraphic circuits involving two or more domestic communications bureaus, or between Japan and the colonies, Manchukuo, China, the Southern Region, or foreign countries") that local operating authorities were supposed to monitor daily, reporting their findings to the Com-

58. Radio Tokyo, March 22, 1944; in Office of Strategic Services, Research and Analysis Branch, comp., *Transportation and Communication in Japan*, 200.

59. November 21, 1944; in Office of Strategic Services, Research and Analysis Branch, comp., *Transportation and Communication in Japan*, 193.

mand Center in Tokyo. Headed by the director of the Operations Bureau, the Command Center was given the extraordinary authority to take emergency actions—such as temporarily altering telegraphic circuits and relay routes—to ensure the smooth operation of telegraphic communications. In addition, all local operating authorities were required to accommodate similar requests from other regional operators even before orders arrived from the Command Center in Tokyo.[60] The conference called on all operators in Japanese-controlled areas of East Asia to set up such command centers as soon as possible, as the MTT had done.[61]

In February 1945, the Board of Communications set up an Emergency Communications Headquarters (Hijō tsūshin honbu) and issued more restrictions to deal with telegraphic services. Headed by the president of the board, the new agency was charged with the planning and speedy execution of all vital wartime communications policies, as well as with general control of emergency communications measures.[62] The board called for "a special handling of essential telegrams and a thoroughgoing curbing of nonurgent ones, to absolutely assure the speedy transmission of essential telegrams between Japan, Manchukuo, and China." In case of a breakdown in the trunk cable lines, vital telegrams would be dispatched by wireless in secret code.[63]

Japanese telecommunications providers on the continent adopted similar measures to cope with the problem of congestion, to little avail. In North China, for example, the Beijing–Tokyo circuit, carrying 1,350 telegrams on average per day, was out of service an average of fifteen hours each day, and the Beijing–Osaka circuit (capacity 1,050 telegrams) was routinely disabled for eight hours. In the spring of 1944, the NCTT launched a two-month campaign aimed at strengthening

60. "Denshin sotsū shirei honbu no setchi," *TDTZ* 4 (November 1944): 12–14.

61. Tanaka Shizuo, "Daitōa denki tsūshin kaigi dai-2-kai kaigi no gaikyō," *TDTZ* 4 (November 1944): 9.

62. Radio Tokyo (Japanese), February 14, 1945; in Office of Strategic Services, Research and Analysis Branch, comp., *Transportation and Communication in Japan*, 189.

63. Radio Tokyo, February 14, 1945, March 3, 1945; in Office of Strategic Services, Research and Analysis Branch, comp., *Transportation and Communication in Japan*, 203.

"communications fighting power."[64] It seemed ineffective, however. In a report submitted to North China's own Decongestion Command Center in late July 1945, a Japanese official in charge of easing the congestion in Beijing cited sabotage by Chinese guerrillas as the main cause of frequent breakdowns in the major lines. Moreover, their equipment was in such fragile condition and suffered from such poor maintenance, he admitted in the confidential report, that operating such equipment was like "whipping a corpse."[65] Occasionally, many of the emergency measures probably compounded the problem of coordination. Chaos reigned when long-used frequencies were discontinued under military orders; in one case this led to the breakdown of direct communication between Inner Mongolia and Shanghai for more than a week.

As a key link in Japan's imperial telecommunications network, the Manchukuo–Korea–Japan route became even more crucial during the latter stage of the Pacific War. As more and more circuits on the Japan–Manchukuo cable were devoted to relaying communications between Japan and North China, construction began on a second cable linking Keijō with northern Manchuria. It reached Wonsan in northeastern Korea by the end of the war. Domestic services in Korea were reduced to meet the need of through traffic as well as air defense. Still, the Korea Strait became the biggest bottleneck in Japan's telecommunications network. This was especially worrisome because the planned new cable linking Japan and Central China, which would have relieved some of this burden, failed to materialize. At the Fourth East Asian Telecommunications Conference in 1942, the MTT called for improving the link between Japan and the continent, citing frequent breakdowns in the Japan–Manchukuo cable, especially its underwater section in the Korean Strait.[66] A second non-loaded submarine cable—with six telephone channels—was placed across the strait in December 1943, but soon ceased functioning. Lack of repair ships further aggravated the problem.

---

64. "Dai-1-ji tsūshin senryoku zōkyō undō senka gaiyō" (September 1944), NCTT Records 2028/1922.

65. "Denshin shirei 20-nichikan no hōkoku" (July 25, 1945), submitted by Officer on Decongestion in Beijing (Oda) to Head of the Decongestion Command Center, NCTT Records 2028/1419.

66. Tōa denki tsūshin jimukyoku, *Tōa denki tsūshin kyōgikai dai-4-kai gijiroku*, 147–49.

In early 1945, congestion and periodic breakdown of the vital telegraphic connections between Manchukuo and Japan became so severe that at one point the MTT resorted to using airplanes to deliver backlogs of telegrams from Manchukuo to Tokyo.[67] Airmail service could hardly become a sufficient substitute, however. After 1944, as Japan lost air supremacy as well as most of its planes, even regular domestic airmail service could no longer be maintained. To ensure this vital link between Japan and the continent, an ultra-shortwave (VHF) connection, promising a total of twelve channels, was established, but its relay station on the island of Tsushima was destroyed in a fire. In a last-minute effort, the Japanese undertook to construct new direct cross-channel wireless facilities without the need of relays. Ironically, the work was completed on August 15, 1945, the very day of the Japanese surrender.[68]

The Final Countdown

The rapidly deteriorating condition of Japan's telecommunications network was already painfully obvious when the Second Greater East Asian Communications Conference was held in Tokyo in late 1944. This was the last time Japanese representing all the telecommunications operators in the imperium gathered under one roof to discuss the fate of Japan's East Asian telecommunications network. Compared to previous gatherings, it was a sober affair. Both the General Assembly and the Technical Section were canceled due to travel difficulties for most representatives; given the pressing demand for technical operations in all areas, it was also difficult to dispatch engineers to Tokyo for another meeting. The entire conference was reduced from the previously announced six days to only two. In his opening speech, Suzuki Kyōichi of the Board of Communications called on all those engaged in telecommunications to renew their efforts to "fully realize the capability of telecommunications in Japan, Manchukuo, China and the rest of Greater East Asia."[69]

---

67. "Fukumeisho" (July 23, 1945), NCTT Records 2028/1419.

68. *DDJS*, 6: 245–57.

69. Tanaka Shizuo, "Daitōa denki tsūshin kaigi dai-2-kai kaigi no gaikyō," *TDTZ* 4 (November 1944): 3–10; "Daitōa denki tsūshin kaigi dai-2-kai kaigi," *TKZ* 435 (December 1944): 12–14. See also Radio Tokyo broadcasts on October 24, 30, and November 4, 1944; in Office of Strategic Services, Research and Analysis Branch, comp., *Transporta-*

The reality must have been starkly clear to those delegates present, who included all senior executives of major telecommunications operators. On November 1, 1944, the last day of the conference, as delegates were preparing the final draft of a revised East Asian Telecommunications Agreement, air-raid sirens twice forced everyone into an underground bomb shelter. This was the first time B-29 bombers had flown over Tokyo from Saipan in the Mariana Islands, although this time it was for reconnaissance only. Although Watanabe Otojirō, who attended the meeting as a senior executive representing the NCTT, claimed that the raid only stiffened everyone's resolve, even he had to admit that it was "too dramatic" an experience to have the revised agreement completed in an underground bomb shelter.[70] As he waxed nostalgic about the autumn foliage and hot springs in Japan, it must have been obvious to him that the end was near.

None of the emergency measures taken by the Japanese during the war could change the fate of its network. Due to deteriorating infrastructure and the loss of skilled workers, even telecommunications at home became unreliable. As one Japanese official revealed in mid-1944, compared with the year before the outbreak of the war in China, the number of telegrams had increased by over 60 percent, but the number of circuits and employees had grown by only over 10 percent. The increase in long-distance telephone calls also outpaced the increase in available circuits. As a result, errors and delays in telecommunications service rose sharply. Several hours of waiting time were required to place even a special emergency long-distance telephone call, whereas some 40 percent of local calls could not be completed due to interruptions.[71] Such was the state of Japan's imperial telecommunications net-

---

tion and Communication in Japan, 191, 201. The 1943 meeting left only a draft record, and the 1944 conference did not leave any. This may be one of the reasons for the confusion, for it apparently slipped the memory of one ex-MOC official, Sasaki Kazuo, who later claimed that the 1944 annual conference never took place. See Sasaki, "Daitōa denki tsūshin taisei no kōzō," in Teishin gaishi kankōkai, *Teishin shiwa* II: 493.

70. Tanaka, "Daitōa denki tsūshin kaigi," *TDTZ* 4 (November 1944): 4; Watanabe Otojirō, "Daitōa denki tsūshin kaigi suisō," *TDTZ* 4 (November 1944): 11–12. When Watanabe later met with former MOC colleague Kajii Takeshi in Tokyo, he was apparently still wedded to the visions of a Greater East Asian telecommunications policy. See Kajii's diary entry of November 23, 1944, in Kajii Takeshi, *Kajii Takeshi ikōshū*, 124.

71. Nagata Yūji, "Kessenka denki tsūshin no setsubi to sutsū," *TKZ* 429 (June 1944): 4.

work at home and abroad as the empire approached its twilight. Japan's imperial telecommunications network disintegrated soon after its formal surrender, although some of the military and diplomatic communications channels remained in operation a little longer.[72]

The final countdown of Japan's empire can perhaps better be measured in terms of Japan's dwindling communications with the outside world, as key allies ceased to exist and last-minute diplomatic negotiations were attempted with increasing desperation. The beginning of the end came in the early morning of April 24, 1945, when Japan's communications link with its wartime ally Germany finally ceased operation. As a result, more than 100 telegrams, including 42 Japanese government telegrams and 33 official telegrams from the German embassy in Tokyo, could not be sent. After that date, Japan's direct communications link with Europe—and with the world outside Asia—was limited to a handful of cities in neutral countries, such as Moscow, Stockholm, Geneva, and Lisbon.[73]

On June 26, 1945, the Japanese Cabinet issued its last major policy statement on telecommunications. The "Policy to Ensure Communications During the Battle in the Home Islands" was a resolution designed to strengthen communications for defense against Allied invasion. Determined to wage a last-ditch effort of home defense, the Army had already begun construction of a new imperial headquarters in the mountains of Nagano prefecture, which became the project of highest priority near the end of the war. Considerable resources were diverted to the project, which included a new communications center for all of Japan. One thousand communications personnel were to be stationed there, with transmitters powerful enough to reach Europe.[74]

---

72. Japanese record shows that on August 17 Tokyo was still capable of communicating with Beijing, Shanghai, Nanjing, Singapore, Makassar, and Saigon. "Taigai tsūshin renraku jōkyō ni kansuru ken" (August 18, 1945), MOC Records II, 665. For instances of such communications, see Louis Allen, *The End of the War in Asia.*

73. "Tai-Doku musen denshin renraku tozetsu ni kansuru ken" (May 3, 1945), MOC Records II-561.

74. Nakamura, "Daihonhei to tsūshin," *DJ* 32 (October 1982). Nakamura notes that Japan emulated the success of Germany by setting up ultra-shortwave multiplex wireless facilities at the front, but never established a direct telephone link between the General Headquarters in Tokyo and commanders in the field, as Germany did; see Nakamura, "Rikugun tsūshinhei yomoyama hanashi," *DJ* 31 (July 1981): 45.

On July 26, 1945, the United States, Britain, and China issued the Potsdam Declaration as an ultimatum to Japan, calling on its government to proclaim "unconditional surrender of all Japanese armed forces" or face "prompt and utter destruction." The declaration stipulated demilitarization and demobilization, punishment of war criminals, removal of obstacles to democratization, material reparations, and territorial reduction in accordance with the 1943 Cairo Declaration, and occupation of Japan until all these objectives had been accomplished. When the Potsdam Declaration was received via the Swiss government, the Japanese government was still counting that the Soviet Union would play a part in bringing about a conditional surrender. Deeply divided over how to respond to the ultimatum, Japanese leaders were particularly concerned about the status of the imperial institution, which was not clarified in the declaration. The Army also objected to many other demands as well. Strong pressure from the military led Prime Minister Suzuki Kantarō to announce that the Cabinet "would offer no comment" (*mokusatsu*), essentially ignoring the Allied demands. Japan's failure to promptly accept the Potsdam Declaration, as well as the Allies' insistence on unconditional surrender, had grave, albeit unforeseen, consequences. In early August, two atomic bombs were dropped on the home islands, and the Soviet Union joined the war against Japan.[75]

By this time the state of Japan's international communications was even more precarious than it had been before. As a final blow to Japan's peace effort, its link with Moscow went dead in the early morning of August 9, the same day the Soviet Union declared war on Japan. At the request of the Ministry of Foreign Affairs, Japan's telegraph office attempted to reopen the Tokyo–Moscow circuit the next day, but to no avail. Desperate Japanese telegraphic operators resorted to announcing their intentions to resume communication via overseas radio broadcasts. On August 11, two telegrams from the Soviet embassy in Tokyo to Soviet Foreign Minister Molotov were sent via Geneva. By late afternoon, the Moscow link had resumed for a few hours, barely long enough for one of the two telegrams to be resent.[76] As Japan maintained only three

75. For a recent reinterpretation on Japan's decision to surrender, see Kiyoshi Hasegawa, *Racing the Enemy: Stalin, Truman, and the Surrender of Japan* (Cambridge, Mass.: Harvard University Press, 2005).

76. "Gaishinka jimu nisshi" (January–December 1945), MOC Records II, 608.

channels of telecommunication outside East Asia, in addition to sending official telegrams through Geneva, it also resorted to overseas shortwave broadcasts to convey its acceptance of the Potsdam Declaration.[77]

AFTERMATH OF EMPIRE

Legacies Abroad

Historian Michael Adas has pointed to the need to abandon grand attempts to draw up balance sheets of imperialism.[78] This is certainly applicable to Japan's techno-imperialism in telecommunications. When the Japanese empire collapsed in August 1945, Japan lost all the infrastructure overseas, including all the telecommunications facilities it had built or taken over in Asia during the course of several decades. Nevertheless, Japan's techno-imperialism in the field of telecommunications had important postwar legacies that not only affected former colonies and occupied areas but influenced Japan proper as well.

No accurate figures exist as to total Japanese investment in telecommunications before and during the Pacific War. A group of former Japanese employees charged with chronicling overseas telecommunications operations not long after the war came up with the information given in Table 13.

Statistics can provide only part of the answer, since not all costs and benefits can be tabulated to satisfy the accountants of imperialist projects. The Korean peninsula provided perhaps the most dramatic demonstration of Japan's imperial legacy. After Korea was divided, the south was placed under U.S. occupation. Japanese technologies slowly gave way to American influence, but the colonial telecommunications infrastructure continued to play a role in post-colonial Korea. An

---

77. Kitayama, *Zoku Taiheiyō sensō mediya shiryō*. According to Japanese records, only the following routes were still working on August 17: Geneva, Stockholm, and Lisbon. Temporary wireless communication between Tokyo and Manila was set up on August 16 to handle occupation-related matters.

78. Michael Adas, "'High' Imperialism and 'New' History," 331–44. For an examination of how a ruthless invader can exploit industrial societies, see Liberman, *Does Conquest Pay?*, 118: "If Japan's expansionism had not resulted in an unwinnable war," the author argues, "the empire would have added greatly to its economic self-sufficiency and ability to wage a protracted war."

Table 13
Estimate of Telecommunications Facilities in Japan's Imperium, 1945

Area	Telegraph offices	Telephone subscribers
Taiwan	178	24,043
Korea	1,048	59,496
Karafuto (southern Sakhalin)	134	6,878
Manchurian Telegraph and Telephone Co. (MTT)	922	108,571
North China Telegraph and Telephone Co. (NCTT)	159	52,786
Inner Mongolia Telecommunications Facility Co. (IMTFC)	60	4,238
Central China Telecommunications Co. (CCTC)	78	15,477
Xiamen Telecommunications Co.	2	813
TOTAL	2,581	272,302

SOURCE: *Gaichi kaigai denki tsūshin shisetsu ni kansuru shūsenji shisetsu genjō, shisan hyōka, oyobi hikitsutsuki jōkyō gaisetsu* (Tokyo: Denki tsūshin kyōkai, 1956).

American survey in the late 1940s found the existing telecommunications network in southern Korea "small and in many respects obsolete by American standards," having suffered from years of undermaintenance. However, it was considered "generally adequate to meet the existing needs."[79] In June 1950, five years after Japan's empire had disintegrated, the Korean peninsula was engulfed in a new conflict. Badly in need of communications facilities, the American forces "discovered" the underground "Mukden cable" running through the peninsula and found it to be a crucial asset. American engineers repaired Japanese cables and repeater stations so that the cable could be put to maximum use as a trunk line for troops in Korea, and as the vital link between General MacArthur's headquarters in Tokyo and U.S. forces on the Korean peninsula. As the great artery of communication, a U.S. Army officer recalled, the Mukden cable was a "God-given gift for the US Signal Corps."[80]

Americans were not the only ones who benefited from Japan's imperial infrastructure. The same Mukden cable also served as a major communications link for North Korea and China, which joined the war in

79. McCune, *Korea Today*, 160–61.

80. Nakamura Fumio, "Chōsen sensō to tsūshin," *DT* 35 (January 1983): 64–65; Huston, *Guns and Butter*, 297–98.

October 1950. Lacking spare parts to repair damage to the cable between Sinuiju and P'yongyang, Chinese engineers resorted to removing sections of the NLC along the former Manchukuo-USSR frontier and using them in Korea.[81] This was the case with the railway linkage to Japan and Manchuria as well. Since Japan had built the railroads in Korea during the colonial period, repair and replacement items could be obtained from the Japanese National Railways and quickly airlifted to Korea, an important advantage to UN troops.[82] During the Chinese Civil War of 1946–49, the railway in the north transported remaining Japanese material to the Chinese Communist forces, contributing to their victory over the Nationalists in Manchuria. And in southern Sakhalin, Japanese technicians were employed to repair the NLC in the area.[83] Thus, in a way no one could have predicted, what remained of Japan's imperial telecommunications and transportation network contributed to Northeast Asia's transition to the world of the Cold War.

Many Japanese technicians involved in telecommunications and other fields stayed on—some by choice, others under pressure. In North China, many Japanese employees of the NCTT, including Asami Shin, former director of the Communications Department, remained. In June 1946, at the request of the Chinese authorities, some 109 former Japanese employees stayed; the others were repatriated to Japan. The outbreak of civil war between the Nationalists and the Communist forces in northeastern China, however, made it increasingly difficult for the Nationalist government to retain these Japanese technicians. By year's end, all but nineteen former Japanese employees had returned to Japan. Seventeen of these returned to Japan in November 1948, but two were kept in service by the Chinese Communists.[84] In Central China, after the takeover by the Chinese (and, in the case of Shanghai Telephone Company, by Americans), most Japanese telegraph operators lost their jobs but Japanese technicians were kept on. Between November 1945 and February 1946, some twenty of them were mobilized to

81. Chen Ende, "Kang-Mei yuan-Chao changxiu dianlan," *Youdian wenshi tongxun* 14 (December 1993): 19. Chen headed the Chinese engineering group that carried out the repair work in North Korea between late 1951 and early 1952.

82. Roy E. Appleman, *South to the Naktong, North to the Yalu* (Washington, D.C.: Office of the Chief of Military History, 1960), 261.

83. *DTJGKS*, 253–54.

84. Hokudenkai, *Kahoku denden jigyō shi*, 256–62.

transfer and install about a thousand automatic exchange telephones from Shanghai to Nanjing. About a hundred Japanese technicians in Hankou were also retained to provide technical assistance to the Chinese government until May 1946.[85]

In Manchuria, a number of employees of the MTT's Investigation Bureau in Harbin were arrested for alleged espionage activities against the Soviet Union, while more than 200 Japanese technicians were retained by the Chinese government to repair damaged communications facilities. One former Japanese employee of the MTT even started a Harbin Wireless School where Chinese and Koreans were taught basic skills—all with the encouragement of the Communist-controlled municipal government—and ran it until his repatriation to Japan in 1953.[86] This employment of large numbers of Japanese technical personnel not only facilitated some technology transfer to the new regime but also had important implications for Japan's relationship with its continental neighbor after the war.[87]

Many Chinese and Koreans who had been employed in these Japanese-dominated communications facilities went on to become the backbone of operations after Japan's surrender, even as their countries were moving into the technological orbit of either the United States or the Soviet Union. Although telegraphic communications had largely been unified in China proper before the war, Japan's wartime consolidation of local telephone operations in China (including Manchuria) no doubt made it easier for postwar Chinese authorities—whether the Nationalists or the Communists—to rebuild a national communications network controlled by the state (with the exception of foreign companies such as Shanghai Telephone Company and the Great Northern Telegraph Company before the 1950s).

---

85. *DDJS*, 6: 473–74.

86. Ibid., 876; Ōtake Eiji, *Hito to nami to*, 115–16. For a generally positive recollection by a Chinese who worked with MTT Japanese technicians in the early postwar period in northeast China, see Ni Xingxin, "Guanyu Riben jishurenyuan zai Dongbei jiefangqu dexing gongzuo de huiyi," *Youdian wenshi* 48 (December 1998): 18–20. Ni resumed contact with some of the Japanese and developed friendships with them decades later, after Japan and China restored diplomatic relations.

87. I discuss this subject and its political implications for early postwar Chinese-Japanese relations in "Resurrecting the Empire? Japanese Technicians in Postwar China."

Influences at Home

As Japan came under Allied occupation, the Supreme Command of the Allied Powers (SCAP) introduced drastic reforms in a wide range of areas, under the twin goal of demilitarization and democratization.[88]

Telecommunications was one of the areas where major changes were introduced. Due to the ITC's extensive ties to the military and its active part in overseas expansion, SCAP considered it "the hub of a small communications *zaibatsu*" (financial clique). With 30 percent of its assets located outside Japan proper and thus considered totally lost, the ITC was in financial distress and was ordered disbanded.[89] MOC officials strove to save its operations on the grounds of their technical superiority, but to no avail.[90] One of the most useful tools of Japan's overseas telecommunications expansion and operation was dismantled.

The technical capability and industrial strength developed during this period were not completely lost in the home islands and would serve as the basis for Japan's postwar recovery despite occasional obstacles.[91] In the field of transmission technology, the Japanese were told to discontinue their research, although they managed to continue it behind the backs of the Americans. In certain areas, such as ionospheric research, Japan was allowed to continue its work even during the Allied occupation, thanks to its obvious lead in the field and the usefulness of its findings. Scientists at Fujikura Wire Works studied captured American cable used in radar in Singapore and began research on plastic conductors in 1943, at the request of the government. Under the guidance of the Research Institute for Physics and Chemistry, Fujikura and Hitachi began test manufacturing by the end of the year.

---

88. For a general history, see John Dower, *Embracing Defeat.*

89. SCAP CCS, "Reasons Why ITC Should Be Absorbed by the Ministry of Communications" (December 27, 1946); SCAP CCS, "ITC's Small Zaibatsu" (January 14, 1947); Box 3188, RG 331, National Archives.

90. See memo "Technical Superiority of ITC," and "Reasons Why the Company Conducting Construction and Maintenance of Wireless Facilities Is to Be Kept When ITC Is Re-arranged." They were prepared by ITC and submitted to SCAP on December 11, 1946, Box 3188, RG 331, National Archives.

91. For an analysis in the chemical industry, see Molony, *Technology and Investment,* 318–19.

Although the war came to a close before their final product was ready, their studies would prove useful for postwar recovery.[92]

Although the hardware that remained from Japan's empire-building was plain to see, what Japan had gained from the experience of building a vast wartime empire was less obvious but equally important.

The human factor was perhaps the most important link between Japan's imperial project and its postwar economic resurgence. To be sure, this link was not completely intact. One estimate put the total number of Japanese who worked in communications in Manchuria and China after 1931 at 25,000 to 30,000.[93] Many never made it back to Japan, either perishing in the last phase of the war or dying in custody in Siberia and elsewhere. The nine Japanese female telephone operators in southern Sakhalin who committed suicide by taking poison before the arrival of the Soviet troops were perhaps the most well-known example. A monument was later built to them in northern Hokkaido years later and would be favored with imperial visits.[94] The majority survived, however, and returned to war-ravaged Japan.

Several top executives were purged. Watanabe Otojirō, returning from seven years in China, served briefly as director of the Telephone and Telegraph Bureau when Matsumae Shigeyoshi was president of the Board of Communications, but was relieved of public duty in early 1947 because of his connection with the NCTT. Since it was difficult to find a qualified replacement, SCAP allowed Watanabe to hold office longer than specified.[95] Nonetheless, the postwar career of one of Japan's most talented telecommunications specialists was unremarkable, as if the scale of post-imperial Japan was too small for him.

Technicians survived better than their administrative counterparts. Although Japan's effort to build telecommunications hegemony in East Asia did cost many lives, many of its "communications men," such as engineers associated with the NLC network, not only survived but

92. Fujikura densen kabushiki kaisha, *88-nen no ayumi*, 233–34.

93. Matsuo Matsutarō, "Tairiku ni katsuyakushita teishinjin (1): Manshū," in Teishin gaishi kankōkai, *Teishin shiwa* II: 213.

94. See Karafuto teiyūkai, *Tsuioku no Karafuto teishin*. Also Matsuo Tatsutarō, "Gaichi hikiage to junshoku hiwa," in Teishin gaishi kankōkai, *Teishin shiwa* III: 221–41.

95. SCAP CCS, "Personnel Changes in the Ministry of Communications" (April 4, 1947), Box 3188, RG 331, National Archives; Teishin dōsōkai, comp., *Senpai ni kiku*, 655–71.

went on to play influential roles in postwar Japan. Shinohara Noboru, Matsumae's less outspoken but equally capable lieutenant, became associate director of the new Science and Technology Agency.[96] Matsumae, who had contributed more than anyone else to the development of NLC technology for the East Asian network, founded the large, multi-campus Tōkai University. Known for its strong engineering departments, the university embodied Matsumae's vision for a future Japan. Several others played key roles in postwar telecommunications (see Table 14).[97]

Thus Japan's empire was lost, but its empire-building experience and techniques were not. In some cases, wartime experience proved quite useful for rebuilding postwar Japan. For example, urban planning in Manchukuo in particular exerted much influence on postwar Japan.[98] Perhaps the most successful large postwar project that had a wartime genesis was the *Shinkansen*, the world-famous bullet train.[99] Dreams during the war became a reality in peace. Although the United States wielded much clout in early postwar Japan, there is evidence that Japan's postwar telecommunications policy was influenced by its wartime experience as well. Although state control of telecommunications had been a firmly established principle in Japan, private involvement in international communications had always been considered an alternative that could be used to overcome political complications. The relative financial and political success of these private operations in the imperium provided firsthand experience to many MOC officials. At the sugges-

---

96. For a Japanese work on the connection between the Science and Technology Agency and the legacies of wartime experience, see Ōyodo Shōichi, *Miyamoto Takenosuke to kagaku gijutsu gyōsei*, 518–28. Shinohara himself later commented that his earlier experience was very helpful in his work as vice minister of the Science and Technology Agency; see Nippon denshin denwa kōsha, *Denki tsūshin jishu gijutsu kaihatsu shi [hanso denwa hen]*, 36.

97. Recollections by Fujikawa in Teishin dōsōkai, comp., *Senpai ni kiku*, 666–67.

98. Writing on the Japanese civil engineers and architects who were active in Manchukuo, Koshizawa notes that "technology is developed through actual work, and technology is transmitted not through publications and documents alone but rather through the movement of people. Through work, pupils are trained who, in turn, form the group of technicians" (Koshizawa Akira, *Manshūkoku shutō keikaku*, 29).

99. For a fascinating study, see Takashi Nishiyama, "War, Peace, and Nonweapons Technology: The Japanese National Railways and Products of Defeat, 1880s–1950s," *Technology and Culture* 48 (April 2007): 286–302.

Table 14
Transwar Personnel Continuities in Japanese Telecommunications

Name	Pre-1945 position & accomplishment	Major postwar position(s)
Fukuda Kon	President, CCTC	Executive Director, KDD
Hanaoka Kaoru	MOC official GNTC negotiations	Director, KDD
Kajii Takeshi	MOC engineer MTT establishment NLC adoption	Founding president, JTTPC Science and Technology Committee
Kobayashi Kōji	NEC engineer NLC development	President, NEC
Matsumae Shigeyoshi	MOC engineer NLC development Director, IRAA	Founder & president, Tōkai University Diet member
Ōhashi Hachirō	President, ITC Telecom Committee	2nd President, JTTPC
Shinohara Noboru	MOC engineer NLC development	Associate Director, Cabinet Science and Technology Agency Science and Technology Committee

NOTES: IRAA: Imperial Rule Assistance Association; JTTPC: Japan Telegraph and Telephone Public Corporation.

tion of Watanabe Otojirō, many promising young MOC employees had been transferred to a national policy company like the NCTT for two years before returning to Tokyo, thus gaining firsthand experience of telecommunications operations in a quasi-private environment. When they returned to Japan after the war and resumed important positions in the government, it became obvious that postwar telecommunications were not going to be the same.[100] Accordingly, in early 1950, the advisory Telegraph and Telephone Revitalization Council proposed that a

100. Later it would become a problem, when calculating years of service, whether to include those in the "national policy companies" such as the MTT. That those in favor finally prevailed, according to one MTT veteran, was partly due to the fact that these companies operated under military orders during times of emergency, although they were private enterprises otherwise. Watanabe Tatsuki, "Mōkyō haken," in Manshū denden tsuiokuroku kankōkai, *Akai sekiyō*, 259.

public enterprise incorporating the advantages of private management would be the most appropriate format for telecommunications in Japan.

During the Upper House deliberations in 1952 over the form of telecommunications in postwar Japan, Endō Shin'ichi, a former MOC bureau director who had served as executive vice president of the MTT, testified as one of nine expert witnesses on the government's proposal to create a public corporation for telecommunications. Endō's experience at the MTT proved especially pertinent. "Although the idea of a 'public corporation' is said to have recently come from America," Endō told Diet members, "in fact in Manchukuo we were already operating according to the concept of a public corporation a long time ago." He was referring not only to the SMR but to the MTT, which had been founded in 1933 as a semi-public entity. Although in appearance the MTT had been a private stock company, Endō explained, "in essence I believe it was what we would call a public corporation." Endō had little doubt about the appropriate form of ownership of telecommunications in postwar Japan, arguing that, "seen from the essence of telegraph and telephone enterprise, public corporation is the most appropriate form. Neither state management nor private management is sound."[101] In the end, the Japan Telegraph and Telephone Public Corporation (Nippon denshin denwa kōsha; JTTPC) was launched in August 1952 to operate domestic telegraph and telephone service under a monopoly it enjoyed until its privatization in 1983.

There was occasional soul-searching on the part of Japan's "communications men" about the fate of Japan's imperial enterprise. An early postwar history of ITC attributed its involvement in government-led expansion in part to the militarism and mass media, which created the fantasy of a continental cable via China to Singapore. Kajii admitted the many failures in Japan's colonial policy and wished Japan's had learned more from Britain, which "sent the best and the brightest" to their overseas colonies.[102] Understandably, many engineers blamed Japan's defeat on a failure of science and technology, vowing not to

101. Excerpts of Endō Shin'ichi's testimony are included in Yamashita Takeshi, *Denshin denwa jigyō ron*, 66–72. The MTT was by no means Japan's only experience with a public corporation in telecommunications. In mid-1942, a public telecommunications corporation had been proposed to operate the pan–East Asian network.

102. *KDTKKS*, 3; Kajii, Preface to Hokudenkai, *Kahoku denden jigyōshi.*

succumb to another "tactless (*mubō*) war." Shortly after Japan's surrender, Matsumae Shigeyoshi, as the president of the Board of Communications, reportedly had a discussion with a young engineer about the fate of "Japan's domestic technology." Although Matsumae still believed in "domestic technology," he now had to agree with the young engineer that Japan must abandon the pursuit of Japanese-style technology that exluded foreign input; only through technical exchange with foreign countries, could Japan develop and export standards universal to the world.[103] And when Kajii Takeshi became the first president of the JTTPC, he immediately ordered the importation of all recent foreign innovations, even at the risk of bankrupting the young company. During his overseas tour in 1953, Kajii was surprised to find that "in foreign countries, they are making great efforts at producing coaxial cables." He was disappointed that Japan was not doing so.[104] After years of technological isolation, Japan apparently had much catching up to do. As one Japanese writer pointed out, Japan's telecommunications manufacturers benefited handsomely from technological transfer from the U.S. military, particularly in the area of FM wireless in the early Cold War years.[105]

Revival

Japan's return to early postwar Asia met with mixed results.[106] Like many Japanese who had worked on telecommunications in wartime China, Matsumae wondered after the war what had happened to the NLC network in Manchuria and China. He remained convinced of China's need for Japanese technology: "To bring these superior circuits back to life, one has to bring vacuum tubes and cables from Japan, or these lines can't be maintained. . . . Since they have to rely on Japan for such products, this is how Japan can expand its exports." Making products that could not be imitated overseas, Matsumae asserted, was

103. *DTJGKS*, 268.

104. Ibid., 40–41.

105. See Nakamura Fumio, "Chōsen sensō to tsūshin," *DJ* 33 (May 1983): 56–57.

106. For instance, Jane Robbins has noted in *Tokyo Calling*, her study of Japan's overseas broadcasting, that its wartime experience helped Japan to become a leader in international broadcasting in postwar Asia.

the task facing the Japanese economy at the present and in the future.[107] In some ways, Japan's technological presence in the pre-1945 era helped re-establish postwar links with Asia, especially after its own economy recovered and took off.

After the People's Republic of China's initial economic recovery, it launched its first Five-Year Plan, with the Soviet Union and its Eastern European allies as China's suppliers of equipment, technology, and funding in areas such as telecommunications. Still, because much of the equipment produced by Japanese companies in Manchuria and North China remained in use in China, spare parts and repairs were necessary. Eager to re-enter the China market, Japanese telecommunications manufacturers managed to obtain information about the conditions of telecommunications facilities there from Japanese repatriates and visitors to the People's Republic. Having ascertained the demand for Japanese telecommunications products, the industry lobbied the Japanese government to relax COCOM (Coordinating Committee for East-West Trade Policy) restriction of exports of certain items to China. In June 1954, special permission was granted to export small quantities of radio parts and vacuum tubes. An executive of Japan's Wireless Communications Industry Association (Nihon musen tsūshin kikai kōgyōkai) even took to the airwaves of NHK's Chinese-language programs to promote Japan's products. As a result, by the end of 1954, Japan received an order for 20,000 vacuum tubes valued at $20,000.[108] The Cold War confrontation, however, prevented Japan from playing a larger role in the technological development in mainland China until the 1970s.

Outside the Communist bloc, the revival and strengthening of ties with Japan proceeded with more success, albeit unevenly. In 1959, at the suggestion of Kondō Giichi, a former MOC official, the Asian Telecommunications Assistance Association (Ajia denki tsūshin kyōryoku kai) was founded. The organization consisted of representatives from government, industry, and Japan's two major telecommunications

---

107. Matsumae, Preface to Matsumae Shigeyoshi ronbunshū kankōkai, *Hatsume e no chōsen*, 44–46.

108. "Tsūshin kiki no Chūkyō muke yushutsu," *Tsūshin kōgyōkai hō* 3 (February 1955): 8–9. For an account of conditions in China after the war by a former Japanese employee of the MTT, see Suzuki Yūji, "Chūgoku ni okeru tsūshin shisetsu to un'ei," *Tsūshin kōgyōkai hō* 3 (February 1955): 12–14.

operators. Most projects involved export of telecommunications equipment to Southeast Asian countries. Also in 1959, Furukawa Electric Company and NEC sent top executives on a tour of Thailand, Cambodia, Laos, and Vietnam. Kajii reported that those who had visited Japan before "welcomed Japan's cooperation because they knew the excellence of Japan's technology."[109] In cooperation with these two manufacturers, the association set up the Japan Communications Assistance Company (Nihon tsūshin kyōroku kaisha) to investigate and design telecommunications facilities overseas. As Japan's economy took off in the 1960s, investment and export would bring much of Asia within Japan's embrace.[110] In Japan's former colony of Taiwan, where 400 km of three-channel non-loaded telephone cable between Taihoku (Taipei) and Takao (Kaohsiung) had been completed by the ITC by the end of 1943, Japanese technology continued to influence telecommunications development well after the war. In the late 1960s, the F-24 coaxial cable made in Japan, with technical assistance provided by the JTTPC was completed between Taipei and Kaohsiung and became a major toll trunk line in Taiwan.[111]

Although the re-established technological links helped launch Japan into postwar recovery and onto the international stage, its history of imperial expansion also made Japan's presence often politically sensitive in Asia. Statements by Japanese officials about the "positive contributions" during Japan's colonial rule certainly did not help. The submarine cable between Japan and Korea, used largely by the U.S. forces stationed in both countries, became a major bone of contention between Tokyo and Seoul. Japan insisted on the mid-point between Pusan and Tsushima as the dividing line for maintenance of the cable, whereas Seoul insisted on the mid-point between Fukuoka and Pusan. In 1958, a Japanese ship repairing the cable was accidentally shelled by the South Korean Navy because it was inside Korean waters as defined by the so-called Syngman Rhee Line. And in a move reminiscent of the 1880s, the Japanese government ended up hiring a GNTC cable ship to do repair work in 1961. The dispute was not resolved until after the

109. *DDJS*, 6: 972–76; Kajii, *Waga hansei*, 514.
110. See Hatch and Yamamura, *Asia in Japan's Embrace.*
111. *DTJGKS*, 250–51.

signing of the Basic Treaty between Japan and the Republic of Korea in 1965, under the Park Chung Hee regime.[112]

Given the new political boundaries, it is not surprising that many of Japan's technological blueprints during its empire-building era failed to turn into reality after the war. The planned underwater railway tunnel between Korea and Japan, not to mention the grand concept of a trans-Asian continental railway, was among the projects that never materialized. As a junior partner in U.S.-dominated East Asia, Japan's role had limits. In 1958, the Japanese government proposed that the International Telecommunications Union extend the world telephone cable network to Southeast Asia, the Far East, and Japan. In addition to Japan, other potential regional members included Singapore, Thailand, the Philippines, and Taiwan. The Japanese government convened several international meetings in the early 1960s to discuss the project, but the plan was eventually abandoned.[113] Although it repaired a few existing submarine cables, Japan was unable to build new cables outside its own territorial waters. Since Japan's new cable ship, built in the late 1960s, would be used only for close-to-shore operations, it was only 1,700 tons, compared to ships at 3,000 tons and above built before the war.[114] Post-imperial geopolitical realities as well as the bitter legacies of empire together made such ambitious overseas projects distant dreams.

On the other hand, the revival of Japan's economic presence in Asia was often accompanied by imperial nostalgia in many recollections and histories compiled after the war. It was not long before Japanese wartime technological projects were recast as well-intentioned developmental efforts in Asia that had been derailed. As one former NCTT executive noted, "Needless to say, since it was during the war, we often had to follow the military's orders along the lines of the so-called national policy. Our hearts, however, were filled with the pure love for telecommunications operations. We put our lives at risk for the sake of peace in East Asia, by restoring abandoned telecommunications facilities on the continent, building housing, lines, equipment, and wireless

---

112. Nippon denshin denwa kōsha, Kaiteisen shisetsu jimusho, *Kaiteisen hyakunen no ayumi*, 631–33, 667–70. For a recent Japanese work from a strong nationalist perspective, see Ishihara, *Kokusai tsūshin no Nihon shi*, 207–9.

113. Hanaoka, *Kaitei densen to Taiheiyō no hyakunen*, 265–68.

114. Nippon denshin denwa kōsha, Kaiteisen shisetsu jimusho, *Kaiteisen hyakunen no ayumi*, 752–64.

receiving centers."[115] A former MOC engineer suggested after the war that although Japan's wartime plans for establishing telecommunications companies in Southeast Asia did not materialize, "As we come to discuss the development policy of underdeveloped countries, since there will be more and more opportunities for Japan's telecommunications technology to enter East Asian countries, even if only part of the wartime plans could be revived, it would be a huge contribution to the development of these newly independent countries."[116] Nakamura Jun'ichi, who served as the Director of Telecommunications Bureau during the Pacific War, appeared to agree wholeheartedly:

> Although we lost the war, it is beyond doubt that the footprints left by the communications men on the continent or in the Southern Region have contributed to the postwar recovery and development of these areas. . . . Development aid to underdeveloped regions for the purpose of peace and prosperity has become an important task for the world as well as for Japan. Our telecommunication and related industries can provide more effective and appropriate assistance as they meet the peaceful and constructive nature of communications. In this sense, our experience in various areas during the war, regardless of the outcome, is worth re-examination today.[117]

By the 1970s people like Matsumae seemed to have recovered their belief in Japanese technological prowess: "The research on non-loaded cable that we spent a long time on is one research, at least, that was ahead of other countries. And such research is not something foreign countries can catch up with, even through [they are] trying. We are convinced that it is through creating such technologies that Japan's economy can be established through Japan's industry." Matsumae was far from alone; techno-nationalism, mixed with traces of imperial nostalgia, regained some lost influence in high-growth Japan.

---

115. Hokudenkai, *Kahoku denden jigyō shi.* In *Kokusai denki tsūshin kabushiki kaisha shi*, published in 1949, the company pled its innocence by blaming the government and the military; see *KDTKKS*, 47.

116. Tsuda Ryūzō, "Taiheiyō sensō toki no gaichi tsūshin," in Teishin gaishi kankōkai, *Teishin shiwa* II: 588–89. For other examples of Japanese business ties to Southeast Asia revived after the war, see Adachi Hiroaki, "Furukawa kei kigyō no Nanpō jigyō tenkai," in Hikita Yasuyuki, *Nanpō kyōeiken*, 476–77.

117. Nakamura Jun'ichi, "'Daitōa tsūshin seisaku' jidai kara Denden kōsha tanjō zen'ya made," in Teishin gaishi kankōkai, *Teishin shiwa* III: 438.

Japan's imperial telecommunications network was once a living creature. It functioned as both the nerve system of government and other institutions as well as the artery of daily life for millions of individuals. By the war's end, however, Japan's far-flung empire resembled a fatally wounded beast, with its brain and nervous system still functioning, albeit at reduced capacity. Before the age of telecommunications, the empire would have disintegrated once crucial transportation lines were severed. Paradoxically, the imperial telecommunications network, as damaged as it was, may have prolonged the life of Japan's empire, for better or for worse.

As this chapter has shown, the total defeat and disintegration of the empire did not wipe the slate of Japan's techno-imperialism clean. In terms of human resources as well as technological linkages, legacies of Japan's empire survived well into the postwar era. If, at the moment of defeat, technological self-doubt was prevalent even among Japan's techno-imperialists, it was gradually replaced by the revival of confidence: once dismissed as inadequate, wartime technological accomplishments at home and in the imperium have now come to be seen as a force of modernization and development and a source of national pride in Japan. As a result, empire-building itself has been recast in the collective memory as well. It is the task of historians to re-establish the link between empire-building and technological development by examining the phenomenon of Japan's techno-imperialism *in toto*. It is this task to which I turn in the Conclusion.

CONCLUSION

Just hours after the broadcast of Emperor Hirohito's speech on August 15, 1945, outgoing Prime Minister Suzuki Kantarō went on the radio, exhorting the Japanese people to rebuild Japan by supporting the *kokutai*—Japan's emperor-centered national polity—and improving the country's scientific knowledge. In a confidential interview with his closest attendants a few months later, the emperor himself attributed Japan's defeat to the country giving too much weight to the fighting spirit while slighting the power of science. The emperor repeated his view a few years later in a letter to the crown prince discussing Japan's defeat.[1] It may not have been a coincidence that both the prime minister and the emperor mentioned science but not technology. In the immediate postwar period, the reputation of technology was more tainted by the war and its destruction, whereas pure science could be cast as a shining embodiment of hope for a new, peace-loving Japan.

As if to echo the Japanese leaders, historians of modern Japan have also viewed its scientific and technological development largely as a failure.[2] Even after Japan's postwar economic "miracle" produced many explanations of its success, historians still tend to disassociate science and technology from war and empire, attributing Japan's spectacular expansion in the 1930s to a runaway military, spiritual fanaticism, or the failure of diplomacy.

This is unfortunate. It is true that fundamental causes of Japan's defeat included its overall national strength and lack of material resources, which had much to do with the level of its science and technology. It

1. The eight-hour interview was published in 1999 as "Shōwa tennō dokuhakuroku," *Bungei shunju* (December 1990). See also Takahashi Hiroshi, *Shōwa tennō hatsugenroku: Taishō 9-nen Shōwa 64-nen no shinjutsu* (Tokyo: Shōgakukan, 1989), 115. I thank Timothy George for bringing this additional source to my attention.

2. An influential work that breaks with this trend is Bartholomew, *The Formation of Science in Japan*.

does not follow, however, that Japan's own scientific and technological achievements did not contribute to its overseas expansion and empire-building in significant ways. Japan's fateful decision to launch simultaneous attacks in the mid-Pacific and in Southeast Asia in December 1941 was certainly a risky gamble, but it was taken *after* a relatively successful decade when Japan harnessed modern technologies in conquering and controlling the vast imperial space in Asia.

TECHNO-IMPERIALISM IN MODERN JAPAN

As a study of Japanese imperialism from the standpoint of telecommunications,[3] this book has sought to restore the important nexus between technology and empire in modern Japan. Several broad characteristics of Japan's techno-imperialism can be drawn.

First, techno-imperialism traces its genesis to at least the Meiji era. As Japan borrowed and adapted advanced military and industrial technologies as well as institutions from the West, the country not only preserved its independence but also quickly gained ascendance as a regional power after defeating China and Russia. The birth of Japan's colonial empire at the turn of the twentieth century would have been inconceivable without a plethora of modern technologies. However, techno-imperialism matured as Japan developed its indigenous scientific and technological expertise and progressed with industrialization, and it gained momentum as Japan embarked on a new phase of continental expansion in the 1930s. Hence, the decade saw the construction not only of an intercontinental telecommunications network but also of gigantic hydraulic dams, high-speed trains, and a modern metropolis in the empire. Indeed, technology and empire had become so intertwined that the period can be appropriately considered as the golden age of Japan's techno-imperialism.

By then technological advances and network expansion had become indispensable to Japan's vision of an integrated imperium in Asia. As Japan sought to build a regional political and economic bloc in Asia, fortified and improved telecommunications links became the "nerve

3. I borrow this expression from Donald Robinson's introduction to Davis and Wilburn, *Railway Imperialism*, 5.

system" facilitating close political, economic, and cultural integration. With its capacity and promise to overcome space and enhance control, telecommunications—perhaps better than other types of technology—thus helps clarify the link between Japan's imperialist dreams and modernist visions. The Greater East Asia Co-Prosperity Sphere was not simply a construct of imperial dominance through force; it was to be enhanced and facilitated by powerful modern technologies—railway, aviation, advanced telecommunications such as long-distance telephones and phototelegraphy—that would strengthen the "bonds" among the regions and peoples in the sphere and fortify Japan's position as its undisputed leader. The new Japanese imperium after the 1930s was, in this sense, a technologically imagined community.

Techno-imperialism is not driven by technology alone; rather, it is better conceived as entire systems of institutions as well as individuals with their aspirations. Japan's communications bureaucracy—which had direct responsibility for telecommunications in Japan as well as its colonies and the informal empire—is a good case to illustrate this point. Lack of financial resources, in large part due to domestic budget constraints, forced the bureaucracy to relinquish some control over construction and maintenance of core hardware in order to attract private funds for new facilities overseas. In occupied Manchuria and China proper, Japan borrowed from earlier experience and found a partial solution by creating nominally private joint ventures that remained under the complete control of Japanese directors and employees. Requirements of bureaucratic or diplomatic "correctness" also favored a less conspicuous presence of Tokyo outside Japan. In Southeast Asia, another semi-governmental company was used to extend and consolidate Japan's control. Still, Japan's communications bureaucrats encountered resistance not only from the Chinese and other foreign interests but also from other Japanese interests as well. Given the institutional basis of technology, a major dimension of techno-imperialism was the political struggle over its control and application.

Japan's techno-imperialism was, to a great extent, a collective endeavor of technology bureaucrats. The elevation of technologies for empire was linked to the political transformation in Japan in the 1930s, as engineers associated with technological developments and applications became energized and active. The Japanese engineers and bureaucrats who strove to bring about Japan's telecommunications hegemony in East Asia might not be imperialists in the conventional sense of the word. Many of them were preoccupied, above all, with ridding Japan of its technological dependence on the advanced industrial countries of the West. At the same time, they were eager to exploit Japan's technological superiority over its "backward" neighbors, beginning with Korea and China. Although engineers and bureaucrats might differ with Japan's military leaders on specific issues of policies, they were never against continental expansion per se. Matsumae, for example, opposed Tōjō's decision to continue the war with the United States because Japan lacked comparable productive capabilities, but he had no objection to working closely with the military to extend Japan's strategic telecommunications network on the Asian continent.

Though certainly far from omnipotent, Japan's techno-imperialism was effective. If integration is understood in the minimalist sense of holding different parts of the empire together, then Japan's imperial telecommunications network did almost too well, for all its problems. In facilitating movements of people, news, and social interactions within a rapidly expanding empire, the communications network almost certainly made a multifaceted contribution to fostering imperial consciousness. Ironically, the success of techno-imperialism in the 1930s also sowed seeds of its own destruction by leading Japanese leaders to overextend themselves. When the war finally came to an end in August 1945, Japan's vital shipping lanes were almost entirely cut off, whereas many of its key communications circuits remained in working order, if only to coordinate the broadcast of the emperor's speech and to send orders of surrender throughout the imperium.

The legacies of Japan' techno-imperialism are mixed. At home, operational experience in the empire was valuable to postwar Japan. It did not, however, produce significant levels of technological transfer to the colonial subjects. The Koreans, Taiwanese, Chinese, and Southeast Asians occupied the lower echelon of Japan's vast wartime telecommunications operations. Perhaps with the exception of a few well-placed Chinese, technology—both engineering and administrative skills—remained largely in Japanese hands. This was intentional since Japan was to assume the leadership role in technology; yet it was also unforeseen due to the relatively short lifespan of the empire. Once

Japan's empire disintegrated, these Asians soon found new sources of technologies, from the Americans, Soviets, or the British. It would not be until decades later that Japan would return, this time as truly a technological giant in the world.

COMMUNICATIONS IMPERIALISM IN PERSPECTIVE

Despite the wartime Japanese rhetoric distinguishing Japan's project of building a "co-prosperity sphere" in Asia from Western colonialism, Japan's experience with communication imperialism was far from unique.[4] In fact, a study of telecommunications as Japan's technology of empire helps place the Japanese experience within the annals of communications imperialism.[5]

To begin with, there were many similarities between Japan's communications imperialism and that of Great Britain, despite the vast differences in the histories, composition, and sizes of their empires. In terms of institutions, Britain's telecommunications hegemony was accomplished for the most part through private business endeavors, although government subsidies also played a huge role. When shortwave began to pose a serious challenge to the existing cable network, the British government was galvanized into combining cable and wireless operations to form a mammoth communications enterprise, Imperial Cables and Wireless, in 1928. The amalgamation of Cables and Wireless provided a model for Japan, where state dominance had been strong since the dawn of telecommunications in the Meiji era. In response to external challenges in the area of telecommunications in the 1920s, the Japanese bureaucracy was able to modify the strict state monopoly and create close state-business collaborations without giving up total control. To better facilitate overseas expansion, Japan resorted to a hybrid private-public form of telecommunications construction and management, beginning

4. Gann, "Western and Japanese Colonialism."

5. In a seminal 1971 essay entitled "The Structure of Imperialism," Johan Galtung first coined "communication imperialism" as a category distinct from economic, political, military, and cultural imperialism. The concept of "communications imperialism" used here is broader than Galtung's, which focused on the exchange of information, and is closer to "railway imperialism," as conceived in Davis and Wilburn, *Railway Imperialism*.

with wireless and culminating in wartime telecommunications national policy companies in occupied areas in Asia. In the end, Japan's experience confirmed Daniel Headrick's observation that "private and government ownership of international telecommunications were never mutually exclusive alternatives but points along a spectrum of increasing government involvement."[6]

There are also parallels in the internal tensions of their respective imperial networks. Japan's military as well as civilian bureaucrats in Tokyo and in the colonies fought bitter battles over control of the expanding communications network. The alienation between Britain and Canada over the Pacific cable in the 1920s bears some resemblance to the friction between Japan and its colonial government in Korea over Japan's new cable project in the late 1930s. The dominions of Australia and New Zealand similarly voiced objections even as they signed Britain's Merger Agreement of 1928, transferring the Pacific cable to a private company over which they had little control.[7] When the dominions attained autonomy in 1931, their governments sought to establish their own direct wireless communications with other members of the Commonwealth, just as the Japanese-controlled telecommunications enterprise in North China sought to do a decade later. It has been suggested that even though the submarine cable did strengthen the British empire through its contribution to defense and commerce, it did less to draw the empire together politically than was thought at the time.[8] The same was true with Japan.

Japan's expansion in the 1930s has often been compared—not the least by the Japanese themselves during the war—to that of Germany, its wartime ally. There are certainly parallels in their communications strategy. As a resurgent Germany under Hitler prepared for war, it built up its domestic telecommunications system. Modernization of the infrastructure proceeded as in Japan, and coaxial cables were used to connect major urban centers—Berlin, Nuremberg, Munich, Hamburg,

6. Daniel R. Headrick, "Public-Private Relations in International Telecommunications Before World War II," in Bella Mody, Johannes M. Bauer, and Joseph D. Straubhaar, eds., *Telecommunications Politics: Ownership and Control of the Informational Highway in Developing Countries* (Mahwah, N.J.: Lawrence Erlbaum Associates, 1995), 31.

7. Barty-King, *Girdle Round the Earth*, 211–12, 223.

8. Robert Boyce, "Submarine Cables as a Factor in Britain's Ascendancy as a World Power, 1850–1914," 99; idem, "Canada and the Pacific Cable Controversy, 1923–1928."

Frankfurt, Vienna, and Hanover. After German forces subjugated the majority of European countries during the war, a "telecommunications geopolitik" emerged step by step. In October 1942, at its apex of domination, Germany organized a convention in Vienna of the European Postal and Telecommunications Union that integrated the respective administrations of the German-dominated countries. To smooth organizational coordination and control, Germany posted special telecommunications attachés at its embassies.[9] There were some differences, however. In contrast to Germany, where the telephone system was transferred from the civil bureaucracy to the exclusive control of the Nazi party, Japanese telecommunications did not undergo such a drastic reorganization at home. A greater degree of continuity prevailed.

What, then, distinguished Japan's quest for telecommunications hegemony in Asia? Japanese imperialism differed from Western imperialism in both space and time. One obvious difference is the fact that Japan's empire was closer to home than were the colonial possessions of most European powers, with the exception of Germany's expansion during World War II. This geographic congruity grew out of concerns for security, but quickly served as justification for closer economic integration. It also made an East Asian telecommunications network based on costly land cables more feasible. Moreover, the racial and cultural affinity Japan had with the peoples in its empire in Northeast Asia often led to a greater emphasis on cultural integration, facilitated by the use of modern communications technology such as facsimile. Such affinity, however, did not produce a greater degree of technology transfer. If technology transfer to the local population under European colonialism was, as Raymond Betts puts it, a "surface affair," Japan's communication imperialism was not fundamentally different.[10]

Another key difference that separates Japanese imperialism from most Western counterparts is its timing: Japan's emergence as a modern industrializing country and an imperial power occurred considerably later than that of Britain and France. Japan built its nascent empire after the advent of intercontinental telecommunications. Perhaps more importantly, Japan acquired nearly all its modern technologies after they

9. Frank Thomas, "The Politics of Growth: The German Telephone System," in Mayntz and Hughes, eds., *The Development of Large Technical Systems*, 179–214.
10. Betts, *Uncertain Dimensions*, 91.

were first developed in the West.[11] As the only non-Western member of the imperialist club, Japan's engineers and officials became increasingly self-conscious and uneasy about its technological dependency on the West. Ultimately, this ambiguity threatened to undermine Japan's claim to a revitalized Asia under its leadership. A distinctive character of Japan's communications imperialism, then, was the sensitivity—at times obsession—on the part of many Japanese to this ideological dimension of technology in its imperial expansion. No Western imperialist power was quite as obsessed with catching up and defining a cultural identity of technology the way modern Japan was. It is here that Japan's communications imperialism comes close to being unique.

In the 60 years since the majorities of colonial empires disintegrated, the rapid development of information technologies (IT)—such as satellite communication and integration of computing and communicating, as in the Internet—has drastically altered the landscape of international communications. Visionaries of the "global village" have predicted that the new and more powerful communications technology will obliterate all barriers of time and space, bringing freedom for all.[12] In reality, the widely shared belief that new information technologies would vastly enhance productivity has already contributed to a dot.com boom and bust in the United States. A communications network based on an open architectural design could very well offer new potential to empower the individual,[13] but ultimately, even the new telecommunications technology does not simply "liberate," as global village prophets would like us to believe. Critical communications scholars have called attention to the "digital divide" and focused on news communication to demonstrate the unequal relations between the advanced core and underdeveloped periphery.[14] It is obvious that today international communications capacities are far from being evenly distributed, and powerful governments and

11. For a recent work that examines this issue from the perspective of "cultural materiality," see Wittner, *Technology and the Culture of Progress in Meiji Japan.*

12. See, e.g., Ithiel de Sola Pool, *Technologies of Freedom*; and idem, *Technologies Without Boundaries.*

13. Janet Abbate, *Invention of the Internet* (Cambridge, Mass.: MIT Press, 1999). Or, as Niall Ferguson puts it, "the technological possibilities of the late nineteenth and early twentieth centuries favored imperialists over the individual troublemaker" (Ferguson, *Empire*, viii).

14. See, e.g., Alleyne, *International Power and International Communications*; and Sussman, *Communication, Technology, and Politics in the Information Age.*

business conglomerates are striving to enhance their control capabilities to gain strategic or economic advantages.

Interestingly, disillusionment with the failed promise of a new world order after the end of the Cold War has produced a re-assessment of bygone empires as well as the lessons for global power.[15]In this context, the Japanese experience with imperial telecommunications provides a timely reminder about the exercise of a crucial dimension of structural power: that an expanding communications system is always fraught with increasing tension from the inside. Japan's quest for telecommunications hegemony in Asia was full of conflicts and contradictions: although the telecommunications network was supposed to foster integration and unity within the imperium, telecommunications policies often turned out to heighten rivalries among different Japanese bureaucracies and exacerbate tensions between the Japanese and non-Japanese. This book thus confirms that control is always contested, not just between the controlling and the controlled, but among all those who seek to wield that control. Ultimately, Japan's experience with telecommunications as its technology of empire confirms "the paradoxical fact," namely, communications technologies simultaneously bring enormous enhancements of control to governments, corporations, consumers, and voters, and a quite new order of chaos and uncontrollability—which brings, in turn, a sense that control is unachievable.[16] Technology can produce unintended consequences, and the technology of empire was no exception.

---

15. See, e.g., Niall Ferguson, *Empire: The Rise and Demise of the British World Order and the Lessons for Global Power* (New York: Basic Books, 2003); and Maier, *Among Empires.* From a critical perspective, Michael Hardt and Antonio Negri use "empire" to describe the "sovereign power that governs the world"; see Michael Hardt and Antonio Negri, *Empire* (Cambridge, Mass.: Harvard University Press, 2000).

16. Mulgan, *Communication and Control*, 2–7.

Reference Matter

Bibliography

ARCHIVAL SOURCES

Communications Museum, Tokyo
Japanese Ministry of Communications Records, 1st Series (MOC Records I)
Diplomatic Record Office, Tokyo
Japanese Ministry of Foreign Affairs Archives (JMFA Archives)
Japan Center for Asian Historical Materials (JACAR), Tokyo
Japan International Telegraph and Telephone Company (KDD), Tokyo
Japanese Ministry of Communications Records, 2nd Series (MOC Records II)
Military History Division Library, Institute of Defense Research, Tokyo
Imperial Japanese Army and Navy Archives, 1868–1945
National Archives and Record Administration, College Park, Md. (National Archives)
RG 331 Supreme Command of the Allied Powers (SCAP)
National Public Record Office, Tokyo
Japanese Cabinet Records, 1868–1945
NHK Broadcasting Culture Research Institute, Tokyo
Japan Broadcasting Corporation (NHK) Records
Second Historical Archives of China, Nanjing
Chinese Ministry of Communications Records, 1911–49 (Chinese MOC Records)
North China Telegraph and Telephone Company Records (NCTT Records)
Shanghai Telecommunications Administrations Archives
Central China Telecommunications Company Records (CCTC Records)
Great Northern Telegraph Company Records (GNTC Records)
Shōwa zaiseishi shiryō microfilm
Japanese Ministry of Finance Records, 1926–45

PERIODICALS

The Bell System Technical Journal. Short Hills, N.J.
Chōsen no teishin jigyō. Keijō.
Chōsen sōtokufu teishin nenpō. Keijō.
Chōsen sōtokufu teishin tōkei yōran. Keijō.
Chōsen teishin ichiran. Keijō.
Chōsen teishin kyōkai zasshi. Keijō.
Daitōa denki tsūshin jimukyoku hō. Tokyo.
Denden. Shinkyō.
Denki tsūshin (*DT*). Tokyo.
Denki tsūshin gyōmu kenkyū. Tokyo.
Denmu kenkyō shiryō. Tokyo.
Denmu nenkan. Tokyo.
Denpa to juken (*DJ*). Tokyo.
Denshin kyōkai zasshi (*DKZ*). Tokyo.
Denwa denshin gakkai zasshi; later renamed *Denki tsūshin gakkai zasshi* (*DTGZ*). Tokyo.
[NCTT] *Eigyō geppō*. Beijing.
Far Eastern Review (*FER*). Manila.
[CCTC] *Gongsibao*. Shanghai.
[MTT] *Gyōmu shiryō*. Shinkyō.
Hokuden. Beijing.
Jiaotong zazhi. Nanjing.
Kachū denki tsūshin (*KDT*). Shanghai.
Kagakushugi kōgyō. Tokyo.
Karafutochō teishin jigyō tōkei yōran. Toyohara.
[NCTT] *Kenkyū zasshi*. Beijing.
[MTT] *Manshū denden tōkei geppō*. Shinkyō.
[MTT] *Manshū denden tōkei nenpō*. Shinkyō.
Manshū hōsō nenkan. Shinkyō.
Taiwan sōtofuku teishin tōkei yōran. Taihoku.
Taiwan teishin. Taihoku.
Takumu nenkan. Tokyo.
Teikoku gikai kizokuin gijiroku. Tokyo.
Teishin jimu gaigyō. Tokyo.
Teishin kyōkai zasshi (*TKZ*). Tokyo.
Teishin no chishiki. Tokyo.
[AT&T] *Telephone & Telegraph Statistics of the World*. New York.
Tōa denki tsūshin zasshi (*TDTZ*). Tokyo.
Tsūshin kōgyōkai hō. Tokyo.

Tsūshin tōkei yōran. Tokyo.
Youdian wenshi tongxun. Beijing.
Youdian wenshi ziliao. Beijing.

PUBLICATIONS IN JAPANESE, CHINESE, AND KOREAN

Amazawa Fujirō, ed. *Gendai Nihon sangyō hattatsu shi*, vol. 22, *rikuun, tsūshin.* Tokyo: Kōjunsha shuppankyoku, 1965.

Amishima Tsuyoshi. *Hatō: denpa to tomoni 50-nen.* Tokyo: Denki tsūshin shinkōkai, 1992.

Bōeichō. Bōei kenkyūjo. Senshibu, comp. *Shiryōshū Nanpō no gunsei.* Tokyo: Asagumo shinbunsha, 1985.

Chiba Masashi. *Kindai kōtsū taikei to shin teikoku no henbō: denshin tetsudō nettowāku no keisei to Chūgoku kokka tōgō no hen'yō.* Tokyo: Nihon keizai Hyōronsha, 2006.

Chōsen chūsatsugun shireibu. *Chōsen bōto tobatsu shi.* Keijō, 1913.

Chōsen gyōsei henshūkyoku. *Chōsen tōchi hiwa.* Keijō, 1937.

Chōsen shiryō kenkyūkai. *Chōsen sōtofuku kansei to sono gyōsei kikō.* Tokyo, 1962.

Chōsen sōtokufu. *Chōsen sangyō keizai chōsakai kaigiroku.* Keijō, 1936.

———. *Chōsen sōran.* Keijō, various years.

———. *Chōsen sōtokufu jikyoku taisaku chōsakai shimon tōshinsho.* Keijō, 1938.

———. *Chōsen sōtokufu tokei nenpō, 1930.* Keijō, 1931.

———. Teishinkyoku. *Chōsen tsūshin jigyō enkakushi.* Keijō, 1938.

———. ———. *Chōsen tsūshin jigyō enkaku shoshi.* Keijō, 1914.

———. ———. *Teishin shūi.* Keijō: Chōsen teishin kyōkai, 1936.

Den Kenjirō denki kankōkai. *Den Kenjirō denki.* Tokyo, 1932.

Denki tsūshin kōgakkō. *Denki tsūshin gairon.* Tokyo, 1944.

Denki tsūshin kyōkai. *Shadan hōjin Denki tsūshin kyōkai 20-nen shi.* Tokyo: Denki tsūshin kyōkai, 1958.

Denshin kyōkai. *Denshin kyōkai 50-nen shi.* Tokyo, 1944.

Dentsū tsūshin shi kankōkai. *Dentsū tsūshin shi.* Tokyo, 1976.

Duban Lu'an shanhou shiyi gongshu bianjichu, comp. *Lu'an shanhou yuebao tekan: youdian.* Beijing, 1923.

Fan Zhongyun, comp. *Shinian lai de Zhongguo.* Shanghai: Shangwu yinshuguan, 1937.

Fujii Nobuyuki. *Terekomu no keizai shi.* Tokyo: Ochanomizu shobō, 1997.

Fujikura densen kabushiki kaisha. *88-nen no ayumi.* Tokyo, 1973.

Fujimura Junichirō. "Jōhō dentatsusha hikyaku no katsudō." In Maruyama Yasunari, ed., *Jōhō to kōtsū.* Tokyo: Chūō kōronsha, 1992.

Fujiwara Kamaashi. *Pekin nijūnen.* Tokyo: Heibonsha, 1959.

Furukawa Takahisa. *Shōwa senchūki no sōgō kokusaku kikan.* Tokyo: Yoshikawa kōbunkan, 1992.

Gaichi denki tsūshin shi hensan iinkai, comp. *Gaichi kaigai denki tsūshin shiryō* (*GKDTS*). 13 vols. Tokyo: Denki tsūshin kyōkai, 1956.

———. *Gaichi kaigai denki tsūshin shisetsu ni kansuru shūsenji shisetsu genjō, shisan hyōka oyobi hikitsutsuki jōkyō gaisetsu.* Tokyo: Denki tsūshin kyōkai, 1956.

Gaimushō. *Nihon gaikō nenpyō narabi shuyō bunsho* (*NGNB*). Tokyo: Hara shobō, 1965–66.

Gaimushō gaikō shiryōkan. *Nihon gaikō bunsho* (*NGB*), various years.

Gaimushō hyakunenshi hensan iinkai. *Gaimushō no hyakunen.* Tokyo: Hara shobō, 1969.

Gijutsu kaihatsu no Shōwa shi. Tokyo: Asahi shimbunsha, 1990.

Hanakada Makoto. *Shōwa 20-nen 8-gatsu.* Tokyo: Shinkō shuppansha, 1973.

Hanaoka Kaoru. *Gaikoku denshin ni tsuite.* Tokyo: Furitsu Tōkyō shōkō shōreikan, 1941.

———. *Kaitei densen to Taiheiyō no hyakunen.* Tokyo: Nittō shuppansha, 1968.

———. *Kokusai denshin jigyō ron.* Tokyo: Kōtsū keizaisha, 1936.

———. *Senzen sengo no memochō.* Tokyo: Shūeisha, 1964.

Harada Katsumasa. "Tetsudō shi kenkyū ni okeru seijishiteki shiten." *Tetsudō shigaku* 12 (1993): 37–41.

Hatano Sumio. *Taiheiyō sensō to Nihon gaikō.* Tokyo: Tōkyō daigaku shuppankai, 1995.

Hikita Yasuyuki. *Nanpō kyōeiken: senji Nihon no Tōnan Ajia keizai shihai.* Tokyo: Taga shuppan, 1995.

———. "Nihon no tai-Chūgoku denki tsūshin jigyō toshi ni tsuite: Manshū jihen ki o chūshin ni." *Rikkyō keizaigaku kenkyū* 41.4 (1988): 1–55.

———. "Nihon no tai-Chūgoku denki tsūshin jigyō toshi ni tsuite: shakkan o chūshin ni." In Sakai Takahito kyōju kanreki kinenkai, ed., *Nihon kindaika no shisō to tenkai*, pp. 347–75. Tokyo: Bunken shuppan, 1988.

Hiraoka Yoichirō. *Rikugun tsūshin no gaikan.* Tokyo, 1958.

Hokudenkai, comp. *Kahoku denden jigyō shi.* Tokyo: Denki tsūshin kyōkai, 1975.

Ikei Masaru. "1930 nendai no masumedia." In Miwa Kimitada, ed., *Saikō Taiheiyō sensō zen'ya*, pp. 141–94. Tokyo: Sōseiki, 1981.

Imaida Kiyonori denki hensankai. *Imaida Kiyonori.* Tokyo: Toppan insatsu, 1943.

Isawa Minoru. "Gijutsusha no ido." In Yamamoto Taketoshi et al., eds., *Iwanami kōza: 'Teikoku' Nihon no gakuchi*, vol. 5, *Teikoku no keizaigaku*, pp. 217–57. Tokyo: Iwanami shoten, 2006.

———. "Senjiki Nihon teikoku ni okeru gijutsusha kyōkyu." In Nakamura Tetsu, ed., *Kindai Higashi Ajia keizei no shiteki tenkai*, pp. 325–58. Tokyo: Nihon hyōronsha, 2007.

Ishihara Fujio. *Kokusai tsūshin no Nihon shi: shokuminchika kaishō e kutō no 99-nen.* Tokyo: Tōkai daigaku shuppankai, 1999.

Ishii Kanji. *Jōhōka to kokka, kigyō.* Tokyo: Yamakawa shuppan, 2002.

———. *Jōhō tsūshin no shakai shi.* Tokyo: Yūhikaku, 1994.

———. "Kindai Nihon no tsūshin shi." *Hanashi* 486 (July 1995).

Ishikawa Junkichi, comp. *Kokka sōdōin shi.* 13 vols. Tokyo: Kokka sōdōin shi kankōkai, 1975–87.

Iwanaga Yūkichi-kun denki hensan iinkai, comp. *Iwanaga Yūkichi kun.* Tokyo, 1941.

Jiaotong shi bianzhuan weiyuanhui. *Jiaotong shi dianzheng bian* (*JSDB*). 2 vols. Nanjing, 1936.

Kafu kaigi teikoku zenken jimusho. *Zai-Shi gaikoku musen denshin mondai.* Tokyo, 1921.

Kahoku denden kurabu. *Tairiku ni okeru tsūshin seisaku o ronzu.* Beijing, 1940.

Kahoku denshin denwa kabushiki kaisha. *Sōgyō 1-shūnen kaikō.* Beijing, 1940.

———. *Sōgyō 5-shūnen.* Beijing, 1943.

Kahoku kōtsu kabushiki kaisha. *Kahoku no tsūshin jigyō.* Beijing, 1942.

Kajii Takeshi. *Waga hansei.* Tokyo: Toppan insatsu, 1968.

Kajii Takeshi tsuitō jigyō iinkai. *Kajii Takeshi ikōshū.* Tokyo: Denki tsūshin kyōkai, 1979.

———. *Kajii Takeshi tsūisōroku.* Tokyo: Denki tsūshin kyōkai, 1977.

Kang Ung. "19-seiki matsu Chōsen no denki tsūshin gijutsu." *Kagakushi kenkyū* II-1991 (30): 161–69.

Kangoku keibi denwa. Kensetsubu. *Kangoku keibi denwa Kensetsubu jigyō gaiyō hōkoku.* Seoul, 1911.

Kantōgun. *Kantōgun tokushu tsūshinbu gyōmu yōhō.* Dalian, 1932.

Karafuto teiyūkai. *Tsūioku no Karafuto teishin.* Sapporo: Karafuto teiyūkai, 1988.

Katayama Kunio. *Kindai Nihon kaiun to Ajia.* Tokyo: Ochanomizu shobō, 1996.

Katō Hidetoshi. *Bunka to komyunikeshon.* Tokyo: Shisakusha, 1977.

Kawahara Hiroshi. *Shōwa seiji shisō kenkyū.* Tokyo: Waseda daigaku shuppanbu, 1979.

Kawanobe Tomiji. "Eikoku koshi to no jizen torikime ni yoru denshinki joyaku: Daihoku denshin kaisha keburu rikuage no mondai." *Kōtsūshi kenkyū* 30 (June 1993): 43–65.

———. *Teregurafu komonjo kō.* Chiba: private edition, 1987.

Kawashima Shin. "'Teikoku' to rajio: Manshūkoku ni okite 'seiji o seikatsu suru' koto." In Yamamoto Taketoshi et al., eds., *Iwanami kōza: 'Teikoku' Nihon no gakuchi*, vol. 4, *Mediya naka no "teikoku,"* pp. 207–42. Tokyo: Iwanami shoten, 2006.

KDD. *Kokusai denki tsūshin hattatsu shi (Nihon hen).* Tokyo: KDD, 1984.

Kikakuin. *Kōtsū denryoku dōin jisshi keikaku kōryō sankō.* Tokyo, 1939.

Kikakuin Daitōa kensetsu shingikai, comp. *Daitōa kensetsu shingikai kankei shiryō: sōkai, bukai, sokkiroku* (*DKSKS*). 6 vols. Reprinted—Tokyo: Ryūkei shosha, 1995.

Kikakuin kenkyūkai. *Kokubō kokka no kōryō*. Tokyo: Shinkigensha, 1941.

Kishi Toshihiko. "Nagasaki Shanhai kan 'Teikokusen' o meguru takokukan kōshō to kigyō tokkyoken no igi." *Kokusai seiji* 146 (November 2006): 21–38.

———. "Nitchū tsūshin mondai no ichidanmen." *Tōyō gakuhō* 83 (2001): 431–55.

———. "Senjika ni okeru tai-Ka denki tsūshin sisutemu no tenkai." *Hokutō Ajia kenkyū* 1 (March 2001): 245–68.

———. "Tianjin sokai denwa mondai o meguru chiiki to kokka kan rigai." In Soda Saburō, ed., *Kindai Chūgoku to Nihon: teikei to tekitai no hanseiki*, pp. 259–90. Tokyo: Ochanomizu shobō, 2001.

———. "Tsūshin tokkyo to kokusai kankei: zai-Ka musen ken'eki o meguru takokukan funsō." In Kishi Toshihiko et al., eds., *Mosakusuru kindai Nitchū kankei: Taiwa to kyōzon no jidai*, pp. 229–48. Tokyo: Tōkyō daigaku shuppankai, 2009.

Kitayama Setsurō. *Rajio Tōkyō*. 3 vols. Tokyo: Tabata shoten, 1990.

Kōain gijutsubu. *Chūshi ni okeru hōsō musen denwa no gaikyō*. Shanghai, 1939.

———. *Shanhai denwa kaisha no gaikyō*. Shanghai, 1939.

Kōain Kachū renrakubu. *Chūshi ni okeru denpō kōryū jōkyō, denwa tsūwa jōkyō chōsa setsumei shiryō*. Shanghai, 1939.

———. *Jihengo ni okeru kyū Kōtsubu densei kikan sesshū keii narabini misesshū bubun no sesshū hōsaku ni kansuru chōsa kenkyū*. Shanghai, 1939.

———. *Tsūshin shisetsu ni kansuru chōsa*. Shanghai, 1942.

Kōain seimubu, comp. *Dai-75-kai Teikoku gikai Shina kankei shitsugi otoshu*. Tokyo, 1940.

Kobayashi Hideo. *Daitōa kyōeiken*. Tokyo: Iwanami shoten, 1988.

———. *Daitōa kyōeiken no keisei to hōkai*. Tokyo: Ochanomizu shobō, 1973.

Kobayashi Kōji. *C & C modan komyunikēshon*. Tokyo: Simul Press, 1985.

Kokaze Hidemasa. *Teikokushugika no Nihon kaiun*. Tokyo: Yamakawa shuppan, 1995.

Kokusai denki tsūshin kabushiki kaisha shi hensan iinkai. *Kokusai denki tsūshin kabushiki kaisha shi* (*KDTKKS*). Tokyo, 1949.

Kokusai denshin denwa kabushiki kaisha shiryō sentā. *Aoi umi o hashiru*. Tokyo, 1969.

Kokusai denwa kabushiki kaisha. *Kokusai denwa kabushiki kaisha jigyō shi*. Tokyo, 1938.

Kokusai jōhō denshin chōsa iinkai. *Jōhō kishu ni kansuru tsūshin keiro*. Tokyo, 1924.

Kokusaku kenkyūkai. *Daitōa kyōeiken keizai kōryū taisei ron*. Tokyo: Nihon hyōronsha, 1944.

Kosaki Masatomi. *Denwa jigyō no un'ei to sono riyō*. Tokyo: Kōtsū keizaisha, 1940.

———. *Denwa riyō ron*. Tokyo: Kōtsū keizaisha, 1936.

Koshizawa Akira. *Manshūkoku shutō keikaku*. Tokyo: Nihon keizai hyōronsha, 1988.

———. *Zhongguo dongbei chengshi jihua shi.* Trans. Huang Shimeng. Taibei: Dajia chubanshe, 1987.

Kubomura Masaji. *Dai-11-gun tsūshintai.* Tokyo: Tosho shuppansha, 1987.

Kuroiwa Toshio. *Gendai gijutsu shi ron.* Tokyo: Tōyō keizai shinpōsha, 1987.

Liang Bo. *Jishu yu diguozhuyi yanjiu: Riben zai Zhongguo de zhimin keyan jigou.* Jinan: Shandong jiaoyu chubanshe, 2006.

Mainichi shinbun shashi hensankai. *Mainichi shinbun 70-nen shi.* Osaka, 1952.

Makino Ryōzō. *Tokubetsu kaikei to natta tsūshin jigyō.* Tokyo: Shakai kyōiku kyōkai, 1934.

Manshū denden tsuiokuroku kankōkai. *Akai sekiyō.* Tokyo, 1967.

Manshū denshin denwa kabushiki kaisha. *Denden.* Shinkyō, 1940.

———. *Denden tokuhon.* Shinkyō, 1939.

———. *Denpō naiyō bunseki tōkei.* Shinkyō, 1940.

———. *Denpō toriatsukau naiyō bunseki tōkei hyō.* Shinkyō, 1940.

———. *Manshū denden 5-shūnen.* Shinkyō, 1939.

———. *Manshū denshin denwa kabushiki kaisha 10-nen shi.* Shinkyō, 1943.

———. *Shain kentōroku.* Shinkyō, 1940.

Manshūkoku shi hensan kankōkai. *Manshūkoku shi: kakuron.* Tokyo: Man-Mō dōhō engokai, 1970.

Manshū teikoku yūsei sōkyoku. *Manshū teikoku yūsei jigyō gaiyō.* Shinkyō, 1942.

Manshū teishin kyōkai. *Kantō teishin 30-nen shi.* Dalian, 1936.

Mantetsu. Harbin jimusho. *Manshū no densei.* 2 vols. Dalian, 1930.

———. Keizai chōsakai. *Ritsuan: Manshū tsūshin jigyō hōsaku.* Dalian, 1936.

Matsumae Shigeyoshi. *Gijutsu no shintaisei.* Tokyo: Taisei yokusai kai, 1941.

———. *Nihon gijutsu ron.* Tokyo: Dai Nippon ōbenkai ōdansha, 1942.

———. *Senji seisan ron.* Tokyo: Ōbunsha, 1943.

Matsumae Shigeyoshi and Nishizaki Tarō. *Denki tsūshin gairon.* Tokyo: Korona sha, 1938.

Matsumae Shigeyoshi ronbunshū kankōkai. *Hatsumei e no chōsen: Matsumae Shigeyoshi ronbunshū.* Tokyo: Tōkai daigaku shuppankai, 1983.

Matsuo Hiroshi. *Denshi rikkoku Nihon o sotateta otoko.* Tokyo: Bungei shunju, 1992.

Matsuo Shosaburō. *Nihonkai chūshin ron.* Tokyo, 1942.

Matsuoka Yuzuru. *Nakayama Ryūji.* Tokomachi: Nakayama Ryūji sensei kenshōkai, 1958.

Mei Shaozu and Song Ganggang, eds. *Bainian dianxin zhu huihuang: Shanghai shi Changtu dianxinju jushi (1881–1997).* Beijing: Zhongguo jihua chubanshe, 1998.

Mitarai Tatsuo, ed. *Minami Jirō.* Tokyo, 1957.

Mogami Seizō. *Man-Shi tsūshin tokuhon.* Tokyo: Kōtsū kenkyūsha, 1940.

Murakami Motoyuki. *Ichi gijutsusha no shōgai.* Tokyo, 1966.

Musoka hōshiki kankōkai. *Gijutsu kaihatsu e no michi: Musoka hōshiki no kaihatsu to shido seishin.* Compiled under the supervision of Matsumae Shigeyoshi. Tokyo: Denki tsūshin kyōkai, 1977.

Mutō Tomio. *Watashi to Manshūkoku*. Tokyo: Bungei shunjū, 1988.

Nagashima Yōichi. "Taihoku denshin kaisha no Nihon shinshutsu to sono haikei." *Nihon rekishi* 367 (October 1995): 77–92.

Naikai Asajirō. *Tsūshin tokubetsu kaikei no umareru made*. Tokyo: Kōtsū keizaisha, 1933.

Nakayama Ryūji. *Nakayama gishi shokan: Shina jijō*. Tokyo, 1916.

———. *Nakayama Ryūji gishi shokanshū*. Tokyo, 1918.

———. *Sensō to denki tsūshin*. Tokyo: Denki tsūshin kyōkai, 1943.

———. *Tatakau denpa*. Tokyo: Kagaku shinkōsha, 1943.

———. *Zhongguo dianzheng yijianshu*. Beijing: Jiaotongbu, 1919.

Namikawa Ryō. *Sunaka naki sensō: sekai denpa sen*. Tokyo: Sankaidō shuppanbu, 1943.

Nichi-Man-Shi kōtsu kondankai. Dai-ichi chōsakai. Jimukyoku, comp. *Tōa kyōei-ken nai kōtsuryō tōkei (1936–1940)*. Tokyo, 1942.

Nihon hōsō kyōkai. *Hōsō 50-nen shi: shiryō hen*. Tokyo: NHK, 1977.

Nihon kagakushi gakkai, comp. *Nihon kagaku gijutsu shi daikei (19): denki tsūshin*. Tokyo: Daiichi hōki shuppan, 1969.

Nihon keieishi kenkyūjo. *Nihon denki kabushiki kaisha 70-nen shi*. Tokyo, 1972.

———. *Oki denki 100-nen no ayumi*. Tokyo, 1981.

———. [Furukawa denki kōgyō.] *Sōgyō 100-nen*. Tokyo, 1991.

Nihon kokusai denshin denwa kabushiki kaisha. *Kokusai denki tsūshin nenbyō shiryō*. Tokyo, 1976.

Nihon musen shi hensan iinkai. *Nihon musen shi*. Tokyo: Denpa kanri iinkai, 1951.

Nihon Taiyō kaitei densen kabushiki kaisha. *Nihon Taiyō kaitei densen kabushiki kaisha shi*. Tokyo, 1970.

Nippon denshin denwa kōsha. *Denpō ryōkin no enkaku*. Tokyo, 1955.

———. *Denshin denwa jigyō shi*. 9 vols. Tokyo: Denki tsūshin kyōkai, 1959.

———. *Jido denwa kōkan 25-nen shi*. 3 vols. Tokyo, 1953.

———. Gijutsushitsu. *Denki tsūshin jishu gijutsu kaihatsu shi*. 4 vols. Tokyo, 1972–79.

———. Kaiteisen shisetsu jimusho. *Kaiteisen hyakunen no ayumi*. Tokyo: Denki tsūshin kyōkai, 1971.

Nishibayashi Tadatoshi. *Nihonjin to terefon*. Tokyo: NTT shuppan, 1990.

———. *Tsūshin to keizai*. Tokyo: Keibunkan, 1957.

Nishimura Shigeo. *Chō Gakuryō: Nitchū no haken to Manshū*. Tokyo: Iwanami shoten, 1996.

Noda keizai kenkyūjo. *Senjika no kokusaku kaisha*. Tokyo, 1940.

Ōe Shinobu et al., eds. *Kindai Nihon to shokuminchi*. 8 vols. Tokyo: Iwanami shoten, 1992–94.

Ōhashi Hachirō denki hensan iinkai. *Ōhashi Hachirō*. Tokyo: Ko Ōhashi Hachirō kinen jigyō iinkai, 1970.

Oka Tadao. *Eikoku o chūshin ni mitaru denki tsūshin hattatsu shi.* Tokyo: Tsūshin chōsakai, 1941.

———. *Taiheiyō iki ni okeru denki tsūshin no kokusaiteki bekken.* Tokyo: Tsūshin chōsakai, 1941.

Okamoto Makiko. *Shokuminchi kanryō no seiji shi.* Tokyo: Sangensha, 2008.

Okita Saburō. *Nihon kanryō jijō.* Tokyo: TBS Buritanika, 1984.

Okumura Katsuko, ed. *Tsuioku Okumura Kiwao.* Tokyo, 1970.

Okumura Kiwao. *Teishin ronsō.* Tokyo: Kōtsū kenkyūsha, 1935.

Ōkurashō. *Nihonjin no kaigai katsudō ni kansuru rekishiteki chōsa.* 21 vols. Tokyo: Ōkurasho kanrikyoku, 1949–50.

Ōtake Eiji. *Hito to nami to: hōsō seikatsu 50-nen no kaikō.* Akita, 1976.

Ōyodo Shōichi. *Gijutsu kanryō no seiji sanka.* Tokyo: Chūkō shinsho, 1995.

———. *Miyamoto Takenosuke to kagaku gijutsu gyōsei.* Tokyo: Tōkai daigaku shuppankai, 1989.

———. *Teishinshō kōmukyoku ni kansuru kagaku gijutsu shiteki kenkyū.* Report to Ministry of Education Research Grant, March 1997.

Qingdao Nihon shōkō kaigisho. *Santōshō nai chōto denwa shoyo gaikoku riyo jogyō.* Qingdao, 1934.

Rikugunsho. *Meiji 37-8-nen senyaku Manshū gunsei shi* (2). Reprinted—ed. Kobayashi Hideo. Tokyo: Yumani shobō, 1999.

Sakai Saburō. *Shōwa kenkyūkai: aru chishikijin shūdan no kiseki.* Tokyo: TBS Buritanika, 1979.

Samejima Sunao. *Moto Gunreibu tsūshin kachō no kaisō.* Tokyo, 1981.

Shibusawa Sei'en kinen zaidan ryūmonsha. *Shibusawa Ei'ichi denki shiryō,* vols. 9, 18, 52. Tokyo, 1963.

Shidehara Kijūrō, *Gaikō gojūnen.* Tokyo: Chūō bunko, 1987.

Shinohara Noboru. *Hitori no kokoro.* Tokyo, 1970.

Shōwa kenkyūkai. *Tōa burroku keizai ni kansuru kenkyū.* Tokyo: Seikatsusha, 1939.

Sugiyama Shin'ya. "Jōhō kakumei." In Nishikawa Shunsaku and Yamamoto Yūzō, eds., *Nihon keizaishi,* vol. 5. Tokyo: Iwanami shoten, 1990.

———. "Jōhō no keizaishi." In Shakai keizai shi gakkai, ed., *Shakai keizai shigaku no katai to tenbō*, pp. 268–76. Tokyo: Yūhikaku, 1992.

———. "Meiji zenki ni okeru yūbin nettowāku: jōhō no keizai shi." *Mita gakkai zasshi* 86.8 (1993): 310–20.

Sumitomo denki kōgyō kabushiki kaisha. *Shashi Sumitomo denki kōgyō kabushiki kaisha.* Osaka, 1961.

Sun Li. *Wan-Qing dianbao jiqi chuanbo guannian.* Shanghai: Shanghai shiji chuban jituan, 2007.

Sunaga Noritake. "Chūgoku no tsūshin shihai to Nichi-Bei kankei: Mitsui-Shuangqiao musen dentai shakkan to Federaru shakkan o megutte." *[Nihon daigaku] Keizai shūshi* 60.4 (1990): 157–87.

———. "Senzenki Nihon no taigai tsūshin jigyō no tokushitsu to tsūshin jishuken: Nihon musen denshin kaisha no setsuritsu o chūshin toshite." *[Nihon daigaku] Keisei ronshū* 17 (1991): 35–54.

Suzuki Jun, ed. *Kōbusho to sono jidai*. Tokyo: Yamakawa shuppansha, 2002.

Suzuki Takeo. *Chōsen no keizai*. Tokyo: Nihon hyōronsha, 1942.

Taehan minguk taeshinbu. *Chon'gi t'ongshin saop palsim nyonsa*. Seoul, 1966.

Taiwan yinhang. Jingji yanjiushi. *Taiwan jiaotong shi*. Taibei, 1955.

Takahashi Hidenao. *Nisshin sensō kaisen katei no kenkyū*. Kobe: Kobe shōka daigaku, Keizai kenkyūjo, 1992.

Takahashi Tatsuo. *Nihon shihonshugi to denshin denwa sangyō*. Tokyo, 1958.

Takahashi Yasutaka. *Nihon shokuminchi tetsudō shiron*. Tokyo: Nihon keizai hyōronsha, 1994.

Takahashi Zenshichi. *Oyatoi gaikokujin (7): tsūshin*. Tokyo: Kajima kenkyūjo shuppankai, 1969.

———. *Terashima Munenori: Nihon denki tsūshin no chichi*. Tokyo: Kokusho kankōkai, 1989.

———. *Tsūshin*. Tokyo: Kondō shuppansha, 1986.

Takeyama Akiko. *Gyo'on hōsō*. Tokyo: Banseisha, 1989.

———. *Sensō to hōsō*. Tokyo: Shakai shisōsha, 1994.

Tanabe Harumichi denki hensankai. *Tanabe Harumichi*. Tokyo: Teishin kyōkai, 1953.

Teishin dōsōkai, comp. *Senpai ni kiku*. Tokyo, 1979.

Teishin gaishi kankōkai. *Teishin shiwa*. 3 vols. Tokyo: Denki tsūshin kyōkai, 1962.

Teishin hakubutsukan. *Teishin hakubutsukan 75-nen shi*. Tokyo: Shin'yūsha, 1972.

Teishin kyōkai. *Teishin kyōkai 75-nen shi*. Tokyo: Teishin kyōkai, 1983.

———. *Yūbin sōgyō dan*. Tokyo, 1936.

Teishin 60-nen shi kankōkai. *Teishin 60-nen shi*. Tokyo, 1930.

Teishinshō. *Chōkyori denki tsūshin kaisen no hoshū no tokuisei narabini korega kikō ni tsuite*. Tokyo, 1937.

———. *Kokusai denki tsūshin kabushiki kaisha kakujugo ni okeru denki tsūshin gijutsu no kenkyū keikaku gaiyō*. Tokyo, 1939.

———. *Musoka kēburu hansō tsūshin hōshiki ni tsuite*. Tokyo, 1939.

———. *Nichi-Man renraku denwa shisetsu sekkei keikaku kōyō*. Tokyo, 1937.

———. *Nippon denshin denwa kōji kabushiki kaisha no gaiyō*. Tokyo, 1939.

———. *Nisshi kan tsūshin mondai no gaiyō*. Tokyo, 1939.

———. *Nisshi tsūshin kaigi gijiroku*. Tokyo, n.d.

———. *Teishin 50-nen ryakushi*. Tokyo, 1921.

———. *Teishin jigyō shi*. 7 vols. Tokyo: Teishin kyōkai, 1940.

———. *Tōa ni okeru denki tsūshin jigyō no gaiyō*. Tokyo, 1939.

———. *Tōa ni okeru denki tsūshin seisaku*. Tokyo, 1939.

Teishinshō denmukyoku. *Denpō kōryō jōkyō ni kansuru chōsa*. Tokyo, 1935.

———. *Nichi-Man renraku musōka keburu no kansei ni saishite*. Tokyo, 1939.

———. *Nisshi tsūshin kaigi gijiroku.* Tokyo, 1931.

Teishinshō Nichi-Man denwa kensetsuka. *Nichi-Man renraku denwa kaisen nokōsei ni tsuite.* Tokyo, 1936.

———. Teishin hakubutsukan. *Kōa teishin hakurankai shi.* Tokyo, 1940.

Tianjin dianxin shiliao. Ed. Zhu Zhenggang. 1990–92.

Tōa denki tsūshin jimukyoku. *Tōa denki tsūshin kyōgikai dai-1-kai kaigi gijiroku.* Tokyo, 1939.

———. *Tōa denki tsūshin kyōgikai dai-2-kai kaigi gijiroku.* Tokyo, 1940.

———. *Tōa denki tsūshin kyōgikai dai-3-kai kaigi gijiroku.* Tokyo, 1941.

———. *Tōa denki tsūshin kyōgikai dai-4-kai kaigi gijiroku.* Tokyo, 1942.

Tōkyō chūō denshinkyoku. *Manshū jihen to denshin jimu toriatsukau gaiyō.* Tokyo, 1935.

———. *Tōkyō chūō denshinkyoku enkaku shi.* Tokyo, 1931.

Tōkyō kokusai tsūshin shinkōkai. *Tsūshin ni kansuru gun-kan-min kondaikai kiroku.* Tokyo, 1943.

Tominaga Yuji. *Kōtsu ni okeru shihonshugi no katten.* Tokyo, 1953.

Tsūshin chōsakai. *Kyū Nankin kokumin seifu no kōtsū to tsūshin jigyō.* Tokyo, 1941.

Tsūshinsha shi kankōkai. *Tsūshinsha shi.* Tokyo, 1958.

Wakai Noboru and Takahashi Yūzō. *Terekomu no yōake.* Tokyo: Denki tsūshin shinkōkai, 1994.

Wang Congzhi and Yun Zeng. *Wuxiandian yu Zhongguo.* Nanjing: Jianshe weiyuanhui, 1931.

Watanabe Masami. *Nihon denshin denwa sōgyō shi.* Tokyo: Hifumi shobō, 1958.

Watanabe Otojirō. *Denki tsūshin kokusaku to denki tsūshin jigyō.* Tokyo: Kōtsū keizaisha, 1943.

Yamada Kōichi. *Mantetsu chōsabu: eikō to zasetsu no yonjūnen.* Tokyo: Nihon keizai shinbunsha, 1977.

Yamamoto Taketoshi et al., eds. *Iwanami kōza: 'Teikoku' Nihon no gakuchi.* 8 vols. Tokyo: Iwanami shoten, 2006.

Yamamoto Tsugu. *Densen shi.* Tokyo: Nihon densen kōgyōkai, 1959.

Yamamoto Yūzō, ed. *Teikoku no kenkyū: Genri, ruikei, kankei.* Nagoya: Nagoya daigaku shuppankai, 2003.

Yamamuro Shin'ichi. *Kimera: Manshūkoku no shozō.* Tokyo: Chūō kōronsha, 1993.

Yamane Nobuhiro. "Kōbusho no gijutsusha yōsei." In Suzuki Jun, ed., *Kōbusho to sono jidai,* 187–228. Tokyo: Yamakawa, 2002.

———. "Meijiki Nihon ni okeru denshinmō no zenkokuka to doboku jigyō no kindaika." *Nenpō Kagaku gijutsu shakai* 7 (1998): 59–85.

———. "Sokuryō chizu no shūseki to kokka zen'iki no hōsoku." *Gendai shisō,* December 1999, 205–25.

Yamashita Takeshi. *Denshin denwa jigyō ron.* Tokyo: Tsūshin bunkasha, 1954.

Yamauchi Masayuki et al., eds. *Teikoku to wa nanika?* Tokyo: Iwanami shoten, 1997.

Yang Daqing. "Senji Nihon no Chūgoku denki tsūshin shihai." *Gunji shigaku*, December 1997, 219–41.

———. "Tsūshin ni yoru chiiki tōgo: senzen no Higashi Ajia ni okeru Nihon no keiken." In Mark Caprio, ed., *Kindai Higashi Ajia no globalization*, pp. 184–219. Tokyo: Akashi shobo, 2006.

Yoshida Hideaki. "Tsūshin kiki kigyō no musen heiki bumon shinshutsu: Nihon denki o chūshin ni." In Shimotani Masahiro, ed., *Senji keizai to Nihon kigyō*, 95–124. Kyoto, 1990.

Yoshimi Shun'ya. *"Koe" no shihonshugi.* Tokyo: Kōdansha, 1995.

Youdianbu. *Zhongguo jindai youdian shi.* Beijing: Zhongguo yudian chubanshe, 1984.

Yūsei daijin kanbō. Hishoka. Kōhōshitsu. *Kyū gaichi ni okeru teishin jijō.* Tokyo, 1964.

Yūseishō. *Yūsei hyakunen shi shiryō.* 30 vols. Tokyo: Yoshikawa kōbunkan, 1968–69.

———. *Zoku teishin jigyō shi.* 10 vols. Tokyo: Maejimakai, 1961.

———. Yūmukyoku. Yūbin jigyō shi hensanshitsu. *Yūbin sōgyō 120-nen no rekishi.* Tokyo: Gyōsei, 1991.

Zhang Jian, ed. *Lao dianhua.* Shanghai: Shanghai guji chubanshe, 2002.

Zhao Zengque. *Zhongguo zhi dianxin shiye.* Shanghai: Shangwu yinshuguan, 1943.

Zhongguo wenhua jianshehui, comp. *Kangzhang qian shinian zhi Zhongguo.* 1937. Reprinted—Hong Kong, 1965.

Zhonghua minguo dianxin zongju. *Zhongguo dianxin jiyao.* Taibei: Zhonghua minguo dianxin zongju, 1971.

Zhongshan Longci (Nakayama Ryūji). *Zhongguo dianzhen yijianshu.* Beijing, 1919.

Zhongyang yanjiuyuan. Jindaishi yanjiusuo, comp. *Zhong-Ri guanxi shiliao: yudian hangyu yanlin jiaoshe, 1912–1916.* Taibei, 1975.

Zhu Huisen, comp. *Dianxin shiliao.* Taibei: Guoshiguan, 1990.

Zoku Nihon musen shi kankōkai. *Zoku Nihon musen shi.* 3 vols. Tokyo, 1972–73.

PUBLICATIONS IN ENGLISH

Adas, Michael. *Dominance by Design: Technological Imperative and America's Civilizing Mission.* Cambridge, Mass.: Harvard University Press, 2006.

———. "'High' Imperialism and 'New' History." In idem, ed., *Islamic and European Expansion: The Forging of a Global Order*, pp. 331–44. Philadelphia: Temple University Press, 1993.

———. *Machines as the Measure of Men: Science, Technology, and Ideologies of Western Dominance.* Ithaca, N.Y.: Cornell University Press, 1989.

Ahvenainen, Jorma. *The Far Eastern Telegraphs: The History of Telegraphic Communications Between the Far East, Europe and America Before the First World War.* Helsinki: Soumalainen Tiedeakatemia, 1981.

Akashi, Yoji. "Bureaucracy and the Japanese Military Administration, with Special Reference to Malaya." In William H. Newell, ed., *Japan in Asia*, pp. 46–82. Singapore: Institute of Southeast Asian Studies, 1981.

Allen, Louis. *The End of the War in Asia.* Brooklyn Heights, N.Y.: Beekman/Esanu Publishers, 1976.

Alleyne, Mark D. *International Power and International Communications.* London: Macmillan, 1995.

Auslin, Michael R. *Negotiating with Imperialism: The Unequal Treaties and the Culture of Japanese Diplomacy.* Cambridge, Mass.: Harvard University Press, 2004.

Baark, Erik. *Lightening Wires: The Telegraph and China's Technological Modernization, 1860–1890.* Westport, Conn.: Greenword Press, 1997.

Barnhart, Michael. *Japan Prepares for Total War: The Search for Economic Security, 1919–1941.* Ithaca, N.Y.: Cornell University Press, 1987.

Bartholomew, James R. *The Formation of Science in Japan.* New Haven, Conn.: Yale University Press, 1989.

———. "Science, Bureaucracy, and Freedom in Meiji and Taisho Japan." In Najita Tetsuo and Victor Koschman, eds., *Conflict in Modern Japanese History*, pp. 295–340. Princeton, N.J.: Princeton University Press, 1982.

Barty-King, Hugh. *Girdle Round the Earth: The Story of Cable and Wireless.* London: Heinemann, 1979.

Beasely, W. G. *Japanese Imperialism, 1894–1945.* Oxford: Oxford University Press, 1987.

Beauchamp, Edward, and Akira Iriye, eds. *Foreign Employees in Meiji Japan.* Boulder, Colo.: Westview, 1990.

Beniger, James R. *The Control Revolution: Technological and Economic Origins of the Information Society.* Cambridge, Mass.: Harvard University Press, 1986.

Berger, Gordon M. *Parties Out of Power in Japan, 1931–1941.* Princeton, N.J.: Princeton University Press, 1977.

———. "The Three-Dimensional Empire: Japanese Attitudes and the New Order in Asia, 1937–1945." *Japan Interpreter* 12.3–4 (Summer 1979): 355–83.

Berland, Jody. "Space at the Margins: Critical Theory and Colonial Space After Innis." In Charles R. Acland and William J. Buxton, eds., *Harold Innis in the Twentieth Century: Reflections and Refractions*, pp. 281–308. Montreal and Kingston: McGill-Queen's University Press, 1999.

Bijker, Wiebe, Thomas Hughes, and Trevor Pinch, eds. *The Social Construction of Technological Systems.* Cambridge, Mass.: MIT Press, 1987.

Boyce, Robert. "Canada and the Pacific Cable Controversy, 1923–28: Forgotten Source of Imperial Alienation." *Journal of Imperial and Commonwealth History* 26.1 (January 1998): 72–92.

———. "Submarine Cables as a Factor in Britain's Ascendancy as a World Power." In Michael North, ed., *Kommunikationsrevolutionen: Die neuen Medien des 16. und 19. Jahrhunderts*, pp. 81–99. Vienna: Bohlau Verlag, 1995.

Bracken, James K., and Christopher H. Sterling. *Telecommunications Research Resources: An Annotated Guide.* Mahwah, N.J.: Lawrence Erlbaum Associates, 1995.

Bright, Charles. *Imperial Telegraphic Communication.* London: P. S. King & Son, 1911.

Brooks, Barbara. *Japan's Imperial Diplomacy: Consuls, Treaty Ports, and War in China, 1895–1938.* Honolulu: University of Hawai'i Press, 2000.

Brooks, John. *Telephone: The First Hundred Years.* New York: Harper & Row, 1975.

Brown, Frank James. *The Cable and Wireless Communications of the World.* London: Pitman & Sons, 1927.

Brunn, Stanley D., and Thomas R. Leinbach, eds. *Collapsing Space and Time: Geographic Aspects of Communication and Information.* London: HarperCollins, 1991.

Cain, Robert J. "Telegraph Cables in the British Empire, 1850–1900." Ph.D. diss., Duke University, 1971.

Cairncross, Frances. *The Death of Distance: How the Communications Revolutions Will Change Our Lives.* Boston: Harvard Business School Press, 1997.

Carey, James W. *Communication as Culture: Essays on Media and Society.* Boston: Unwin Hyman, 1989.

Chandler, Alfred D., and James Cortada, eds. *A Nation Transformed by Information: How Information Has Shaped the United States from Colonial Times to the Present.* Oxford: Oxford University Press, 2000.

Chen, Edward I-te. "The Attempt to Integrate the Empire: Legal Perspectives." In Ramon H. Myers and Mark R. Peattie, eds., *The Japanese Colonial Empire, 1895*–1945, pp. 240–74. Princeton, N.J.: Princeton University Press, 1984.

Chida, Tomohei, and Peter N. Davies. *The Japanese Shipping and Shipbuilding Industries: A History of Their Modern Growth.* London: Athlone Press, 1990.

Choucri, Nazli, Robert C. North, and Susumu Yamakage. *The Challenge of Japan Before World War II and After: A Study of National Growth and Expansion.* London: Routledge, 1992.

Christopher, A. J. *The British Empire at Its Zenith.* London: Croom Helm, 1988.

Chu, Chia-hua. *China's Postal and Other Communications Services.* London: Kegan Paul, Trench, Trubner & Co., 1937.

Coates, Vary T., and Bernard Finn. *A Retrospective Technology Assessment. Submarine Telegraphy: The Transatlantic Cable of 1866.* San Francisco: San Francisco Press, 1979.

Codding, George A. *The International Telecommunication Union: An Experiment In International Cooperation*. Leiden: E. J. Brill, 1952.

Cohen, Jerome B. *Japan's Economy in War and Reconstruction*. Minneapolis: University of Minnesota Press, 1949.

Cook, Taya, and Theodore Cook, eds. *Japan at War*. New York: New Press, 1996.

Coox, Alvin D. *Nomonhan: Japan Against Russia, 1939*. Stanford, Calif.: Stanford University Press, 1985.

Crowley, James B. "Intellectuals as Visionaries of the New Asian Order." In *Dilemma of Growth in Prewar Japan*, ed. James W. Morley, pp. 319–73. Princeton, N.J.: Princeton University Press, 1971.

———. *Japan's Quest for Autonomy: National Security and Foreign Policy, 1930–1938*. Princeton, N.J.: Princeton University Press, 1966.

Cusumano, Michael. "'Scientific Industry': Strategy, Technology, and Management in the Riken Industrial Group, 1917–1945." In William Wray, ed., *Managing Industrial Enterprise: Cases from Japan's Prewar Experience*, pp. 269–315. Cambridge, Mass.: Harvard University, Council on East Asian Studies, 1990.

Davis, Clarence B., and Kenneth E. Wilburn Jr., eds. *Railway Imperialism*. New York: Greenwood, 1991.

Davis, R. E. G. *Airlines of Asia Since 1920*. London: Putnam, 1997.

de Cogan, Donard. "International Cable Telegraph and the Austro-Hungarian Monarchy." *Blätter für Technikgeschichte* 50 (1988–90): 7–44.

Dilts, Marion May. *The Telephone in a Changing World*. New York: Longmans, Green, 1941.

Dorpalen, Andreas. *The World of General Haushofer: Geopolitics in Action*. New York: Farrar & Rinehart, 1942.

Dower, John W. *Empire and Aftermath: Yoshida Shigeru and the Japanese Experience, 1878–1954*. Cambridge, Mass.: Harvard University, Council on East Asian Studies, 1979.

———. *Japan in War and Peace*. New York: New Press, 1995.

Drea, Edward. *MacArthur's Ultra*. Kansas City: University of Kansas Press, 1990.

Duara, Prasenjit. *Sovereignty and Authenticity: Manchukuo and the East Asian* Modern. Lanham, Md.: Rowman & Littlefield, 2003.

Duch, Raymond M. *Privatizing the Economy: Telecommunications Policy in Comparative Perspective*. Ann Arbor: University of Michigan Press, 1991.

Dunsheath, Percy. *A History of Electrical Engineering*. London: Faber and Faber, 1962.

Duus, Peter. *The Abacus and the Sword: The Japanese Penetration of Korea, 1895–1910*. Berkeley: University of California Press, 1995.

Duus, Peter, ed. *The Cambridge History of Japan*, vol. 6, *The Twentieth Century*. Cambridge, Eng.: Cambridge University Press, 1989.

Duus, Peter; Ramon H. Myers; and Mark R. Peattie, eds. *The Japanese Informal Empire in China, 1895–1937*. Princeton, N.J.: Princeton University Press, 1989.

———. *The Japanese Wartime Empire, 1931–1945*. Princeton, N.J.: Princeton University Press, 1996.

Eckert, Carter. *Offsprings of Empire: The Koch'ang Kims and the Colonial Origins of Korean Capitalism, 1876–1945*. Seattle: University of Washington Press, 1991.

Ericson, Steven J. *The Sound of the Whistle: Railroads and the State in Meiji Japan*. Cambridge, Mass.: Harvard University, Council on East Asian Studies, 1996.

Fagen, M. D., ed. *A History of Engineering and Science in the Bell System: Transmission Technology (1925–1975)*, vol. 1. N.p.: Bell Telephone Laboratories, 1978.

Finn, Bernard, and Daqing Yang, eds. *Communication Under the Seas: The Evolving Cable Network and Its Implications*. Cambridge, Mass.: MIT Press, 2009.

Fletcher, William Miles, III. *The Search for a New Order: Intellectuals and Fascism in Prewar Japan*. Chapel Hill, N.C.: University of North Carolina Press, 1982.

Flichy, Patrice. *A History of Modern Communications*. Los Angeles: Sage Publications, 1995.

Foreign Affairs Association of Japan. *The Japan Year Book, 1943–44*. Tokyo, 1943. Reprinted—U.S. International Departmental Committee for the Acquisition of Foreign Publications.

Forman, Harrison. *Changing China*. New York: Crown Publishers, 1948.

Fransman, Martin. *Japan's Computer and Communications Industry*. Oxford: Oxford University Press, 1995.

Galtung, Johan. "Structure of Imperialism." *Journal of Peace Research* 8.2 (1971): 81–117.

Gerr, Stanley. *Japanese-English Glossary of Electrical and Electrocommunication Terms*. New York: G. E. Stechert & Co., 1944.

Goldhamer, Herbert. *The Social Effects of Communication Technology*. Santa Monica, Calif.: Rand Corporation, 1970.

Gould, Peter. "Dynamic Structure of Geographic Space." In Stanley D. Brunn and Thomas R. Leinbach, eds., *Collapsing Space and Time: Geographic Aspects of Communication and Information*, pp. 3–30. New York: HarperCollins, 1991.

Griset, Pascal. "Technical Systems and Strategy: Intercontinental Telecommunications in the First Quarter of the Twentieth Century." In Olivier Coutard, ed., *The Governance of Large Technical Systems*, pp. 58–72. New York: Routledge, 1999.

Harvey, David. *The Condition of Postmodernity*. London: Blackwell, 1990.

Hata Ikuhiko. "The Army's Move into Northern Indochina." In James William Morley, ed., *The Fateful Choice: Japan's Advance into Southeast Asia, 1939–1941*, pp. 155–208. New York: Columbia University Press, 1980.

———. "Continental Expansion, 1905–1941." Trans. Alvin D. Coox. In Peter Duus, ed., *Cambridge History of Japan*, vol. 6, *The Twentieth Century*, pp. 271–314. Cambridge, Eng.: Cambridge University Press, 1989.

Hatano, Sumio. "The Japanese Navy and the Development of Southward Expansion." In Shin'ya Sugiyama and Milagros Guerrero, eds., *International Commercial Rivalry in Southeast Asia in the Interwar Period*, pp. 97–108. New Haven, Conn.: Yale Southeast Asia Studies, 1994.

Hatch, Walter, and Kozo Yamamura. *Asia in Japan's Embrace: Building a Regional Production Alliance*. Cambridge, Eng.: Cambridge University Press, 1996.

Headrick, Daniel. *The Invisible Weapon: Telecommunication and International Politics, 1851–1945*. New York: Oxford University Press, 1992.

———. *The Tentacles of Progress: Technology Transfer in the Age of Imperialism, 1850–1940*. New York: Oxford University Press, 1988.

———. *The Tools of Empire: Technology and European Imperialism in the Nineteenth Century*. New York: Oxford University Press, 1981.

Hoover, Edgar M. *The Location of Economic Activity*. New York: McGraw-Hill, 1948.

Hoskins, Halford L. *British Routes to India*. London, 1928.

Hudson, Heather E. *When Telephones Reach the Village: The Role of Telecommunications in Rule Development*. Norwood, N.J.: Ablex Publishing, 1984.

Hughes, Thomas P. *Networks of Power: Electrification in Western Society*. Baltimore, Md.: Johns Hopkins University Press, 1983.

Hugill, Peter. *Global Communications Since 1844: Geopolitics and Technology*. Baltimore, Md.: Johns Hopkins University Press, 1999.

Huston, James A. *Guns and Butter, Powder and Rice: U.S. Army Logistics in the Korean War*. Selinsgrove, Ga.: Susquehanna University Press, 1989.

Inaba, Chiharu. "International Telecommunications During the Russo-Japanese War." *The Studies* (Tōyō Eiwa College), no. 32 (1993): 25–41.

Innis, Harold A. *Empire and Communication*. Ed. David Godfrey. Victoria, B.C.: Press Porcepic, 1986.

Investigation Section, Minister's Secretariat, comp. *Outline of Postal Services of Japan*. Tokyo: Ministry of Postal Services, 1956.

Iriye, Akira. "Chang Hsueh-liang and the Japanese." *Journal of Asian Studies* 20.1 (1960): 33–43.

———. "The Failure of Economic Expansion, 1918–1931." In Bernard S. Silberman and H. D. Harootunian, eds., *Japan in Crisis: Essays on Taisho Democracy*, pp. 237–69. Princeton, N.J.: Princeton University Press, 1974.

———. *Power and Culture*. Cambridge, Mass.: Harvard University Press, 1980.

Jacob, Philip E., and James V. Toscano, eds. *The Integration of Political Communities*. Philadelphia: J. B. Lippincott, 1964.

Jennings, John M. *The Opium Empire: Japanese Imperialism and Drug Trafficking in Asia, 1895–1945*. Westport, Conn.: Praeger, 1997.

Johnson, Chalmers. *Japan's Public Policy Companies.* Palo Alto, Calif.: AEI-Hoover, 1978.

———. *MITI and the Japanese Miracle.* Stanford: Stanford University Press, 1982.

Jolly, W. P. *Marconi.* New York: Stein and Day, 1972.

Kahn, B. Winston. "Doihara Kenji and the North China Autonomy Movement." In Alvin Coox and Hilary Conroy, eds., *China and Japan Since World War I*, pp. 177–210. Santa Barbara, Calif.: ABC-Clio, 1974.

Kasza, Gregory J. *The State and the Media in Japan, 1918–1945.* Berkeley: University of California Press, 1988.

Kennedy, P. M. "Imperial Cable Communications and Strategy, 1879–1914." *English Historical Review* 86 (1971): 728–52.

Kern, Stephen. *The Culture of Time and Space, 1880–1918.* Cambridge, Mass.: Harvard University Press, 1983.

Koh, B. C. *Japan's Administrative Elites.* Berkeley: University of California Press, 1989.

Kudo, Akira. *German-Japanese Business Relations.* New York: Routledge, 1997.

Kushner, Barak. *The Thought War: Japanese Imperial Propaganda.* Honolulu: University of Hawai'i Press, 2006.

Larsen, Kirk. *Tradition, Treaties, and Trade: Qing Imperialism and Chosŏn Korea, 1850–1910.* Cambridge, Mass.: Harvard University Asia Center, 2008.

Lebra, Joyce C. "Postwar Perspectives on Japan's Greater East Asia Co-Prosperity Sphere." Harmon Memorial Lecture in Military History at the U.S. Air Force Academy, 1991.

Lebra, Joyce C., ed. *Japan's Greater East Asia Co-Prosperity Sphere in World War II.* Kuala Lumpur: Oxford University Press, 1975.

Li, Lincoln. *The Japanese Army in North China, 1937–1941: Problems of Political and Economic Control.* Tokyo: Oxford University Press, 1975.

Liberman, Peter. *Does Conquest Pay? The Exploitation of Occupied Industrial Societies.* Princeton, N.J.: Princeton University Press, 1996.

Low, Morris F. "The Japan That Can Say No: The Rise of Techno-Nationalism and Its Impact on Technological Change." In Robert Fox, ed., *Technological Change: Methods and Themes in the History of Technology*, pp. 201–14. Amsterdam: Harwood Academic Publishers, 1996.

———. *Science and the Building of a New Japan.* New York: Palgrave Macmillan, 2005.

Low, Morris, ed. *Building a Modern Japan: Science, Technology, and Medicine in the Meiji Era and Beyond.* New York: Palgrave Macmillan, 2005.

MacMurray, John V. A. *Treaties and Agreements with or Concerning China, 1894–1919.* New York, 1921.

Mance, Osborne. *International Telecommunications.* London: Oxford University Press, 1944.

Marvin, Carolyn. *When the Old Technologies Were New: Thinking About Electric Communication in the Late Nineteenth Century.* New York: Oxford University, 1988.

Matsumae, Shigeyoshi. *The Life and Struggles of a Fighter for Peace.* Tokyo: Tokai University Press, 1988.

———. *The Second World War: A Tragedy for Japan.* Tokyo: Tokai University Press, 1981.

———. *Study on the Long Distance Communication System by Non-Loaded Cable.* Tokyo: Corona, 1938.

Matsusaka, Tak Yoshihisa. "Japanese Empire," in William M. Tsutsui, ed., *A Companion to Japanese History*, pp. 225–40. Malden, Mass.: Blackwell, 2007.

———. "Japanese Imperialism and the South Manchuria Railway Company, 1904–1914." Ph.D. diss., Harvard University, 1993.

———. *The Making of Japanese Manchuria, 1904–1932.* Cambridge, Mass.: Harvard University Asia Center, 2000.

———. "Managing Occupied Manchuria, 1931–1934." In Peter Duus, Ramon H. Myers, and Mark R. Peattie, eds., *The Japanese Wartime Empire, 1931–1945*, pp. 97–135. Princeton, N.J.: Princeton University Press, 1996.

Mayntz, Renate, and Thomas P. Hughes, eds. *The Development of Large Technical Systems.* Boulder, Colo.: Westview, 1988.

Mayo, Marlene, ed. *The Emergence of Imperial Japan: Self-Defense or Calculated Aggression?* New York: D. C. Heath, 1970.

McCormack, Gavan. *Chang Tso-lin in Northeast China, 1911–1928: China, Japan, and the Manchurian Idea.* Berkeley: University of California Press, 1977.

McCune, George. *Korea Today.* Cambridge, Mass.: Harvard University Press, 1950.

Melody, William H., et al., eds. *Culture, Communication and Dependency: The Tradition of H. A. Innis.* Norwood, N.J.: Ablex Publishing, 1981.

Mills, J. Saxon. *The Press and Communications of the Empire.* London: W. Collins Sons, 1924.

Mimura, Janis. "Technocratic Visions of Empire: Technology Bureaucrats and the 'New Order for Science and Technology.'" In Harald Fuess, ed., *The Japanese Empire in East Asia and Its Postwar Legacies*, pp. 97–118. Munich: Iudicium, 1998.

Ministry of Communications (Japan). *Outline of the History of Telegraphs in Japan.* Tokyo, 1892.

———. *A Short History of the Post and Telegraph Services in Japan.* Tokyo, 1902.

Miwa, Kimitaka. "Japanese Policies and Concepts for a Regional Order in Asia, 1938–1940." In James W. White, ed., *The Ambivalence of Nationalism: Modern Japan Between East and West*, pp. 133–56. Lanham, Md.: University Press of America, 1990.

Mizuno Hiromi. *Science for the Empire: Scientific Nationalism.* Stanford: Stanford University Press, 2009.

Molony, Barbara. *Technology and Investment: The Prewar Japanese Chemical Industry.* Cambridge, Mass.: Harvard University, Council on East Asian Studies, 1990.

Mommsen, Wolfgang J., and Jurgen Osterhammel, eds. *Imperialism and After: Continuities and Discontinuities.* London: Allen & Unwin, 1986.

Morley, James William, ed. *The China Quagmire: Japan's Expansion on the Asian Continent, 1933–1941.* New York: Columbia University Press, 1983.

———. *The Dilemma of Growth in Prewar Japan.* Princeton, N.J.: Princeton University Press, 1971.

———. *The Fateful Choice: Japan's Advance into Southeast Asia, 1939–1941.* New York: Columbia University Press, 1980.

———. *Japan Erupts: The London Naval Conference and the Manchurian Incident, 1928–1932.* New York: Columbia University Press, 1984.

Morris-Suzuki, Tessa. *The Technological Transformation of Modern Japan: From the Seventeenth to the Twentieth Century.* Cambridge, Eng.: Cambridge University Press, 1994.

Mosher, John. *Japanese Post Offices in China.* Lawrence, Mass.: Quarterman Publications, 1978.

Mulgan, M. J. *Communications and Control.* Cambridge, Eng.: Polity Press, 1991.

Myers, Ramon H. "Creating a Modern Enclave Economy: The Economic Integration of Japan, Manchuria, and North China, 1932–1945." In Peter Duus, Ramon H. Myers, and Mark R. Peattie, eds., *The Japanese Wartime Empire, 1931–1945*, pp. 136–70. Princeton, N.J.: Princeton University Press, 1996.

Myers, Ramon H., and Mark R. Peattie, eds. *The Japanese Colonial Empire, 1895–1945.* Princeton, N.J.: Princeton University Press, 1984.

Nagaoka Shinjirō. "The Drive into Southern Indochina and Thailand." In James William Morley, ed., *The Fateful Choice: Japan's Advance into Southeast Asia, 1939–1941*, pp. 209–41. New York: Columbia University Press, 1980.

———. "Economic Demands on the Dutch East Indies." In James William Morley, ed., *The Fateful Choice: Japan's Advance into Southeast Asia, 1939–1941*, pp. 125–54. New York: Columbia University Press, 1980.

Nakamura Takafusa. "Japan's Economic Thrust into North China, 1933–1938: Formation of the North China Development Corporation." In Akira Iriye, ed., *The Chinese and the Japanese: Essays in Political and Cultural Interactions*, pp. 249–50. Princeton, N.J.: Princeton University Press, 1980.

———. "The Japanese War Economy as a 'Planned Economy.'" In Erich Pauer, ed., *Japan's War Economy*, pp. 9–22. London: Routledge, 1999.

Namikawa Ryō. "Japanese Overseas Broadcasting: A Personal View." In K. R. M. Short, ed., *Film & Radio Propaganda in World War II*, pp. 319–33. Knoxville: University of Tennessee Press, 1983.

Nickles, David. *Under the Wire: How Telegraph Changed Diplomacy* Cambridge, Mass.: Harvard University Press, 2007.

Nish, Ian. "Some Thoughts on Japanese Expansion." In Hans J. Mommsen and Jurgen Osterhammel, eds., *Imperialism and After: Continuities and Discontinuities.* London: Allen & Unwin, 1986.

Numnonda, Thamsook. *Thailand and the Japanese Presence, 1941–45.* Singapore: Institute of Southeast Asian Studies, 1977.

Odagiri, Hiroyuki, and Akira Goto. *Technology and Industrial Development in Japan.* Oxford: Clarendon Press, 1996.

Office of Strategic Services. Research and Analysis Branch, comp. *Transportation and Communication in Japan.* FCC intercepts of shortwave broadcasts (except those of a purely military nature) from Radio Tokyo and affiliated stations, from December 1941, to April 1, 1945 and from OSS sources. Honolulu, May 12, 1945.

Ogata, Sadako. *Defiance in Manchuria: The Making of Japanese Foreign Policy, 1931–1932.* Berkeley: University of California Press, 1964.

Okochi, Akio, and Hoshimi Uchida, eds. *Development and Diffusion of Technology: Electrical and Chemical Industries.* Tokyo: Tokyo University Press, 1980.

O'Neill, E. F., ed. *A History of Engineering and Science in the Bell System: Transmission Technology (1925–1975).* N.p.: AT&T Bell Telephone Laboratories, 1978.

Partner, Simon. *Assembled in Japan: Electrical Goods and the Making of the Japanese Consumer.* Berkeley: University of California Press, 1999.

Pauer, Erich, ed. *Japan's War Economy.* London: Routledge, 1999.

———. *Papers on the History of Industry and Technology in Japan II: From the Meiji-Period to Postwar Japan.* Marburg: Marburger Japan-Reihe, 1995.

Peattie, Mark R. *Ishiwara Kanji and Japan's Confrontation with the West.* Princeton, N.J.: Princeton University Press, 1975.

———. "The Japanese Colonial Empire, 1895–1945." In Peter Duus, ed., *Cambridge History of Japan*, vol. 6, *The Twentieth Century*, pp. 217–70. Cambridge, Eng.: Cambridge University Press, 1989.

———. *Nan'yō: The Rise and Fall of the Japanese in Micronesia, 1885–1945.* Honolulu: University of Hawai'i Press, 1988.

Pickworth, George. "Germany's Imperial Wireless System." *Electronics World + Wireless World* 99 (May 1993): 427–33.

Pool, Ithiel de Sola. *Forecasting the Telephone: A Retrospective Technology Assessment.* Norwood, N.J.: Ablex, 1983.

———. *Technologies of Freedom.* Cambridge, Mass.: Harvard University Press, 1983.

———. *Technologies Without Boundaries* Cambridge, Mass.: Harvard University Press, 1990.

Pool, Ithiel de Sola, ed. *The Social Impact of the Telephone.* Cambridge, Mass.: MIT Press, 1977.

Pyle, Kenneth. "The Technology of Japanese Nationalism: The Local Improvement Movement, 1900–1918." *Journal of Asian Studies* 33.1 (November 1973): 51–65.

Read, Donald. *The Power of News: The History of Reuters.* Oxford: Oxford University Press, 1992.

Reynolds, E. Bruce. *Thailand and Japan's Southern Advance, 1941–1945.* New York: St. Martin's Press, 1995.

Rice, Richard. "Economic Mobilization in Wartime Japan: Business, Bureaucracy, and Military in Conflict." *Journal of Asian Studies* 38 (August 1979): 698–702.

Robbins, Jane M. J. *Tokyo Calling: Japanese Overseas Radio Broadcasting, 1937–1945.* Florence: European Press Academic Publishing, 2002.

Robertson, Eric. *The Japanese File: Pre-War Japanese Penetration in Southeast Asia.* Hong Kong: Heinemann Asia, 1979.

Sakamoto, Mamoru. *A Lion Aroused: Conversations with Shigeyoshi Matsumae.* Tokyo: Tokai University Press, 1986.

Samuels, Richard J. *"Rich Nation, Strong Army": National Security and Technological Transformation of Japan.* Ithaca, N.Y.: Cornell University Press, 1994.

Schreiner, George A. *Cables and Wireless and Their Role in the Foreign Relations of the United States.* Boston, 1924.

Shimizu, Hajime. "Southeast Asia as a Regional Concept in Modern Japan: An Analysis of Geography Textbooks." In Saya Shiraishi and Takashi Shiraishi, eds., *The Japanese in Colonial Southeast Asia*, pp. 21–61. Ithaca, N.Y.: Cornell University Southeast Asia Program, 1993.

Shiraishi, Saya, and Takashi Shiraishi, eds. *The Japanese in Colonial Southeast Asia.* Ithaca, N.Y.: Cornell University Southeast Asia Program, 1993.

Siemens, Georg. *History of the House of Siemens*, vol. 2, *The Era of World Wars.* Trans. A. F. Rodger and Lawrence N. Hole. Freiburg: Karl Alber, 1957.

Snyder, Jack. *Myth of Empire: Domestic Politics and International Ambitions.* Ithaca, N.Y.: Cornell University Press, 1991.

Spaulding, Robert M. "The Bureaucracy as a Political Force, 1920–1945." In James William Morley, ed., *Dilemma of Growth in Prewar Japan*, pp. 33–80. Princeton, N.J.: Princeton University Press, 1971.

———. *Imperial Japan's Higher Civil Service Examination.* Princeton, N.J.: Princeton University Press, 1967.

Summerton, Jane, ed. *Changing Large Technical Systems.* Boulder, Colo.: Westview Press, 1994.

Sussman, Gerald. *Communication, Technology, and Politics in the Information Age.* Thousand Oaks, Calif.: Sage Publications, 1997.

Thompson, John B. *The Media and Modernity: A Social Theory of the Media.* Stanford, Calif.: Stanford University Press, 1995.

Tribolet, Leslie B. *The International Aspects of Electric Communications in the Pacific Area.* Baltimore, Md.: Johns Hopkins Press, 1929.

United States, Department of State. *Foreign Relations of the United States (FRUS).* Washington, D.C.: U.S. Government Printing Office, various years.

Ward, Robert S. *Asia for the Asiatics.* Chicago: University of Chicago Press, 1945.

Wasserman, Neil H. *From Invention to Innovation: Long-Distance Telephone Transmission at the Turn of the Century.* Baltimore, Md.: Johns Hopkins University Press, 1985.

Watanabe, Masao. *The Japanese and Western Science.* Philadelphia: University of Pennsylvania Press, 1990.

Westney, D. Eleanor. "Building the National Communications System: Adopting and Adapting Western Organizational Models in Meiji Japan." In James W. White, Michio Umegaki, and Thomas R. H. Havens, eds., *The Ambivalence of Nationalism: Modern Japan Between East and West*, pp. 39–59. Lanham, Md.: University Press of America, 1990.

Willoughby, Westel Woodbury. *China at the Conference; A Report.* Baltimore, Md.: Johns Hopkins University Press, 1922.

———. *Foreign Rights and Interests in China.* Rev. and enl. ed. Baltimore, Md.: Johns Hopkins University Press, 1927.

Winkler, Jonathan Reed. *Nexus: Strategic Communications and American Security in World War I.* Cambridge, Mass.: Harvard University Press, 2008.

Winks, Robin W., ed. *The Oxford History of the British Empire: Historiography.* Oxford: Oxford University Press, 1999.

Winseck, Dwayne R., and Robert M. Pike. *Communication and Empire: Media, Markets, and Globalization, 1860–1930.* Durham, N.C.: Duke University Press, 2007.

Wittner, David G. *Technology and the Culture of Progress in Meiji Japan.* New York: Routledge, 2008.

Wray, William. *Mitsubishi and the NYK, 1870–1914.* Cambridge, Mass.: Harvard University, Council on East Asian Studies, 1984.

Yamamura, Kozo. "Japan's Deus ex Machina: Western Technology in the 1920s." *Journal of Japanese Studies* 12.1 (1985): 65–95.

———. "Success Ill-gotten? The Role of Meiji Militarism in Japan's Technological Progress." *Journal of Economic History* 37.1 (March 1977): 113–35.

Yamamuro Shin'ichi. *Manchuria Under Japanese Domination.* Trans. Joshua A. Fogel. Philadelphia: University of Pennsylvania Press, 2006.

Yang, Daqing. "Colonial Korea in Japan's Imperial Telecommunications Network." In *Modernity, Domination, and Identity in Colonial Korea*, ed. Gi-Wook Shin and Michael Robinson, pp. 161–90. Cambridge, Mass.: Harvard University Asia Center, 1999.

———. "Resurrecting the Empire? Japanese Technicians in Postwar China." In Harald Fuess, ed., *Japanese Empire in East Asia and Its Postwar Legacies*, pp. 185–205. Munich: Iudicium, 1998.

Yardley, Herbert. *The American Black Chamber*. Indianapolis, 1931.

Yoshihashi, Takehiko. *Conspiracy at Mukden: The Rise of the Japanese Military*. New Haven, Conn.: Yale University Press, 1963.

Young, Louise A. *Japan's Total Empire: Manchuria and the Culture of Wartime Imperialism*. Berkeley: University of California Press, 1998.

Young, Peter. *Person to Person: The International Impact of the Telephone*. Cambridge, Eng.: Granta Editions, 1991.

Index

Harvard East Asian Monographs
(*out-of-print)

*1. Liang Fang-chung, *The Single-Whip Method of Taxation in China*
*2. Harold C. Hinton, *The Grain Tribute System of China, 1845–1911*
3. Ellsworth C. Carlson, *The Kaiping Mines, 1877–1912*
*4. Chao Kuo-chün, *Agrarian Policies of Mainland China: A Documentary Study, 1949–1956*
*5. Edgar Snow, *Random Notes on Red China, 1936–1945*
*6. Edwin George Beal, Jr., *The Origin of Likin, 1835–1864*
7. Chao Kuo-chün, *Economic Planning and Organization in Mainland China: A Documentary Study, 1949–1957*
*8. John K. Fairbank, *Ching Documents: An Introductory Syllabus*
*9. Helen Yin and Yi-chang Yin, *Economic Statistics of Mainland China, 1949–1957*
10. Wolfgang Franke, *The Reform and Abolition of the Traditional Chinese Examination System*
11. Albert Feuerwerker and S. Cheng, *Chinese Communist Studies of Modern Chinese History*
12. C. John Stanley, *Late Ching Finance: Hu Kuang-yung as an Innovator*
13. S. M. Meng, *The Tsungli Yamen: Its Organization and Functions*
*14. Ssu-yü Teng, *Historiography of the Taiping Rebellion*
15. Chun-Jo Liu, *Controversies in Modern Chinese Intellectual History: An Analytic Bibliography of Periodical Articles, Mainly of the May Fourth and Post-May Fourth Era*
*16. Edward J. M. Rhoads, *The Chinese Red Army, 1927–1963: An Annotated Bibliography*
*17. Andrew J. Nathan, *A History of the China International Famine Relief Commission*
*18. Frank H. H. King (ed.) and Prescott Clarke, *A Research Guide to China-Coast Newspapers, 1822–1911*
*19. Ellis Joffe, *Party and Army: Professionalism and Political Control in the Chinese Officer Corps, 1949–1964*
*20. Toshio G. Tsukahira, *Feudal Control in Tokugawa Japan: The Sankin Kōtai System*
*21. Kwang-Ching Liu, ed., *American Missionaries in China: Papers from Harvard Seminars*
*22. George Moseley, *A Sino-Soviet Cultural Frontier: The Ili Kazakh Autonomous Chou*

23. Carl F. Nathan, *Plague Prevention and Politics in Manchuria, 1910–1931*
*24. Adrian Arthur Bennett, *John Fryer: The Introduction of Western Science and Technology into Nineteenth-Century China*
*25. Donald J. Friedman, *The Road from Isolation: The Campaign of the American Committee for Non-Participation in Japanese Aggression, 1938–1941*
*26. Edward LeFevour, *Western Enterprise in Late Ching China: A Selective Survey of Jardine, Matheson and Company's Operations, 1842–1895*
27. Charles Neuhauser, *Third World Politics: China and the Afro-Asian People's Solidarity Organization, 1957–1967*
*28. Kungtu C. Sun, assisted by Ralph W. Huenemann, *The Economic Development of Manchuria in the First Half of the Twentieth Century*
*29. Shahid Javed Burki, *A Study of Chinese Communes, 1965*
30. John Carter Vincent, *The Extraterritorial System in China: Final Phase*
31. Madeleine Chi, *China Diplomacy, 1914–1918*
*32. Clifton Jackson Phillips, *Protestant America and the Pagan World: The First Half Century of the American Board of Commissioners for Foreign Missions, 1810–1860*
*33. James Pusey, *Wu Han: Attacking the Present Through the Past*
*34. Ying-wan Cheng, *Postal Communication in China and Its Modernization, 1860–1896*
35. Tuvia Blumenthal, *Saving in Postwar Japan*
36. Peter Frost, *The Bakumatsu Currency Crisis*
37. Stephen C. Lockwood, *Augustine Heard and Company, 1858–1862*
38. Robert R. Campbell, *James Duncan Campbell: A Memoir by His Son*
39. Jerome Alan Cohen, ed., *The Dynamics of China's Foreign Relations*
40. V. V. Vishnyakova-Akimova, *Two Years in Revolutionary China, 1925–1927*, trans. Steven L. Levine
41. Meron Medzini, *French Policy in Japan During the Closing Years of the Tokugawa Regime*
42. Ezra Vogel, Margie Sargent, Vivienne B. Shue, Thomas Jay Mathews, and Deborah S. Davis, *The Cultural Revolution in the Provinces*
43. Sidney A. Forsythe, *An American Missionary Community in China, 1895–1905*
*44. Benjamin I. Schwartz, ed., *Reflections on the May Fourth Movement.: A Symposium*
*45. Ching Young Choe, *The Rule of the Taewŏngun, 1864–1873: Restoration in Yi Korea*
46. W. P. J. Hall, *A Bibliographical Guide to Japanese Research on the Chinese Economy, 1958–1970*
47. Jack J. Gerson, *Horatio Nelson Lay and Sino-British Relations, 1854–1864*
48. Paul Richard Bohr, *Famine and the Missionary: Timothy Richard as Relief Administrator and Advocate of National Reform*
49. Endymion Wilkinson, *The History of Imperial China: A Research Guide*
50. Britten Dean, *China and Great Britain: The Diplomacy of Commercial Relations, 1860–1864*
51. Ellsworth C. Carlson, *The Foochow Missionaries, 1847–1880*
52. Yeh-chien Wang, *An Estimate of the Land-Tax Collection in China, 1753 and 1908*
53. Richard M. Pfeffer, *Understanding Business Contracts in China, 1949–1963*

*54. Han-sheng Chuan and Richard Kraus, *Mid-Ching Rice Markets and Trade: An Essay in Price History*
55. Ranbir Vohra, *Lao She and the Chinese Revolution*
56. Liang-lin Hsiao, *China's Foreign Trade Statistics, 1864–1949*
*57. Lee-hsia Hsu Ting, *Government Control of the Press in Modern China, 1900–1949*
*58. Edward W. Wagner, *The Literati Purges: Political Conflict in Early Yi Korea*
*59. Joungwon A. Kim, *Divided Korea: The Politics of Development, 1945–1972*
60. Noriko Kamachi, John K. Fairbank, and Chūzō Ichiko, *Japanese Studies of Modern China Since 1953: A Bibliographical Guide to Historical and Social-Science Research on the Nineteenth and Twentieth Centuries, Supplementary Volume for 1953–1969*
61. Donald A. Gibbs and Yun-chen Li, *A Bibliography of Studies and Translations of Modern Chinese Literature, 1918–1942*
62. Robert H. Silin, *Leadership and Values: The Organization of Large-Scale Taiwanese Enterprises*
63. David Pong, *A Critical Guide to the Kwangtung Provincial Archives Deposited at the Public Record Office of London*
*64. Fred W. Drake, *China Charts the World: Hsu Chi-yü and His Geography of 1848*
*65. William A. Brown and Urgrunge Onon, translators and annotators, *History of the Mongolian People's Republic*
66. Edward L. Farmer, *Early Ming Government: The Evolution of Dual Capitals*
*67. Ralph C. Croizier, *Koxinga and Chinese Nationalism: History, Myth, and the Hero*
*68. William J. Tyler, tr., *The Psychological World of Natsume Sōseki*, by Doi Takeo
69. Eric Widmer, *The Russian Ecclesiastical Mission in Peking During the Eighteenth Century*
*70. Charlton M. Lewis, *Prologue to the Chinese Revolution: The Transformation of Ideas and Institutions in Hunan Province, 1891–1907*
71. Preston Torbert, *The Ching Imperial Household Department: A Study of Its Organization and Principal Functions, 1662–1796*
72. Paul A. Cohen and John E. Schrecker, eds., *Reform in Nineteenth-Century China*
73. Jon Sigurdson, *Rural Industrialism in China*
74. Kang Chao, *The Development of Cotton Textile Production in China*
75. Valentin Rabe, *The Home Base of American China Missions, 1880–1920*
*76. Sarasin Viraphol, *Tribute and Profit: Sino-Siamese Trade, 1652–1853*
77. Ch'i-ch'ing Hsiao, *The Military Establishment of the Yuan Dynasty*
78. Meishi Tsai, *Contemporary Chinese Novels and Short Stories, 1949–1974: An Annotated Bibliography*
*79. Wellington K. K. Chan, *Merchants, Mandarins and Modern Enterprise in Late Ching China*
80. Endymion Wilkinson, *Landlord and Labor in Late Imperial China: Case Studies from Shandong by Jing Su and Luo Lun*
*81. Barry Keenan, *The Dewey Experiment in China: Educational Reform and Political Power in the Early Republic*
*82. George A. Hayden, *Crime and Punishment in Medieval Chinese Drama: Three Judge Pao Plays*

*83. Sang-Chul Suh, *Growth and Structural Changes in the Korean Economy, 1910–1940*
84. J. W. Dower, *Empire and Aftermath: Yoshida Shigeru and the Japanese Experience, 1878–1954*
85. Martin Collcutt, *Five Mountains: The Rinzai Zen Monastic Institution in Medieval Japan*
86. Kwang Suk Kim and Michael Roemer, *Growth and Structural Transformation*
87. Anne O. Krueger, *The Developmental Role of the Foreign Sector and Aid*
*88. Edwin S. Mills and Byung-Nak Song, *Urbanization and Urban Problems*
89. Sung Hwan Ban, Pal Yong Moon, and Dwight H. Perkins, *Rural Development*
*90. Noel F. McGinn, Donald R. Snodgrass, Yung Bong Kim, Shin-Bok Kim, and Quee-Young Kim, *Education and Development in Korea*
*91. Leroy P. Jones and Il SaKong, *Government, Business, and Entrepreneurship in Economic Development: The Korean Case*
92. Edward S. Mason, Dwight H. Perkins, Kwang Suk Kim, David C. Cole, Mahn Je Kim et al., *The Economic and Social Modernization of the Republic of Korea*
93. Robert Repetto, Tai Hwan Kwon, Son-Ung Kim, Dae Young Kim, John E. Sloboda, and Peter J. Donaldson, *Economic Development, Population Policy, and Demographic Transition in the Republic of Korea*
94. Parks M. Coble, Jr., *The Shanghai Capitalists and the Nationalist Government, 1927–1937*
95. Noriko Kamachi, *Reform in China: Huang Tsun-hsien and the Japanese Model*
96. Richard Wich, *Sino-Soviet Crisis Politics: A Study of Political Change and Communication*
97. Lillian M. Li, *China's Silk Trade: Traditional Industry in the Modern World, 1842–1937*
98. R. David Arkush, *Fei Xiaotong and Sociology in Revolutionary China*
*99. Kenneth Alan Grossberg, *Japan's Renaissance: The Politics of the Muromachi Bakufu*
100. James Reeve Pusey, *China and Charles Darwin*
101. Hoyt Cleveland Tillman, *Utilitarian Confucianism: Chen Liang's Challenge to Chu Hsi*
102. Thomas A. Stanley, *Ōsugi Sakae, Anarchist in Taishō Japan: The Creativity of the Ego*
103. Jonathan K. Ocko, *Bureaucratic Reform in Provincial China: Ting Jih-ch'ang in Restoration Kiangsu, 1867–1870*
104. James Reed, *The Missionary Mind and American East Asia Policy, 1911–1915*
105. Neil L. Waters, *Japan's Local Pragmatists: The Transition from Bakumatsu to Meiji in the Kawasaki Region*
106. David C. Cole and Yung Chul Park, *Financial Development in Korea, 1945–1978*
107. Roy Bahl, Chuk Kyo Kim, and Chong Kee Park, *Public Finances During the Korean Modernization Process*
108. William D. Wray, *Mitsubishi and the N.Y.K, 1870–1914: Business Strategy in the Japanese Shipping Industry*
109. Ralph William Huenemann, *The Dragon and the Iron Horse: The Economics of Railroads in China, 1876–1937*
*110. Benjamin A. Elman, *From Philosophy to Philology: Intellectual and Social Aspects of Change in Late Imperial China*
111. Jane Kate Leonard, *Wei Yüan and China's Rediscovery of the Maritime World*

112. Luke S. K. Kwong, *A Mosaic of the Hundred Days:. Personalities, Politics, and Ideas of 1898*
*113. John E. Wills, Jr., *Embassies and Illusions: Dutch and Portuguese Envoys to K'ang-hsi, 1666–1687*
114. Joshua A. Fogel, *Politics and Sinology: The Case of Naitō Konan (1866–1934)*
*115. Jeffrey C. Kinkley, ed., *After Mao: Chinese Literature and Society, 1978–1981*
116. C. Andrew Gerstle, *Circles of Fantasy: Convention in the Plays of Chikamatsu*
117. Andrew Gordon, *The Evolution of Labor Relations in Japan: Heavy Industry, 1853–1955*
*118. Daniel K. Gardner, *Chu Hsi and the "Ta Hsueh": Neo-Confucian Reflection on the Confucian Canon*
119. Christine Guth Kanda, *Shinzō: Hachiman Imagery and Its Development*
*120. Robert Borgen, *Sugawara no Michizane and the Early Heian Court*
121. Chang-tai Hung, *Going to the People: Chinese Intellectual and Folk Literature, 1918–1937*
*122. Michael A. Cusumano, *The Japanese Automobile Industry: Technology and Management at Nissan and Toyota*
123. Richard von Glahn, *The Country of Streams and Grottoes: Expansion, Settlement, and the Civilizing of the Sichuan Frontier in Song Times*
124. Steven D. Carter, *The Road to Komatsubara: A Classical Reading of the Renga Hyakuin*
125. Katherine F. Bruner, John K. Fairbank, and Richard T. Smith, *Entering China's Service: Robert Hart's Journals, 1854–1863*
126. Bob Tadashi Wakabayashi, *Anti-Foreignism and Western Learning in Early-Modern Japan: The "New Theses" of 1825*
127. Atsuko Hirai, *Individualism and Socialism: The Life and Thought of Kawai Eijirō (1891–1944)*
128. Ellen Widmer, *The Margins of Utopia: "Shui-hu hou-chuan" and the Literature of Ming Loyalism*
129. R. Kent Guy, *The Emperor's Four Treasuries: Scholars and the State in the Late Chien-lung Era*
130. Peter C. Perdue, *Exhausting the Earth: State and Peasant in Hunan, 1500–1850*
131. Susan Chan Egan, *A Latterday Confucian: Reminiscences of William Hung (1893–1980)*
132. James T. C. Liu, *China Turning Inward: Intellectual-Political Changes in the Early Twelfth Century*
*133. Paul A. Cohen, *Between Tradition and Modernity: Wang T'ao and Reform in Late Ching China*
134. Kate Wildman Nakai, *Shogunal Politics: Arai Hakuseki and the Premises of Tokugawa Rule*
*135. Parks M. Coble, *Facing Japan: Chinese Politics and Japanese Imperialism, 1931–1937*
136. Jon L. Saari, *Legacies of Childhood: Growing Up Chinese in a Time of Crisis, 1890–1920*
137. Susan Downing Videen, *Tales of Heichū*
138. Heinz Morioka and Miyoko Sasaki, *Rakugo: The Popular Narrative Art of Japan*
139. Joshua A. Fogel, *Nakae Ushikichi in China: The Mourning of Spirit*

140. Alexander Barton Woodside, *Vietnam and the Chinese Model: A Comparative Study of Vietnamese and Chinese Government in the First Half of the Nineteenth Century*
*141. George Elison, *Deus Destroyed: The Image of Christianity in Early Modern Japan*
142. William D. Wray, ed., *Managing Industrial Enterprise: Cases from Japan's Prewar Experience*
*143. T'ung-tsu Ch'ü, *Local Government in China Under the Ching*
144. Marie Anchordoguy, *Computers, Inc.: Japan's Challenge to IBM*
145. Barbara Molony, *Technology and Investment: The Prewar Japanese Chemical Industry*
146. Mary Elizabeth Berry, *Hideyoshi*
147. Laura E. Hein, *Fueling Growth: The Energy Revolution and Economic Policy in Postwar Japan*
148. Wen-hsin Yeh, *The Alienated Academy: Culture and Politics in Republican China, 1919–1937*
149. Dru C. Gladney, *Muslim Chinese: Ethnic Nationalism in the People's Republic*
150. Merle Goldman and Paul A. Cohen, eds., *Ideas Across Cultures: Essays on Chinese Thought in Honor of Benjamin L Schwartz*
151. James M. Polachek, *The Inner Opium War*
152. Gail Lee Bernstein, *Japanese Marxist: A Portrait of Kawakami Hajime, 1879–1946*
*153. Lloyd E. Eastman, *The Abortive Revolution: China Under Nationalist Rule, 1927–1937*
154. Mark Mason, *American Multinationals and Japan: The Political Economy of Japanese Capital Controls, 1899–1980*
155. Richard J. Smith, John K. Fairbank, and Katherine F. Bruner, *Robert Hart and China's Early Modernization: His Journals, 1863–1866*
156. George J. Tanabe, Jr., *Myōe the Dreamkeeper: Fantasy and Knowledge in Kamakura Buddhism*
157. William Wayne Farris, *Heavenly Warriors: The Evolution of Japan's Military, 500–1300*
158. Yu-ming Shaw, *An American Missionary in China: John Leighton Stuart and Chinese-American Relations*
159. James B. Palais, *Politics and Policy in Traditional Korea*
*160. Douglas Reynolds, *China, 1898–1912: The Xinzheng Revolution and Japan*
161. Roger R. Thompson, *China's Local Councils in the Age of Constitutional Reform, 1898–1911*
162. William Johnston, *The Modern Epidemic: History of Tuberculosis in Japan*
163. Constantine Nomikos Vaporis, *Breaking Barriers: Travel and the State in Early Modern Japan*
164. Irmela Hijiya-Kirschnereit, *Rituals of Self-Revelation: Shishōsetsu as Literary Genre and Socio-Cultural Phenomenon*
165. James C. Baxter, *The Meiji Unification Through the Lens of Ishikawa Prefecture*
166. Thomas R. H. Havens, *Architects of Affluence: The Tsutsumi Family and the Seibu-Saison Enterprises in Twentieth-Century Japan*
167. Anthony Hood Chambers, *The Secret Window: Ideal Worlds in Tanizaki's Fiction*
168. Steven J. Ericson, *The Sound of the Whistle: Railroads and the State in Meiji Japan*
169. Andrew Edmund Goble, *Kenmu: Go-Daigo's Revolution*

170. Denise Potrzeba Lett, *In Pursuit of Status: The Making of South Korea's "New" Urban Middle Class*
171. Mimi Hall Yiengpruksawan, *Hiraizumi: Buddhist Art and Regional Politics in Twelfth-Century Japan*
172. Charles Shirō Inouye, *The Similitude of Blossoms: A Critical Biography of Izumi Kyōka (1873–1939), Japanese Novelist and Playwright*
173. Aviad E. Raz, *Riding the Black Ship: Japan and Tokyo Disneyland*
174. Deborah J. Milly, *Poverty, Equality, and Growth: The Politics of Economic Need in Postwar Japan*
175. See Heng Teow, *Japan's Cultural Policy Toward China, 1918–1931: A Comparative Perspective*
176. Michael A. Fuller, *An Introduction to Literary Chinese*
177. Frederick R. Dickinson, *War and National Reinvention: Japan in the Great War, 1914–1919*
178. John Solt, *Shredding the Tapestry of Meaning: The Poetry and Poetics of Kitasono Katue (1902–1978)*
179. Edward Pratt, *Japan's Protoindustrial Elite: The Economic Foundations of the Gōnō*
180. Atsuko Sakaki, *Recontextualizing Texts: Narrative Performance in Modern Japanese Fiction*
181. Soon-Won Park, *Colonial Industrialization and Labor in Korea: The Onoda Cement Factory*
182. JaHyun Kim Haboush and Martina Deuchler, *Culture and the State in Late Chosŏn Korea*
183. John W. Chaffee, *Branches of Heaven: A History of the Imperial Clan of Sung China*
184. Gi-Wook Shin and Michael Robinson, eds., *Colonial Modernity in Korea*
185. Nam-lin Hur, *Prayer and Play in Late Tokugawa Japan: Asakusa Sensōji and Edo Society*
186. Kristin Stapleton, *Civilizing Chengdu: Chinese Urban Reform, 1895–1937*
187. Hyung Il Pai, *Constructing "Korean" Origins: A Critical Review of Archaeology, Historiography, and Racial Myth in Korean State-Formation Theories*
188. Brian D. Ruppert, *Jewel in the Ashes: Buddha Relics and Power in Early Medieval Japan*
189. Susan Daruvala, *Zhou Zuoren and an Alternative Chinese Response to Modernity*
*190. James Z. Lee, *The Political Economy of a Frontier: Southwest China, 1250–1850*
191. Kerry Smith, *A Time of Crisis: Japan, the Great Depression, and Rural Revitalization*
192. Michael Lewis, *Becoming Apart: National Power and Local Politics in Toyama, 1868–1945*
193. William C. Kirby, Man-houng Lin, James Chin Shih, and David A. Pietz, eds., *State and Economy in Republican China: A Handbook for Scholars*
194. Timothy S. George, *Minamata: Pollution and the Struggle for Democracy in Postwar Japan*
195. Billy K. L. So, *Prosperity, Region, and Institutions in Maritime China: The South Fukien Pattern, 946–1368*
196. Yoshihisa Tak Matsusaka, *The Making of Japanese Manchuria, 1904–1932*

197. Maram Epstein, *Competing Discourses: Orthodoxy, Authenticity, and Engendered Meanings in Late Imperial Chinese Fiction*
198. Curtis J. Milhaupt, J. Mark Ramseyer, and Michael K. Young, eds. and comps., *Japanese Law in Context: Readings in Society, the Economy, and Politics*
199. Haruo Iguchi, *Unfinished Business: Ayukawa Yoshisuke and U.S.-Japan Relations, 1937–1952*
200. Scott Pearce, Audrey Spiro, and Patricia Ebrey, *Culture and Power in the Reconstitution of the Chinese Realm, 200–600*
201. Terry Kawashima, *Writing Margins: The Textual Construction of Gender in Heian and Kamakura Japan*
202. Martin W. Huang, *Desire and Fictional Narrative in Late Imperial China*
203. Robert S. Ross and Jiang Changbin, eds., *Re-examining the Cold War: U.S.-China Diplomacy, 1954–1973*
204. Guanhua Wang, *In Search of Justice: The 1905–1906 Chinese Anti-American Boycott*
205. David Schaberg, *A Patterned Past: Form and Thought in Early Chinese Historiography*
206. Christine Yano, *Tears of Longing: Nostalgia and the Nation in Japanese Popular Song*
207. Milena Doleželová-Velingerová and Oldřich Král, with Graham Sanders, eds., *The Appropriation of Cultural Capital: China's May Fourth Project*
208. Robert N. Huey, *The Making of 'Shinkokinshū'*
209. Lee Butler, *Emperor and Aristocracy in Japan, 1467–1680: Resilience and Renewal*
210. Suzanne Ogden, *Inklings of Democracy in China*
211. Kenneth J. Ruoff, *The People's Emperor: Democracy and the Japanese Monarchy, 1945–1995*
212. Haun Saussy, *Great Walls of Discourse and Other Adventures in Cultural China*
213. Aviad E. Raz, *Emotions at Work: Normative Control, Organizations, and Culture in Japan and America*
214. Rebecca E. Karl and Peter Zarrow, eds., *Rethinking the 1898 Reform Period: Political and Cultural Change in Late Qing China*
215. Kevin O'Rourke, *The Book of Korean Shijo*
216. Ezra F. Vogel, ed., *The Golden Age of the U.S.-China-Japan Triangle, 1972–1989*
217. Thomas A. Wilson, ed., *On Sacred Grounds: Culture, Society, Politics, and the Formation of the Cult of Confucius*
218. Donald S. Sutton, *Steps of Perfection: Exorcistic Performers and Chinese Religion in Twentieth-Century Taiwan*
219. Daqing Yang, *Technology of Empire: Telecommunications and Japanese Expansion in Asia, 1883–1945*